Lecture Notes in Computer Science 7941

Commenced Publication in 1973
Founding and Former Series Editors:
Gerhard Goos, Juris Hartmanis, and Jan van Leeuwen

Advanced Research in Computing and Software Science

Subline of Lectures Notes in Computer Science

Masahito Hasegawa (Ed.)

Typed Lambda Calculi and Applications

11th International Conference, TLCA 2013
Eindhoven, The Netherlands, June 26-28, 2013
Proceedings

Springer

Volume Editor

Masahito Hasegawa
Kyoto University, Research Institute for Mathematical Sciences
Kitashirakawa-Oiwake-cho, Sakyo-ku, Kyoto 606-8502, Japan
E-mail: hassei@kurims.kyoto-u.ac.jp

ISSN 0302-9743 e-ISSN 1611-3349
ISBN 978-3-642-38945-0 e-ISBN 978-3-642-38946-7
DOI 10.1007/978-3-642-38946-7
Springer Heidelberg Dordrecht London New York

Library of Congress Control Number: 2013940266

CR Subject Classification (1998): D.1.6, D.3.2, F.3, F.4.1, F.4, I.2.3

LNCS Sublibrary: SL 1 – Theoretical Computer Science and General Issues

Typesetting: Camera-ready by author, data conversion by Scientific Publishing Services, Chennai, India

Printed on acid-free paper

Springer is part of Springer Science+Business Media (www.springer.com)

Preface

This volume contains the papers of the 11th International Conference on Typed Lambda Calculi and Applications (TLCA 2013), which was held during June 26–28, 2013, in Eindhoven, The Netherlands. TLCA 2013 was part of the 7th International Conference on Rewriting, Deduction, and Programming (RDP 2013), together with the 24th International Conference on Rewriting Techniques and Applications (RTA 2013), the Workshop on Control Operators and their Semantics (COS 2013), the Workshop on Haskell And Rewriting Techniques (HART 2013), the 11th International Workshop on Reduction Strategies in Rewriting and Programming (WRS 2013), the 27th International Workshop on Unification (UNIF 2013), the Second International Workshop on Confluence (IWC 2013), the Workshop on Infinitary Rewriting (WIR 2013), and the annual meeting of the IFIP Working Group 1.6 on Term Rewriting.

The TLCA series of conferences serves as a forum for presenting original research results that are broadly relevant to the theory and applications of lambda calculus. Previous TLCA conferences were held in Utrecht (1993), Edinburgh (1995), Nancy (1997), L'Aquila (1999), Kraków (2001), Valencia (2003), Nara (2005), Paris (2007), Brasília (2009), and Novi Sad (2011).

A total of 15 papers were accepted out of 41 submissions for presentation at TLCA 2013 and for inclusion in the proceedings. I would like to thank everyone who submitted a paper and to express my regret that many interesting papers could not be included. Each submitted paper was reviewed by at least three members of the Program Committee, who were assisted in their work by 59 external reviewers. I thank the members of the Program Committee and the external reviewers for their review work, as well as Andrei Voronkov for providing the EasyChair system that proved invaluable throughout the review process and the preparation of this volume.

In addition to the contributed papers, the TLCA 2013 program contained invited talks by Simon Peyton-Jones (joint with RTA 2013), Hugo Herbelin, and Damiano Mazza. This volume contains abstracts for the talks by Peyton-Jones and Herbelin, and an extended abstract for the talk by Mazza. An extended abstract for the talk by Peyton-Jones appears in the proceedings of RTA 2013.

Many people helped to make TLCA 2013 a success. I would like to thank the Conference Chair Herman Geuvers, RDP 2013 Chair Hans Zantema, and the Organizing Committee, the local organizing team, TLCA Publicity Chair Luca Paolini, and the TLCA Steering Committee. The financial support from the Netherlands Organisation for Scientific Research (NWO) is gratefully acknowledged.

April 2013 Masahito Hasegawa

Organization

Program Chair

Masahito Hasegawa	Kyoto University, Japan

Program Committee

Andreas Abel	LMU München, Germany
Patrick Baillot	CNRS and ENS Lyon, France
Nick Benton	Microsoft Research Cambridge, UK
Lars Birkedal	University of Aarhus, Denmark
Herman Geuvers	Radboud University Nijmegen and Eindhoven University of Technology, The Netherlands
Masahito Hasegawa	Kyoto University, Japan
Naoki Kobayashi	University of Tokyo, Japan
Paul-André Melliès	CNRS and Universite Paris Diderot, France
Thomas Streicher	TU Darmstadt, Germany
Lorenzo Tortora de Falco	Università Roma Tre, Italy
Stephanie Weirich	University of Pennsylvania, USA

Conference Chair

Herman Geuvers	Radboud University Nijmegen and Eindhoven University of Technology, The Netherlands

Organizing Committee

Sacha Claessens	Eindhoven University of Technology, The Netherlands
Herman Geuvers	Radboud University Nijmegen and Eindhoven University of Technology, The Netherlands
Jürgen Giesl	RWTH Aachen University, Germany
Erik de Vink	Eindhoven University of Technology, The Netherlands
Hans Zantema	Eindhoven University of Technology, The Netherlands

TLCA Steering Committee Chair

Paweł Urzyczyn	University of Warsaw, Poland

TLCA Steering Committee

Samson Abramsky	University of Oxford, UK
Henk Barendregt	Radboud University Nijmegen, The Netherlands
Pierre-Louis Curien	CNRS and Université Paris Diderot, France
Mariangiola Dezani-Ciancaglini	University of Turin, Italy
Martin Hofmann	LMU München, Germany
Luke Ong	University of Oxford, UK
Simona Ronchi Della Rocca	University of Turin, Italy
Paweł Urzyczyn	University of Warsaw, Poland

TLCA Honorary Advisor

Roger Hindley	Swansea University, UK

TLCA Publicity Chair

Luca Paoloni	University of Turin, Italy

Additional Reviewers

Beniamino Accattoli
Kenichi Asai
Federico Aschieri
Emmanuel Beffara
Stefano Berardi
Alexis Bernadet
Aleš Bizjak
Valentin Blot
Pierre Bourreau
Christopher Broadbent
Chris Casinghino
Iliano Cervesato
Pierre Clairambault
Ugo dal Lago
Ugo de'Liguoro
Henry DeYoung
Harley Eades III
José Espírito Santo
Andrzej Filinski
Dan Ghica
Stéphane Graham-Lengrand
Nicolas Guenot
Stefano Guerrini
Daniel Gustafsson
Ichiro Hasuo
Hugo Herbelin
Daniel Hirschkoff
Naohiko Hoshino
Barry Jay
Delia Kesner
Robbert Krebbers
Neelakantan Krishnaswami
James Laird
Marc Lasson
Davorin Lesnik
Daniel R. Licata
Luigi Liquori
Giulio Manzonetto
Ralph Matthes
Damiano Mazza
Rasmus Møgelberg
Andrzej Murawski
Koji Nakazawa
Georg Neis

Fredrik Nordvall Forsberg
Michele Pagani
Luca Paolini
Tomas Petricek
Jakob Rehof
Bernhard Reus
Colin Riba
Claudio Sacerdoti Coen
Christian Sattler
Peter Selinger
Paula Severi
Vilhelm Sjöberg
Takeshi Tsukada
Nikos Tzevelekos
Noam Zeilberger

Table of Contents

Contributed Papers

Type-Directed Compilation in the Wild: Haskell and Core

Simon Peyton-Jones
Microsoft Research, Cambridge, UK

Academic papers often describe typed calculi, but it is rare to find one in a production compiler. Indeed, I think the Glasgow Haskell Compiler (GHC) may be the only production compiler in the world that really has a remorselessly strongly-typed intermediate language, informally called "Core", or (when writing academic papers) the more respectable-sounding "System FC".

As real compilers go, GHC's Core language is tiny: it is a slight extension of System F, with letrec, data types, and case expressions. Yet all of Haskell (now a bit of a monster) gets translated into it. In the last few years we have added one new feature to Core, namely typed (but erasable) coercions that witness type equalities. This single addition has opened the door to a range of source-language extensions, such as GADTs and type families.

In this talk I'll describe Core, and how it has affected GHC's development over the last two decades, concentrating particularly on recent developments. I'll also mention the role of user-written rewrite rules as a compiler extension mechanism. Overall, I will try to give a sense of the ways in which the work of the typed lambda-calculi and rewriting communities has influenced at least one real compiler.

M. Hasegawa (Ed.): TLCA 2013, LNCS 7941, p. 1, 2013.

Proving with Side Effects

Hugo Herbelin

INRIA Paris-Rocquencourt, πr^2 Team
PPS Lab, Univ. Paris Diderot, Paris, France

Control operators, such as *callcc*, are effectful constructions. But control operators relate to classical logic along the proofs-as-programs correspondence, so, in this sense, classical reasoning is an effectful form of reasoning.

The effects of control operators can be interpreted in a purely functional language by working in the continuation monad. Otherwise said, control operators provide "direct-style" programming for computational behaviours already implicitly available in purely functional languages. In logic, this corresponds to interpreting classical reasoning within minimal logic via the double-negation translation.

In programming, there are various kinds of effectful constructions whose effects can be captured in a purely functional way through appropriate monads. In the talk, we focus on memory assignment and investigate its strength in the context of proving. In particular, we look at memory assignment as a direct-style presentation of reasoning through Kripke/Cohen forcing translations, getting close to similar ideas from Krivine.

M. Hasegawa (Ed.): TLCA 2013, LNCS 7941, p. 2, 2013.

Non-linearity as the Metric Completion of Linearity

Damiano Mazza

CNRS, UMR 7030, LIPN, Université Paris 13, Sorbonne Paris Cité
Damiano.Mazza@lipn.univ-paris13.fr

Abstract. We summarize some recent results showing how the lambda-calculus may be obtained by considering the metric completion (with respect to a suitable notion of distance) of a space of affine lambda-terms, i.e., lambda-terms in which abstractions bind variables appearing at most once. This formalizes the intuitive idea that multiplicative additive linear logic is "dense" in full linear logic (in fact, a proof-theoretic version of the above-mentioned construction is also possible). We argue that thinking of non-linearity as the "limit" of linearity gives an interesting point of view on well-known properties of the lambda-calculus and its relationship to computational complexity (through lambda-calculi whose normalization is time-bounded).

1 Linearity and Approximations

The concept of linearity in logic and computer science, introduced over two decades ago [12], has now entered firmly into the "toolbox" of proof theorists and functional programming language theorists. It is present, in one way or another, in a broad range of contexts, such as: denotational semantics [11], games semantics [22] and categorical semantics [8]; computational interpretations of classical logic [18,9]; optimal implementation of functional programming languages [3,19]; the theory of explicit substitutions [2]; higher-order languages for probabilistic [10] and quantum computation [24]; typing systems for polynomial-time [4], non-size-increasing [14] and resource-aware computation [17]; and even concurrency theory [6,15].

Technically, linearity imposes a severe restriction on the behavior of programs: data must be accessed exactly once. Its cousin *affinity*, which is more relevant for the purposes of this text, slightly relaxes the constraint: although data may be discarded, it may nevertheless be accessed at most once. In any case, linearity and affinity forbid re-use, forcing the programmer to explicitly keep track of how many copies of a given piece of information are needed in order to perform a computation.

How can general, non-linear computation be performed in an affine setting? In other words, how can a persistent memory be simulated by a volatile memory? The intuitive answer is clear: one persistent memory cell, accessible arbitrarily many times, may be perfectly simulated by infinitely many volatile memory cells, each accessible only once. Of course, if only a finite memory is available, then

M. Hasegawa (Ed.): TLCA 2013, LNCS 7941, pp. 3–14, 2013.

only an imperfect simulation will be possible in general. However, the important point is that affine computation may approximate non-linear computation to an arbitrary degree of precision.

2 A Polyadic Affine Lambda-Calculus

Let us see how the above intuition may be formalized. Consider the fragment

$$A, B ::= X \mid (A_1 \,\&\, 1) \otimes \cdots \otimes (A_n \,\&\, 1) \multimap B$$

of multiplicative additive linear logic (if $n = 0$, then the premise of the implication is the logical constant 1). The proofs of this simple logical system correspond to (simply-typed) terms of the following language:

$$t, u ::= x \mid \lambda x_1 \ldots x_n.t \mid t\langle u_1, \ldots, u_n\rangle,$$

with the requirement that variables appear at most once in terms. In other words, we have a "multilinear", or polyadic affine λ-calculus.

The reduction of a simply-typed non-linear λ-term such as $M = (\lambda x.Nxx)I$ may be "linearized" as

$$[\![M]\!] = (\lambda x_0 x_1.[\![N]\!]\langle x_0\rangle\langle x_1\rangle)\langle [\![I]\!], [\![I]\!]\rangle,$$

in which we see how the duplication of the subterm I by the head redex of M forces us to explicitly introduce two copies of $[\![I]\!]$ (the linearization of I). This is of course very naive: if M duplicates I again (for instance, if $N = \lambda y.zyy$), we will be forced to include additional copies of $[\![I]\!]$ in $[\![M]\!]$ and it would be hard in general to statically determine exactly how many are necessary (we would essentially need to normalize M).

We are thus naturally led to consider an *infinitary* calculus. The rigorous manipulation of infinity requires some form of topology, which will actually be the key to a satisfactory formalization of the above intuition: we will be able to say that affine terms approximate non-linear terms to an arbitrary degree of precision in a clear technical sense, that of metric spaces.

Our first step is to switch to an untyped framework, so that our analysis will be valid in the most general terms. To this extent, we introduce a term $\bot$ in the language, which is used to solve possible mismatches between the arity of abstractions and applications: when reducing $(\lambda x_0 x_1.t)\langle u\rangle$, the sequence in the outer application in not "long enough", so the term $\bot$ will be substituted to x_1.

We also switch from variables to explicit *occurrences*, which is to realize that the affine (or linear) λ-calculus is, in a way, a calculus of occurrences. This, although not technically necessary (and not done in [20]) will simplify the exposition.

So, our definition of (untyped) polyadic affine λ-calculus is the following:

$$t, u ::= \bot \mid x_i \mid \lambda x.t \mid t\mathbf{u},$$

where:
- in x_i, $i \in \mathbb{N}$ is a unique identifier of the occurrence of x, *i.e.*, we require that if x_i, x_j appear in the same term, then $i \neq j$;
- abstractions bind *variables*, *i.e.*, $\lambda x.t$ binds every free occurrence of the form x_i in t (free and bound occurrences are defined as usual);
- $\mathbf{u}$ is a finite sequence of terms. Actually, since we have $\perp$, it is technically simpler to say that $\mathbf{u}$ is a function from $\mathbb{N}$ to terms which is almost everywhere equal to $\perp$.

As usual, terms are always considered up to α-equivalence.

The most important point is how we define reduction:

$$(\lambda x.t)\mathbf{u} \quad \to \quad t[\mathbf{u}/x],$$

where the notation $t[\mathbf{u}/x]$ means that we substitute $\mathbf{u}(i)$ to the at most unique free occurrence x_i in t. We call the set of terms defined above $\Lambda_{\mathrm{p}}^{\mathrm{aff}}$. The superscript reminds us that the calculus is affine, whereas the subscript stands for "polyadic".

The calculus $\Lambda_{\mathrm{p}}^{\mathrm{aff}}$ is strongly confluent (*i.e.*, reduction in at most one step, denoted by $\to^{=}$, enjoys the diamond property) and strongly normalizing. Both properties are immediate consequences of affinity: redexes cannot be duplicated, the theory of residues is trivial and local confluence is achieved in at most one step; moreover, the size of terms strictly decreases during reduction.

3 A Metric Space of Terms and Its Infinitary Completion

Let us now define a function $\Lambda_{\mathrm{p}}^{\mathrm{aff}} \times \Lambda_{\mathrm{p}}^{\mathrm{aff}} \to [0,1]$, by induction on the first argument:

$$d(\perp, t') = \begin{cases} 0 & \text{if } t' = \perp \\ 1 & \text{otherwise} \end{cases}$$

$$d(x_i, t') = \begin{cases} 0 & \text{if } t' = x_i \\ 1 & \text{otherwise} \end{cases}$$

$$d(\lambda x.t_1, t') = \begin{cases} d(t_1, t_1') & \text{if } t' = \lambda x.t_1' \\ 1 & \text{otherwise} \end{cases}$$

$$d(t_1\mathbf{u}, t') = \begin{cases} \max\left(d(t_1, t_1'), \sup_{i\in\mathbb{N}} 2^{-i-1} d(\mathbf{u}(i), \mathbf{u}'(i))\right) & \text{if } t' = t_1'\mathbf{u}' \\ 1 & \text{otherwise} \end{cases}$$

Note that, in the abstraction case, we implicitly used α-equivalence to force the variables abstracted in t_1 and t_1' to coincide. This small nuisance could be avoided by resorting to de Bruijn's notation [5] but, except for the following two paragraphs, we prefer to stick to the usual notation, for better readability.

One may check that d is a bounded ultrametric on $\Lambda_{\mathrm{p}}^{\mathrm{aff}}$, *i.e.*, it is a bounded (by 1) metric which further satisfies $d(t, t'') \leq \max(d(t, t'), d(t', t''))$ for all $t, t', t'' \in \Lambda_{\mathrm{p}}^{\mathrm{aff}}$ (a stronger version of the triangle inequality). A more in-depth analysis of d reveals the following. Consider the poset $\mathbb{N}^*$ of finite sequences of integers, ordered by the prefix relation. A *tree* is, as usual, a downward-closed subset of $\mathbb{N}^*$ (note that non-well-founded and infinitely branching trees are both allowed). Let $\Sigma = \{\perp, \lambda, @\} \cup \mathbb{N}^2$, and let $f : \mathbb{N}^* \to \Sigma$. We define $\operatorname{supp} f = \{a \in \mathbb{N}^* \mid f(a) \neq \perp\}$. We may see the of terms of $\Lambda_{\mathrm{p}}^{\mathrm{aff}}$ (in de Bruijn notation) as finite labeled trees, *i.e.*, as functions t from $\mathbb{N}^*$ (arbitrary integers are needed because applications have arbitrarily large width) to Σ (de Bruijn indices must be *pairs* of integers: one for identifying the abstraction, one for identifying the occurrence), such that $\operatorname{supp} t$ is a finite tree.

Now, if we endow Σ with the discrete uniformity, the ultrametric d may be seen to yield the uniformity of uniform convergence on finitely branching (but possibly infinite) trees. In this uniformity, a sequence of terms $(t_n)_{n\in\mathbb{N}}$ (which are particular functions) converges to t if, for every finitely branching tree $\tau \subseteq \mathbb{N}^*$, there exists $k \in \mathbb{N}$ such that, whenever $n \geq k$, we have $t_n(a) = t(a)$ for all $a \in \tau$. In other words, t_n eventually coincides with t on every finitely branching tree.

Let us look at an example, using the metric d. Let

$$\Delta_n = \lambda x.x_0\langle x_1, \ldots, x_n\rangle$$

(when we write a sequence $\mathbf{u}$ as $\langle u_0, \ldots, u_{n-1}\rangle$ we mean that $\mathbf{u}(i) = u_i$ for $0 \leq i < n$ and $\mathbf{u}(i) = \perp$ for $i \geq n$). We invite the reader to check that, for all $n \in \mathbb{N}$ and $p > 0$, $d(\Delta_n, \Delta_{n+p}) = 2^{-n-1}$, so the sequence is Cauchy.[1] And yet, no term of $\Lambda_{\mathrm{p}}^{\mathrm{aff}}$ may be the limit of $(\Delta_n)_{n\in\mathbb{N}}$, because the sequence is obviously tending to the infinitary term

$$\Delta = \lambda x.x_0\langle x_1, x_2, \ldots\rangle$$

(eventually, Δ_n coincides with Δ on every finitely branching tree).

The above example proves that the metric space $(\Lambda_{\mathrm{p}}^{\mathrm{aff}}, d)$ is not complete. We denote its completion by $\Lambda_{\infty}^{\mathrm{aff}}$. Its terms may no longer be defined inductively, because they may have infinite height. However, they are well-founded, *i.e.*, as trees, they contain no infinite branch from their root. In terms of the above description of terms, $t \in \Lambda_{\infty}^{\mathrm{aff}}$ iff, as a function $t : \mathbb{N}^* \to \Sigma$, $\operatorname{supp} t$ is a well-founded tree. This means that the strict subterm relation $t \sqsubset t'$ is well-founded, so we may still reason by induction on $\Lambda_{\infty}^{\mathrm{aff}}$, in stark contrast with usual infinitary λ-calculi [16]. This is a consequence of the notion of (uniform) convergence induced by d: since a sequence $(t_n)_{n\in\mathbb{N}}$ tending to t must eventually coincide with t on every finitely branching tree, it coincides in particular on infinite trees, which, by König's lemma, must be non-well-founded. But if $(t_n)_{n\in\mathbb{N}}$ is a sequence of $\Lambda_{\mathrm{p}}^{\mathrm{aff}}$, every t_n is finite and in particular well-founded, so it cannot coincide with t on a non-well-founded tree unless t is also well-founded.

On the other hand, finitely high but infinitely wide terms such as Δ are the typical inhabitants of $\Lambda_{\infty}^{\mathrm{aff}} \setminus \Lambda_{\mathrm{p}}^{\mathrm{aff}}$. In fact, in [20] we defined the metric so that only terms of finite height are added to the completion (it is enough to consider the ultrametric $\max(d, \rho)$, where ρ is the discrete pseudometric such that $\rho(t, t') = 0$ as soon as t, t' have the same height, and $\rho(t, t') = 1$ otherwise), on the grounds

[1] Since d is an ultrametric, it is actually enough to check this for $p = 1$ only.

that these are the most interesting ones and are easier to manipulate (we may apply induction on the height even in the infinitary case). However, in this exposition we prefer to bring forth the more natural and topologically better behaved metric d.

Reduction in $\Lambda_\infty^{\mathrm{aff}}$ is defined just as in $\Lambda_{\mathrm{p}}^{\mathrm{aff}}$:

$$(\lambda x.t)\mathbf{u} \quad \to \quad t[\mathbf{u}/x],$$

except that now it may be necessary to perform infinitely many (linear) substitutions, because we may have that x_i is free in t for infinitely many $i \in \mathbb{N}$. We would like to observe that, from a topological point of view, this obvious definition is actually the only possible one. Indeed, it is possible to show, in a sense that we do not make precise here, that reduction as defined above is continuous on $\Lambda_\infty^{\mathrm{aff}}$.[2] Since a continuous function is entirely determined by its behavior on a dense subset like $\Lambda_{\mathrm{p}}^{\mathrm{aff}}$, there is really no other topologically sound way of extending reduction to infinitary terms.

In spite of the presence of infinitary terms, reduction is strongly confluent, because the calculus is still affine, *i.e.*, it is a "calculus of occurrences", in which no subterm is duplicated during reduction. In spite of this, infinitary terms may not normalize. This is easily seen by considering the term

$$\Omega = \Delta\langle\Delta, \Delta, \ldots\rangle,$$

which reduces to itself. Indeed, Δ takes a possibly infinite list, extracts the head (which is $\perp$ if the list is empty) and applies it to the rest of the list. If the list we feed to Δ is made up of infinitely many copies of Δ itself, we obviously loop.

This example gives us the opportunity to see concretely, in a simple but already meaningful case, how affine terms approximate non-linear terms. Of course, technically speaking, the term Ω above is still affine. However, it behaves exactly like its namesake term in the usual λ-calculus (indeed, we will see that it corresponds to it in a precise sense), so we may consider it to be an example of non-linear term. Consider now the finite terms

$$\Omega_n = \Delta_n\langle\overbrace{\Delta_n, \ldots, \Delta_n}^{n \text{ times}}\rangle.$$

We invite again the reader to check that $d(\Omega_n, \Omega) = 2^{-n-1}$, so that $\lim \Omega_n = \Omega$. Hence, Ω_n is supposed to approximate Ω better and better, as n grows. In the case of Ω, there is not much to approximate except divergence; and in fact, $\Omega_n \to^* \perp\langle\rangle$ in $n+1$ steps, *i.e.*, the reduction of Ω_n is longer and longer, approximating the diverging behavior of Ω.

---

[2] We are alluding to Proposition 8 of [20]. Unfortunately, we made a mistake in that paper and Proposition 8 is actually false for the metric used therein. The result does hold for the metric d considered here, which is why we said above that it is "topologically better behaved". The mistake luckily does not affect the main results of [20], in which Proposition 8 plays no role.

4 Uniformity and the Isomorphism with the Usual Lambda-Calculus

There are far too many terms (a continuum of them) in $\Lambda_\infty^{\text{aff}}$ for it be directly in correspondence with the usual λ-calculus. We might say that $\Lambda_\infty^{\text{aff}}$ is a *non-uniform* λ-calculus, in the same sense as non-uniform families of circuits: if we accept $\Lambda_\infty^{\text{aff}}$ as a computational model, every function on $\mathbb{N}$ becomes computable, with respect to any standard encoding of natural numbers. To retrieve the λ-calculus, we need to introduce some notion of uniformity.

Definition 1 (Uniformity). *We define $\approx$ to be the smallest partial equivalence relation on $\Lambda_\infty^{\text{aff}}$ such that:*

- $x_i \approx x_j$ *for every variable x and $i, j \in \mathbb{N}$;*
- *if $t \approx t'$, then $\lambda x.t \approx \lambda x.t'$ for every variable x;*
- *if $t \approx t'$ and $\mathbf{u}, \mathbf{u}'$ are such that, for all $i, i' \in \mathbb{N}$, $\mathbf{u}(i) \approx \mathbf{u}'(i')$, then $t\mathbf{u} \approx t'\mathbf{u}'$.*

A term t is uniform *if $t \approx t$. We denote by $\Lambda_\infty^{\text{u}}$ the set of uniform terms.*

Intuitively, $\approx$ equates terms that "look alike" under any possible permutation of the terms appearing in its application sequences. In particular, it equates all occurrences of the same variable: while it is important that we distinguish two occurrences of x by naming one of them x_i and the other x_j (with $i \neq j$), it does not matter which is assigned i and which j.

A term u is uniform if $u \neq \bot$ and if u "looks like itself" even if we permute some of its subterms in application sequences. For instance, any term containing a finite application, such as $z_0\langle x_0 \rangle$, cannot be uniform, because $\langle x_0 \rangle = \langle x_0, \bot \rangle$ and $z_0\langle x_0, \bot \rangle$ and $z_0\langle \bot, x_0 \rangle$ do not "look alike" (indeed, $x_0 \not\approx \bot$). On the other hand, terms like Δ and Ω are uniform (but not Δ_n or Ω_n: by the above remark, a finite approximation of a uniform term containing an application can never be uniform). Note that, if $t\mathbf{u}$ is uniform, then every $\mathbf{u}(i)$ has the same height, that of $\mathbf{u}(0)$. Hence, uniform terms all have finite height. This is why we said above that the terms of finite height are "the most interesting ones".

The set $\Lambda_\infty^{\text{u}}$ is not closed under reduction: in $t = x_0\langle u, u, \ldots \rangle$, with u closed, uniform and such that $u \to u'$, the reduct $t \to x_0\langle u', u, \ldots \rangle$ is in general not uniform, because u' has no reason to "look like" u. The solution is obvious: we must reduce *all* of the copies of u at the same time:

Definition 2 (Infinitary reduction). *We define the relations $\Rightarrow_k$ on $\Lambda_\infty^{\text{u}}$, with $k \in \mathbb{N}$, as the smallest relations satisfying:*

- $(\lambda x.t)\mathbf{u} \Rightarrow_0 t[\mathbf{u}/x]$;
- *if $t \Rightarrow_k t'$, then $\lambda x.t \Rightarrow_k \lambda x.t'$;*
- *if $t \Rightarrow_k t'$, then $t\mathbf{u} \Rightarrow_k t'\mathbf{u}$;*
- *if $t\mathbf{u} \in \Lambda_\infty^{\text{u}}$ and $\mathbf{u}(0) \Rightarrow_k u_0'$, by uniformity the "same" reduction may be performed in all $\mathbf{u}(i)$, $i \in \mathbb{N}$, obtaining the term u_i'. If we define $\mathbf{u}'(i) = u_i'$ for all $i \in \mathbb{N}$, then $t\mathbf{u} \Rightarrow_{k+1} t\mathbf{u}'$.*

We denote by $\Rightarrow$ the union of all $\Rightarrow_k$, for $k \in \mathbb{N}$.

Note that $\Rightarrow_k$ is infinitary iff $k > 0$. Indeed, $\Rightarrow_0$ is head reduction,[3] which corresponds to a single reduction step (which may of course perform infinitely many substitutions, but this is not what we mean by "infinitary". Rather, we mean that infinitely many reductions steps are performed together).

Proposition 1. *Let $t \in \Lambda^{\mathrm{u}}_{\infty}$. Then:*

- *$t \Rightarrow t'$ implies $t' \in \Lambda^{\mathrm{u}}_{\infty}$;*
- *furthermore, for all $u \approx t$, $u \Rightarrow u' \approx t'$.*

Proposition 1 asserts that uniform terms are stable under $\Rightarrow$ and that such a rewriting relation is compatible with the equivalence classes of $\approx$. Therefore, the set $\Lambda^{\mathrm{aff}}_{\infty}/\approx$ may be endowed with the (one-step) reduction relation $\Rightarrow$. It turns out that this is exactly the usual, non-linear λ-calculus. In the following, we write Λ for the set of usual λ-terms and $\to_\beta$ for usual β-reduction.

Theorem 1 (Isomorphism). *We have*

$$(\Lambda^{\mathrm{aff}}_{\infty}/\approx, \Rightarrow) \quad \cong \quad (\Lambda, \to_\beta),$$

in the Curry-Howard sense, i.e., there exist two maps

$$(\!|\cdot|\!) : \Lambda^{\mathrm{u}}_{\infty} \to \Lambda \qquad\qquad [\![\cdot]\!] : \Lambda \to \Lambda^{\mathrm{u}}_{\infty}$$

such that, for all $M \in \Lambda$ and $t \in \Lambda^{\mathrm{u}}_{\infty}$:

1. *$(\!|[\![M]\!]|\!) = M$;*
2. *$[\![(\!|t|\!)]\!] \approx t$;*
3. *$M \to_\beta M'$ implies $[\![M]\!] \Rightarrow t' \approx [\![M']\!]$;*
4. *$t \Rightarrow t'$ implies $(\!|t|\!) \to_\beta (\!|t'|\!)$.*

The two maps of the isomorphism are both defined by induction. For what concerns $(\!|\cdot|\!)$, we have:

$$(\!|x_i|\!) = x \quad \text{(for all } i \in \mathbb{N}),$$
$$(\!|\lambda x.t|\!) = \lambda x.(\!|t|\!),$$
$$(\!|t\mathbf{u}|\!) = (\!|t|\!)(\!|\mathbf{u}(0)|\!).$$

For what concerns the other direction, we first fix a bijective function $\ulcorner\cdot\urcorner : \mathbb{N}^* \to \mathbb{N}$ to encode finite sequences of integers as integers. Then, we define a family of parametric maps $[\![\cdot]\!]_a$, with $a \in \mathbb{N}^*$, as follows:

$$[\![x]\!]_a = x_{\ulcorner a \urcorner}$$
$$[\![\lambda x.M]\!]_a = \lambda x.[\![M]\!]_a$$
$$[\![MN]\!]_a = [\![M]\!]_{a0}\langle [\![N]\!]_{a1}, [\![N]\!]_{a2}, [\![N]\!]_{a3}, \ldots\rangle$$

One can prove that, for any $a, a' \in \mathbb{N}^*$ and any $t \in \Lambda^{\mathrm{u}}_{\infty}$, we actually have $[\![t]\!]_a \in \Lambda^{\mathrm{u}}_{\infty}$ and $[\![t]\!]_a \approx [\![t]\!]_{a'}$. Of course, Theorem 1 holds for any choice of $a \in \mathbb{N}^*$, that is why we simply write $[\![\cdot]\!]$. We let the reader check that the uniform infinitary terms Δ and Ω introduced above are (modulo $\approx$) the images through $[\![\cdot]\!]$ of their well-known namesake λ-terms.

[3] It is actually *spinal* reduction, but the distinction is inessential.

5 The Proof-Theoretic Perspective

As already mentioned, our idea of obtaining the λ-calculus through a metric completion process has proof-theoretic roots, in particular in linear logic. In fact, the above constructions may be reformulated using proofs instead of λ-terms. In [21], we show how a fully-complete model of polarized multiplicative exponential linear logic may be built as a metric completion of a model of the sole multiplicative fragment. Roughly speaking, we take objects which are very much related to the *designs* of Girard's ludics [13], introduce a metric completely analogous to the one given here, and construct the model in the completed space. What we obtain closely resembles Abramsky, Jagadeesan and Malacaria's formulation of games semantics [1].

Recently, Melliès and Tabareau [23] used a similar idea to provide an explicit formula for constructing the free commutative comonoid in certain symmetric monoidal categories. This offers a categorical viewpoint on our work, and yields some potentially interesting remarks.

Melliès and Tabareau's construction starts with a symmetric monoidal category $(\mathcal{C}, \otimes, 1)$ with finite products, which we denote by $A \,\&\, B$. We define $\dagger A = A \,\&\, 1$, the free co-pointed object on A, with its canonical projection $\pi^A : \dagger A \longrightarrow 1$. We also inductively define $A^{\otimes 0} = 1$, $A^{\otimes n+1} = A^{\otimes n} \otimes A$.

Using the symmetry of $\mathcal{C}$, for every $n \in \mathbb{N}$ we may build $n!$ parallel isomorphisms $\sigma_i^{A,n} : (\dagger A)^{\otimes n} \longrightarrow (\dagger A)^{\otimes n}$. We define $A^{\leq n}$ to be the equalizer, if it exists, of $\sigma_1^{A,n}, \ldots, \sigma_{n!}^{A,n}$.

Now, by the universal property of equalizers on the morphism π^A, we know that there is a canonical projection $\pi_n^A : A^{\leq n+1} \longrightarrow A^{\leq n}$, for all $n \in \mathbb{N}$. Then, we define $!A$ to be the limit, if it exists, of the diagram

$$A^{\leq 0} \xleftarrow{\pi_0^A} A^{\leq 1} \xleftarrow{\pi_1^A} A^{\leq 2} \xleftarrow{\pi_2^A} \cdots$$

Melliès and Tabareau's result is the following:

Proposition 2 ([23]). *If the equalizers and the projective limit considered above exist in $\mathcal{C}$ and if these limits commute with the tensor product of $\mathcal{C}$, then, for every object A of $\mathcal{C}$, $!A$ is the free commutative comonoid on A.*

It is known that, in a $*$-autonomous category with finite products, the existence of the free commutative comonoid on every object yields a denotational model of full linear logic (a result due to Lafont, see Melliès's survey in [8]). Therefore, Proposition 2 provides a way of building, under certain conditions, models of full linear logic starting from models of its multiplicative additive fragment.

The conditions required by Proposition 2 are however not anodyne. In fact, Tasson showed [23] how the construction fails in a well known model of linear logic, Ehrhard's finiteness spaces [11]. In this model, although all the required limits exist, the projective limit does not commute with the tensor product.

Our approach seems to offer an alternative construction to that of Melliès and Tabareau's, in which the two main steps for building the free comonoid are reversed: first one computes a projective limit, then one equalizes. This follows our procedure for recovering the λ-calculus: we first complete the space $\Lambda_{\mathrm{p}}^{\mathrm{aff}}$ to obtain $\Lambda_{\infty}^{\mathrm{aff}}$, then we introduce uniformity and obtain the λ-calculus as the quotient $\Lambda_{\infty}^{\mathrm{aff}}/\approx$.

More in detail, we start by defining $p_n^A : (\dagger A)^{\otimes n+1} \longrightarrow (\dagger A)^{\otimes n}$ as the morphism obtained by composing $id_{(\dagger A)^{\otimes n}} \otimes \pi^A$ with the iso $(\dagger A)^{\otimes n} \otimes 1 \cong (\dagger A)^{\otimes n}$. Then, we define ∇A as the limit (if it exists) of the diagram

$$(\dagger A)^{\otimes 0} \xleftarrow{p_0^A} (\dagger A)^{\otimes 1} \xleftarrow{p_1^A} (\dagger A)^{\otimes 2} \xleftarrow{p_2^A} \cdots$$

At this point, if we suppose that the above limit commutes with the tensor, *i.e.*, that $\nabla A \otimes \nabla A$ is the limit of the diagram

$$(\dagger A)^{\otimes 0} \otimes (\dagger A)^{\otimes 0} \xleftarrow{p_0^A \otimes p_0^A} (\dagger A)^{\otimes 1} \otimes (\dagger A)^{\otimes 1} \xleftarrow{p_1^A \otimes p_1^A} (\dagger A)^{\otimes 2} \otimes (\dagger A)^{\otimes 2} \xleftarrow{p_2^A \otimes p_2^A} \cdots,$$

then it is not hard to see that ∇A is also a cone for the second diagram, and that $\nabla A \otimes \nabla A$ is a cone for the first. Therefore, we have two canonical morphisms $\varphi : \nabla A \longrightarrow \nabla A \otimes \nabla A$ and $\psi : \nabla A \otimes \nabla A \longrightarrow \nabla A$. Using these and the symmetry of $\mathcal{C}$, we build infinitely many endomorphisms of ∇A, of the form $\nabla A \longrightarrow (\nabla A)^{\otimes n} \longrightarrow (\nabla A)^{\otimes n} \longrightarrow \nabla A$. We define $!A$ to be the equalizer (if it exists) of all these endomorphisms.

If we apply this construction to the category of finiteness spaces, $!A$ actually turns out to be the free commutative comonoid on A. Whether this is this just a coincidence or whether a suitable rephrasing of Proposition 2 holds is currently unknown and is doubtlessly an interesting topic of further research.

6 Complexity-Bounded Calculi

We add purely linear terms to our syntax, *i.e.*, we consider a denumerably infinite set of linear variables, disjoint from the set of usual variables and ranged over by $a, b, c, \ldots$, and we modify the grammar defining $\Lambda_{\mathrm{p}}^{\mathrm{aff}}$ as follows:

$$t, u ::= \bot \mid x_i \mid \lambda x.t \mid t\mathbf{u} \mid a \mid \ell a.t \mid tu.$$

Furthermore, we require that:

- occurrences of variables (x_i) and linear variables (a) both appear at most once in terms;
- in $\ell a.t$, which is a linear abstraction, the variable a must appear free in t;
- in $t\mathbf{u}$, no $\mathbf{u}(i)$ contains free linear variables, for $i \in \mathbb{N}$.

We denote by $\ell\Lambda_{\mathrm{p}}^{\mathrm{aff}}$ the set of terms thus obtained.

Proof-theoretically, this calculus corresponds to allowing simple linear implication in the fragment of multiplicative additive linear logic we consider:

$$A, B ::= X \mid A \multimap B \mid (A_1 \,\&\, 1) \otimes \cdots \otimes (A_n \,\&\, 1) \multimap B.$$

Reduction in $\ell\Lambda_{\mathrm{p}}^{\mathrm{aff}}$ is defined by adding a purely linear β-reduction rule besides the one already present in $\Lambda_{\mathrm{p}}^{\mathrm{aff}}$:

$$(\ell a.t)u \to t[u/a],$$
$$(\lambda x.t)\mathbf{u} \to t[\mathbf{u}/x].$$

Note that the absence of types produces "clashes", *i.e.*, terms of the form $(\ell a.t)\mathbf{u}$ or $(\lambda x.t)u$, which look like redexes (especially the latter...) but are not reduced. This is unproblematic for our purposes.

The ultrametric d on $\ell\Lambda_{\mathrm{p}}^{\mathrm{aff}}$ is defined just as in Sect. 3 for the inductive cases already present in $\Lambda_{\mathrm{p}}^{\mathrm{aff}}$, and is trivially extended to the other cases:

$$d(a,t') = \begin{cases} 0 & \text{if } t' = a \\ 1 & \text{otherwise} \end{cases}$$

$$d(\ell a.t_1, t') = \begin{cases} d(t_1, t_1') & \text{if } t' = \ell a.t_1' \\ 1 & \text{otherwise} \end{cases}$$

$$d(t_1 u, t') = \begin{cases} \max\left(d(t_1, t_1'), d(u, u')\right) & \text{if } t' = t_1' u' \\ 1 & \text{otherwise} \end{cases}$$

We denote by $\ell\Lambda_{\infty}^{\mathrm{aff}}$ the completion of $\ell\Lambda_{\mathrm{p}}^{\mathrm{aff}}$ with respect to d.

The partial equivalence relation $\approx$ is extended to $\ell\Lambda_{\infty}^{\mathrm{aff}}$ in the obvious way: $a \approx a$ for every linear variable a; if $t \approx t'$, then $\ell a.t \approx \ell a.t'$; if $t \approx t'$ and $u \approx u'$, then $tu \approx t'u'$. Hence, a term $t \in \ell\Lambda_{\infty}^{\mathrm{aff}}$ is uniform if $t \approx t$. Infinitary reduction is also extended to the uniform terms of $\ell\Lambda_{\infty}^{\mathrm{aff}}$ in the obvious way (the index of $\Rightarrow_k$ does not increase when reducing inside the argument of a linear application).

Of course, $\ell\Lambda_{\infty}^{\mathrm{aff}}$ brings nothing really new with respect to $\Lambda_{\infty}^{\mathrm{aff}}$. In particular, if we are only interested in the λ-calculus, purely linear terms are useless. They become interesting when we restrict the space of finite terms, *i.e.*, the approximations we are allowed to use.

Definition 3 (Depth, stratified term). *The depth of a free occurrence of variable x_i in a term $t \in \ell\Lambda_{\mathrm{p}}^{\mathrm{aff}}$, denoted by $\delta_{x_i}(t)$, is defined by induction on t:*

- $\delta_{x_i}(x_i) = 0$;
- $\delta_{x_i}(\lambda y.t_1) = \delta_{x_i}(\ell a.t_1) = \delta_{x_i}(t_1)$;
- *if $t = t_1\mathbf{u}$, then x_i is free in $\mathbf{u}(p)$ for some $p \in \mathbb{N}$, and we set $\delta_{x_i}(t) = \delta_{x_i}(\mathbf{u}(p)) + 1$;*
- *similarly, if $t = t_1 t_2$, then x_i must be free in t_p for $p \in \{1, 2\}$, and we set $\delta_{x_i}(t) = \delta_{x_i}(t_p)$.*

A term $t \in \ell\Lambda_{\mathrm{p}}^{\mathrm{aff}}$ is stratified *if:*

- *whenever x_i is free in t, $\delta_{x_i}(t) = 1$;*
- *for every subterm of t of the form $\lambda x.u$ and for every $i \in \mathbb{N}$ such that x_i is free in u, $\delta_{x_i}(u) = 1$.*

We denote by $\ell\Lambda_{\mathrm{p}}^{\mathrm{s}}$ the set of all stratified terms.

The definition of stratified term clarifies why we need to consider purely linear terms: in their absence, the only stratified applications would be of the form $\bot\mathbf{u}$, *i.e.*, head variables are excluded, because their depth is always 0.

As a subset of $\ell\Lambda_{\mathrm{p}}^{\mathrm{aff}}$, $\ell\Lambda_{\mathrm{p}}^{\mathrm{s}}$ is also a metric space, with the same ultrametric d. However, $\ell\Lambda_{\mathrm{p}}^{\mathrm{s}}$ is not dense in $\ell\Lambda_{\infty}^{\mathrm{aff}}$. In fact, its completion, which is equal to its topological closure as a subset of $\ell\Lambda_{\infty}^{\mathrm{aff}}$ and which we denote by $\ell\Lambda_{\infty}^{\mathrm{s}}$, is strictly smaller. We may see this by considering the term Δ introduced in Sect. 3. In order for any $t \in \ell\Lambda_{\mathrm{p}}^{\mathrm{aff}}$ to be such that $d(t, \Delta) < 1$, we must have $t = \lambda x.x_0\mathbf{u}$, which is not stratified. Hence, no sequence in $\ell\Lambda_{\mathrm{p}}^{\mathrm{s}}$ ever tends to Δ, and this term is not present in $\ell\Lambda_{\infty}^{\mathrm{s}}$. Similarly, $\Omega \notin \ell\Lambda_{\infty}^{\mathrm{s}}$.

The above example is interesting because it excludes the most obvious source of divergence in $\ell\Lambda_{\infty}^{\mathrm{aff}}$. In fact, $\ell\Lambda_{\infty}^{\mathrm{s}}/\approx$ is actually an elementary λ-calculus, in the same sense as that of [7]. When suitably typed in a system/logic containing a type $\mathbf{N}$ corresponding to natural numbers, the terms of type $\mathbf{N} \to \mathbf{N}$ represent exactly the elementary functions, which are those computable by a Turing machine in time bounded by a tower of exponentials of fixed height.

We believe that a polytime λ-calculus may be obtained by *considering another metric* on $\ell\Lambda_{\mathrm{p}}^{\mathrm{s}}$. That is, the approximations are the same, but they do not have the same meaning. To give an analogy (which is purely suggestive, not technical), we may consider the standard sequence spaces used in analysis. The set c_{00} of infinite sequences of real numbers which are almost everywhere null (hence virtually finite) may be endowed with many different metrics, according to which the completion only contains sequences which tend to 0. However, the rate at which they are allowed to vanish is different: any rate (c_0), strictly more than the linear inverse (ℓ^1), strictly more than the inverse square ($\ell^{\frac{1}{2}}$)...

At the moment, we have a metric such that, when we complete $\ell\Lambda_{\mathrm{p}}^{\mathrm{s}}$ with respect to it and consider uniform terms, we seem to obtain a space of terms roughly corresponding to a poly-time λ-calculus such as the one of [25]. Although we have no precise results yet, this research direction looks promising and is definitely worth further investigation. In particular, thanks to non-uniform terms, this might lead to a λ-calculus characterization of the class $\mathbf{P}/_{\mathrm{poly}}$.

Acknowledgments. This summary is mostly based on [20] and the journal version [21] (under review), which benefited from the partial support of ANR projects COMPLICE (08-BLAN-0211-01), PANDA (09-BLAN-0169-02) and LOGOI (10-BLAN-0213-02).

References

1. Abramsky, S., Jagadeesan, R., Malacaria, P.: Full abstraction for PCF. Inform. Comput. 163(2), 409–470 (2000)
2. Accattoli, B., Kesner, D.: Preservation of strong normalisation modulo permutations for the structural lambda-calculus. Logical Methods in Computer Science 8(1) (2012)
3. Asperti, A., Guerrini, S.: The Optimal Implementation of Functional Programming Languages. Cambridge University Press (1998)
4. Baillot, P., Terui, K.: Light types for polynomial time computation in lambda calculus. Inf. Comput. 207(1), 41–62 (2009)
5. de Bruijn, N.G.: Lambda calculus notation with nameless dummies. Indagat. Math. 34, 381–392 (1972)
6. Caires, L., Pfenning, F.: Session types as intuitionistic linear propositions. In: Gastin, P., Laroussinie, F. (eds.) CONCUR 2010. LNCS, vol. 6269, pp. 222–236. Springer, Heidelberg (2010)

7. Coppola, P., Martini, S.: Typing lambda terms in elementary logic with linear constraints. In: Abramsky, S. (ed.) TLCA 2001. LNCS, vol. 2044, p. 76. Springer, Heidelberg (2001)
8. Curien, P.L., Herbelin, H., Krivine, J.L., Melliès, P.A.: Interactive Models of Computation and Program Behavior. AMS (2010)
9. Curien, P.-L., Munch-Maccagnoni, G.: The duality of computation under focus. In: Calude, C.S., Sassone, V. (eds.) TCS 2010. IFIP AICT, vol. 323, pp. 165–181. Springer, Heidelberg (2010)
10. Danos, V., Ehrhard, T.: Probabilistic coherence spaces as a model of higher-order probabilistic computation. Inf. Comput. 209(6), 966–991 (2011)
11. Ehrhard, T.: Finiteness spaces. Mathematical Structures in Computer Science 15(4), 615–646 (2005)
12. Girard, J.Y.: Linear logic. Theor. Comput. Sci. 50(1), 1–102 (1987)
13. Girard, J.Y.: Locus solum. Math. Struct. Comput. Sci. 11(3), 301–506 (2001)
14. Hofmann, M.: Linear Types and Non-Size-Increasing Polynomial Time Computation. Inform. Comput. 183(1), 57–85 (2003)
15. Honda, K., Laurent, O.: An exact correspondence between a typed pi-calculus and polarised proof-nets. Theor. Comput. Sci. 411(22-24), 2223–2238 (2010)
16. Kennaway, R., Klop, J.W., Sleep, R., de Vries, F.J.: Infinitary lambda calculus. Theor. Comput. Sci. 175(1), 93–125 (1997)
17. Lago, U.D., Gaboardi, M.: Linear dependent types and relative completeness. Logical Methods in Computer Science 8(4) (2011)
18. Laurent, O., Regnier, L.: About translations of classical logic into polarized linear logic. In: Proceedings of LICS, pp. 11–20. IEEE Computer Society (2003)
19. Mackie, I.: Efficient lambda-evaluation with interaction nets. In: van Oostrom, V. (ed.) RTA 2004. LNCS, vol. 3091, pp. 155–169. Springer, Heidelberg (2004)
20. Mazza, D.: An infinitary affine lambda-calculus isomorphic to the full lambda-calculus. In: Dershowitz, N. (ed.) Proceedings of LICS, pp. 471–480. IEEE Computer Society (2012)
21. Mazza, D.: Non-linearity as the metric completion of linearity (submitted, 2013), `http://lipn.univ-paris13.fr/~mazza/?page=pub`
22. Melliès, P.A., Tabareau, N.: Resource modalities in tensor logic. Ann. Pure Appl. Logic 161(5), 632–653 (2010)
23. Melliès, P.-A., Tabareau, N., Tasson, C.: An explicit formula for the free exponential modality of linear logic. In: Albers, S., Marchetti-Spaccamela, A., Matias, Y., Nikoletseas, S., Thomas, W. (eds.) ICALP 2009, Part II. LNCS, vol. 5556, pp. 247–260. Springer, Heidelberg (2009)
24. Selinger, P., Valiron, B.: A lambda calculus for quantum computation with classical control. Mathematical Structures in Computer Science 16(3), 527–552 (2006)
25. Terui, K.: Light affine calculus and polytime strong normalization. In: Proceedings of LICS, pp. 209–220. IEEE Computer Society (2001)

System F_i

A Higher-Order Polymorphic λ-Calculus with Erasable Term-Indices

Ki Yung Ahn[1], Tim Sheard[1], Marcelo Fiore[2], and Andrew M. Pitts[2]

[1] Portland State University, Portland, Oregon, USA*
{kya,sheard}@cs.pdx.edu
[2] University of Cambridge, Cambridge, UK
{Marcelo.Fiore,Andrew.Pitts}@cl.cam.ac.uk

Abstract. We introduce a foundational lambda calculus, System F_i, for studying programming languages with term-indexed datatypes – higher-kinded datatypes whose indices range over data such as natural numbers or lists. System F_i is an extension of System F_ω that introduces the minimal features needed to support term-indexing. We show that System F_i provides a theory for analysing programs with term-indexed types and also argue that it constitutes a basis for the design of logically-sound light-weight dependent programming languages. We establish erasure properties of F_i-types that capture the idea that term-indices are discardable in that they are irrelevant for computation. Index erasure projects typing in System F_i to typing in System F_ω. So, System F_i inherits strong normalization and logical consistency from System F_ω.

Keywords: term-indexed data types, generalized algebraic data types, higher-order polymorphism, type-constructor polymorphism, higher-kinded types, impredicative encoding, strong normalization, logical consistency.

1 Introduction

We are interested in the use of indexed types to state and maintain program properties. A type parameter (like `Int` in `(List Int)`) usually tells us something about data stored in values of that type. A type-index (like `3` in `(Vector Int 3)`) states an inductive property of values with that type. For example, values of type `(Vector Int 3)` have three elements.

Indexed types come in two flavors: *type-indexed* and *term-indexed* types.

An example of type-indexing is a definition of a *representation type* [8] using GADTs in Haskell:

```
data TypeRep t where
  RepInt  :: TypeRep Int
  RepBool :: TypeRep Bool
  RepPair :: TypeRep a -> TypeRep b -> TypeRep (a,b)
```

* Supported by NSF grant 0910500.

M. Hasegawa (Ed.): TLCA 2013, LNCS 7941, pp. 15–30, 2013.

Here, a value of type `(TypeRep t)` is isomorphic in shape with the type-index `t`. For example, `(RepPair RepInt RepBool) :: TypeRep (Int,Bool)`.

An example of *Term-indices* are datatypes with indices ranging over data structures, such as natural numbers (like `Z`, `(S Z)`) or lists (like `Nil` or `(Cons Z Nil)`). A classic example of a term-index is the second parameter to the length-indexed list type `Vec` (as in `(Vec Int (S Z))`).

In languages such as Haskell[1] or OCaml [10], which support GADTs with only type-indexing, term-indices are simulated (or faked) by reflecting data at the type-level with uninhabited type constructors. For example,

```
data S n
data Z
data Vec t n where
  Cons :: a -> Vec a n -> Vec a (S n)
  Nil  :: Vec a Z
```

This simulation comes with a number of problems. First, there is no way to say that types such as `(S Int)` are ill-formed, and second the costs associated with duplicating the constructors of data to be used as term-indices. Nevertheless, GADTs with "faked" term-indices have become extremely popular as a light-weight, type-based mechanism to raise the confidence of users that software systems maintain important properties.

Our approach in this direction is to design a new foundational calculus, System F_i, for functional programming languages with term-indexed datatypes. In a nutshell, System F_i is obtained by minimally extending System F_ω with type-indexed kinds. Notably, this yields a logical calculus that is expressive enough to embed non-dependent *term-indexed datatypes* and their eliminators. Our contributions in this development are as follows.

- Identifying the features that are needed in a higher-order polymorphic λ-calculus to embed term-indexed datatypes (Sect. 2), in isolation from other features normally associated with such calculi (e.g., general recursion, large elimination, dependent types).
- The design of the calculus, System F_i (Sect. 4), and its use to study properties of languages with term-indexed datatypes, including the embedding of term-indexed datatypes into the calculus (Sect. 6) using Church or Mendler style encodings, and proofs about these encodings. For instance, one can use System F_i to prove that the Mendler-style eliminators for GADTs [3] are normalizing.
- Showing that System F_i enjoys a simple erasure property (Sect. 5.2) and inherits meta-theoretic results, strong normalization and logical consistency, from F_ω (Sect. 5.3).

2 Motivation: From System F_ω to System F_i, and Back

It is well known that datatypes can be embedded into polymorphic lambda calculi by means of functional encodings [5].

[1] See Sect. 7 for a very recent GHC extension, which enable true term-indices.

In System F, one can embed *regular datatypes*, like homogeneous lists:

Haskell: `data List a = Cons a (List a) | Nil`
System F: $\texttt{List}\ A \triangleq \forall X.(A \to X \to X) \to X \to X$
$\texttt{Cons} \triangleq \lambda w.\lambda x.\lambda y.\lambda z.\,y\ w\ (x\ y\ z),\ \texttt{Nil} \triangleq \lambda y.\lambda z.z$

In such regular datatypes, constructors have algebraic structure that directly translates into polymorphic operations on abstract types as encapsulated by universal quantification over types (of kind $*$).

In the more expressive System F_ω (where one can abstract over type constructors of any kind), one can encode more general *type-indexed datatypes* that go beyond the regular datatypes. For example, one can embed powerlists with heterogeneous elements in which an element of type `a` is followed by an element of the product type `(a,a)`:

Haskell: `data Powl a = PCons a (Powl(a,a)) | PNil`
`-- PCons 1 (PCons (2,3) (PCons ((3,4),(1,2)) PNil)) :: Powl Int`
System F_ω: $\texttt{Powl} \triangleq \lambda A^{*}.\forall X^{*\to *}.(A \to X(A \times A) \to XA) \to XA \to XA$

Note the non-regular occurrence (`Powl(a,a)`) in the definition of (`Powl a`), and the use of universal quantification over higher-order kinds ($\forall X^{*\to *}$). The term encodings for `PCons` and `PNil` are exactly the same as the term encodings for `Cons` and `Nil`, but have different types.

What about term-indexed datatypes? What extensions to System F_ω are needed to embed term-indices as well as type-indices? Our answer is System F_i.

In a functional language supporting term-indexed datatypes, we envisage that the classic example of homogeneous length-indexed lists would be defined along the following lines (in Nax[2]-like syntax):

```
data Nat  = S Nat | Z
data Vec : * -> Nat -> * where
  VCons  : a -> Vec a {i} -> Vec a {S i}
  VNil   : Vec a {Z}
```

Here the type constructor `Vec` is defined to admit parameterisation by both type and term-indices. For instance, the type (`Vec (List Nat) {S (S Z)}`) is that of two-dimensional vectors of natural numbers. By design, our syntax directly reflects the difference between type and term-indexing by enclosing the latter in curly braces. We also make this distinction in System F_i, where it is useful within the type system to guarantee the static nature of term-indexing.

The encoding of the vector datatype in System F_i is as follows:

$$\texttt{Vec} \triangleq \lambda A^{*}.\lambda i^{\texttt{Nat}}.\forall X^{\texttt{Nat}\to *}.(\forall j^{\texttt{Nat}}.A \to X\{j\} \to X\{\texttt{S}\ j\}) \to X\{\texttt{Z}\} \to X\{i\}$$

where `Nat`, `Z`, and `S` respectively encode the natural number type and its two constructors, zero and successor. Again, the term encodings for `VCons` and `VNil` are exactly the same as the encodings for `Cons` and `Nil`, but have different types.

[2] We are developing a language called Nax whose theory is based on System F_i.

Without going into the details of the formalism, which are given in the next section, one sees that such a calculus incorporating term-indexing structure needs four additional constructs (see Fig. 1 for the highlighted extended syntax).

1. Type-indexed kinding ($A \to \kappa$), as in ($\mathtt{Nat}{\to}{*}$) in the example above, where the compile-time nature of term-indexing will be reflected in the typing rules, enforcing that A be a closed type (rule (Ri) in Fig. 2).
2. Term-index abstraction $\lambda i^A.F$ (as $\lambda i^{\mathtt{Nat}}.\cdots$ in the example above) for constructing (or introducing) term-indexed kinds (rule (λi) in Fig. 2).
3. Term-index application $F\{s\}$ (as $X\{\mathtt{Z}\}$, $X\{j\}$, and $X\{\mathtt{S}\ j\}$ in the example above) for destructing (or eliminating) term-indexed kinds, where the compile-time nature of indexing will be reflected in the typing rules, enforceing that the index be statically typed (rule ($@i$) in Fig. 2) .
4. Term-index polymorphism $\forall i^A.B$ (as $\forall j^{\mathtt{Nat}}.\cdots$ in the example above) where the compile-time nature of polymorphic term-indexing will be reflected in the typing rules enforcing that the variable i be static of closed type A (rule ($\forall Ii$) in Fig. 2).

As described above, System F_i maintains a clear-cut separation between type-indexing and term-indexing. This adds a level of abstraction to System F_ω and yields types that in addition to parametric polymorphism also keep track of inductive invariants using term-indices. All term-index information can be erased, since it is only used at compile-time. It is possible to project any well-typed System F_i term into a well-typed System F_ω term. For instance, the erasure of the F_i-type `Vec` is the F_ω-type `List`. This is established in Sect. 5 and used to deduce the strong normalization of System F_i.

3 Why Term-Indexed Calculi? (Rather Than Dependent Types)

We claim that a moderate extension to the polymorphic calculus (F_ω) is a better candidate than a dependently typed calculus for the basis of a practical programming system. We hope to design a unified system for programming as well as reasoning. Language designs based on indexed types can benefit from existing compiler technology and type inference algorithms for functional programming languages. In addition, theories for term-indexd datatypes are simpler than theories for full-fledged dependent datatypes, because term-indexd datatypes can be encoded as functions (using Church-like encodings).

The implementation technology for functional programming languages based on polymorphic calculi is quite mature. The industrial strength Glasgow Haskell Compiler (GHC), whose intermediate core language is an extension of F_ω, is used by thousands every day. Our term-indexed calculus F_i is closely related to F_ω by an index-erasure property. The hope is that a language implementation based on F_i can benefit from these technologies. We have built a language implementation of these ideas, which we call Nax.

Type inference algorithms for functional programming languages are often based on certain restrictions of the Curry-style polymorphic lambda calculi. These restrictions are designed to avoid higher-order unification during type inference. We have developed a conservative extension of Hindley–Milner type inference for Nax. This was possible because Nax is based on a restricted F_i. Dependently typed languages, on the other hand, are often based on bidirectional type checking, which requires annotations on top level definitions, rather than Hindley–Milner-style type inference.

In dependent type theories, datatypes are usually introduced as primitive constructs (with axioms), rather than as functional encodings (e.g., Church encodings). One can give functional encodings for datatypes in a dependent type theory, but one soon realizes that the induction principles (or, dependent eliminators) for those datatypes cannot be derived within the pure dependent calculi [11]. So, dependently typed reasoning systems support datatypes as primitives. For instance, Coq is based on Calculus of Inductive Constructions, which extends Calculus of Constructions [7] with dependent datatypes and their induction principles.

In contrast, in polymorphic type theories, all imaginable datatypes within the calculi have functional encodings (e.g., Church encodings). For instance, F_ω need not introduce datatypes as primitive constructs, since F_ω can embed all these datatypes, including non-regular recursive datatypes with type indices.

Another reason to use F_i is to extend the application of Mendler-style recursion schemes, which are well-understood in the context of polymorphic lambda calculi like F_ω. Researchers have thought about (though not published)[3] Mendler-style primitive recursion over dependently-typed functions over positive datatypes (i.e., datatypes that have a map), but not for negative (or, mixed-variant) datatypes. In System F_i, we can embed Mendler-style recursion schemes, (just as we embedded them in F_ω) that are also well-defined for negative datatypes.

4 System F_i

System F_i is a higher-order polymorphic lambda calculus designed to extend System F_ω by the inclusion of term-indices. The syntax and rules of System F_i are described in Figs. 1, 2 and 3. The extensions new to System F_i, which are not originally part of System F_ω, are highlighted by grey boxes. Eliding all the grey boxes from Figs. 1, 2 and 3, one obtains a version of System F_ω with Curry-style terms and the typing context separated into two parts (type-level context Δ and term-level context Γ).

We assume readers to be familiar with System F_ω and focus on describing the new constructs of F_i, which appear in grey boxes.

Kinds (Fig. 1). The key extension to F_ω is the addition of term-indexed arrow kinds of the form $A \to \kappa$. This allows type constructors to have terms as indices. The rest of the development of F_i flows naturally from this single extension.

[3] Tarmo Uustalu described this on a whiteboard when we met with him at the University of Cambridge in 2011.

Syntax:	Term Variables		$x, y, z, \ldots, i, j, k, \ldots$
	Type Constructor Variables		$X, Y, Z, \ldots$
	Sort		$\square$
	Kinds	κ	$::= * \mid \kappa \to \kappa \mid A \to \kappa$
	Type Constructors	A, B, F, G	$::= X \mid A \to B \mid \lambda X^\kappa.F \mid F\,G \mid \forall X^\kappa.B$
			$\mid \lambda i^A.F \mid F\,\{s\} \mid \forall i^A.B$
	Terms	r, s, t	$::= x \mid \lambda x.t \mid r\,s$
	Typing Contexts	Δ	$::= \cdot \mid \Delta, X^\kappa \mid \Delta, i^A$
		Γ	$::= \cdot \mid \Gamma, x : A$

Reduction: $\boxed{t \rightsquigarrow t'}$

$$\frac{}{(\lambda x.t)\,s \rightsquigarrow t[s/x]} \qquad \frac{t \rightsquigarrow t'}{\lambda x.t \rightsquigarrow \lambda x.t'} \qquad \frac{r \rightsquigarrow r'}{r\,s \rightsquigarrow r'\,s} \qquad \frac{s \rightsquigarrow s'}{r\,s \rightsquigarrow r\,s'}$$

Fig. 1. Syntax and Reduction rules of F_i

Sorting (Fig. 2). The formation of indexed arrow kinds is governed by the sorting rule (*Ri*). The rule (*Ri*) specifies that an indexed arrow kind $A \to \kappa$ is well-sorted when A has kind $*$ under the empty type-level context $(\cdot)$ and κ is well-sorted. Requiring A to be well-kinded under the empty type-level context avoids dependent kinds (i.e., kinds depending on type-level or value-level bindings). That is, A should be a closed type of kind $*$, which does not contain any free type variables or index variables. For example, $(\mathit{List}\,X \to *)$ is not a well-sorted kind since X appears free, while $((\forall X^*.\,\mathit{List}\,X) \to *)$ is a well-sorted kind.

Typing contexts (Fig. 1). Typing contexts are split into two parts. Type level contexts (Δ) for type-level (static) bindings, and term-level contexts (Γ) for term-level (dynamic) bindings. A new form of index variable binding (i^A) can appear in type-level contexts in addition to the traditional type variable bindings (X^κ). There is only one form of term-level binding $(x : A)$ that appears in term-level contexts. Note, both x and i represent the same syntactic category of "Type Variables". The distinction between x and i is only a convention for the sake of readability.

Well-formed typing contexts (Fig. 2). A type-level context Δ is well-formed if (1) it is either empty, or (2) extended by a type variable binding X^κ whose kind κ is well-sorted under Δ, or (3) extended by an index binding i^A whose type A is well-kinded under the empty type-level context at kind $*$. This restriction is similar to the one that occurs in the sorting rule (*Ri*) for term-indexed arrow kinds (see the paragraph *Sorting*). The consequence of this is that, in typing contexts and in sorts, A must be a closed type (not a type constructor!) without free variables.

Well-formed typing contexts:

$$\boxed{\vdash \Delta} \quad \frac{}{\vdash \cdot} \quad \frac{\vdash \Delta \quad \vdash \kappa : \Box}{\vdash \Delta, X^{\kappa}}\,(X \notin \mathsf{dom}(\Delta)) \quad \frac{\vdash \Delta \quad \cdot \vdash A : *}{\vdash \Delta, i^{A}}\,(i \notin \mathsf{dom}(\Delta))$$

$$\boxed{\Delta \vdash \Gamma} \quad \frac{\vdash \Delta}{\Delta \vdash \cdot} \quad \frac{\Delta \vdash \Gamma \quad \Delta \vdash A : *}{\Delta \vdash \Gamma, x : A}\,(x \notin \mathsf{dom}(\Gamma))$$

Sorting: $\boxed{\vdash \kappa : \Box}$

$$(A)\,\frac{}{\vdash * : \Box} \quad (R)\,\frac{\vdash \kappa : \Box \quad \vdash \kappa' : \Box}{\vdash \kappa \to \kappa' : \Box} \quad (Ri)\,\frac{\cdot \vdash A : * \quad \vdash \kappa : \Box}{\vdash A \to \kappa : \Box}$$

Kinding: $\boxed{\Delta \vdash F : \kappa}$

$$(Var)\,\frac{X^{\kappa} \in \Delta \quad \vdash \Delta}{\Delta \vdash X : \kappa} \quad (\to)\,\frac{\Delta \vdash A : * \quad \Delta \vdash B : *}{\Delta \vdash A \to B : *}$$

$$(\lambda)\,\frac{\vdash \kappa : \Box \quad \Delta, X^{\kappa} \vdash F : \kappa'}{\Delta \vdash \lambda X^{\kappa}.F : \kappa \to \kappa'} \quad (\lambda i)\,\frac{\cdot \vdash A : * \quad \Delta, i^{A} \vdash F : \kappa}{\Delta \vdash \lambda i^{A}.F : A \to \kappa}$$

$$(@)\,\frac{\Delta \vdash F : \kappa \to \kappa' \quad \Delta \vdash G : \kappa}{\Delta \vdash F\,G : \kappa'} \quad (@i)\,\frac{\Delta \vdash F : A \to \kappa \quad \Delta; \cdot \vdash s : A}{\Delta \vdash F\,\{s\} : \kappa}$$

$$(\forall)\,\frac{\vdash \kappa : \Box \quad \Delta, X^{\kappa} \vdash B : *}{\Delta \vdash \forall X^{\kappa}.B : *} \quad (\forall i)\,\frac{\cdot \vdash A : * \quad \Delta, i^{A} \vdash B : *}{\Delta \vdash \forall i^{A}.B : *}$$

$$(Conv)\,\frac{\Delta \vdash A : \kappa \quad \Delta \vdash \kappa = \kappa' : \Box}{\Delta \vdash A : \kappa'}$$

Typing: $\boxed{\Delta; \Gamma \vdash t : A}$

$$(:)\,\frac{(x : A) \in \Gamma \quad \Delta \vdash \Gamma}{\Delta; \Gamma \vdash x : A} \quad (:i)\,\frac{i^{A} \in \Delta \quad \Delta \vdash \Gamma}{\Delta; \Gamma \vdash i : A}$$

$$(\to I)\,\frac{\Delta \vdash A : * \quad \Delta; \Gamma, x : A \vdash t : B}{\Delta; \Gamma \vdash \lambda x.t : A \to B} \quad (\to E)\,\frac{\Delta; \Gamma \vdash r : A \to B \quad \Delta; \Gamma \vdash s : A}{\Delta; \Gamma \vdash r\,s : B}$$

$$(\forall I)\,\frac{\vdash \kappa : \Box \quad \Delta, X^{\kappa}; \Gamma \vdash t : B}{\Delta; \Gamma \vdash t : \forall X^{\kappa}.B}\,(X \notin \mathrm{FV}(\Gamma)) \quad (\forall E)\,\frac{\Delta; \Gamma \vdash t : \forall X^{\kappa}.B \quad \Delta \vdash G : \kappa}{\Delta; \Gamma \vdash t : B[G/X]}$$

$$(\forall Ii)\,\frac{\cdot \vdash A : * \quad \Delta, i^{A}; \Gamma \vdash t : B}{\Delta; \Gamma \vdash t : \forall i^{A}.B}\,\begin{pmatrix} i \notin \mathrm{FV}(t), \\ i \notin \mathrm{FV}(\Gamma) \end{pmatrix} \quad (\forall Ei)\,\frac{\Delta; \Gamma \vdash t : \forall i^{A}.B \quad \Delta; \cdot \vdash s : A}{\Delta; \Gamma \vdash t : B[s/i]}$$

$$(=)\,\frac{\Delta; \Gamma \vdash t : A \quad \Delta \vdash A = B : *}{\Delta; \Gamma \vdash t : B}$$

Fig. 2. Well-formedness, Sorting, Kinding, and Typing rules of F_i

Kind equality: $\boxed{\vdash \kappa = \kappa' : \square}$

$$\frac{\cdot \vdash A = A' : * \qquad \vdash \kappa = \kappa' : \square}{\vdash A \to \kappa = A' \to \kappa' : \square}$$

Type constructor equality: $\boxed{\Delta \vdash F = F' : \kappa}$

$$\frac{\Delta, X^{\kappa} \vdash F : \kappa' \qquad \Delta \vdash G : \kappa}{\Delta \vdash (\lambda X^{\kappa}.F)\, G = F[G/X] : \kappa'} \qquad \frac{\Delta, i^{A} \vdash F : \kappa \qquad \Delta; \cdot \vdash s : A}{\Delta \vdash (\lambda i^{A}.F)\, \{s\} = F[s/i] : \kappa}$$

$$\frac{\Delta \vdash F = F' : A \to \kappa \qquad \Delta; \cdot \vdash s = s' : A}{\Delta \vdash F\, \{s\} = F'\, \{s'\} : \kappa}$$

Term equality: $\boxed{\Delta; \Gamma \vdash t = t' : A}$

$$\frac{\Delta; \Gamma, x : A \vdash t : B \qquad \Delta; \Gamma \vdash s : A}{\Delta; \Gamma \vdash (\lambda x.t)\, s = t[s/x] : B}$$

Fig. 3. Equality rules of F_i (only the key rules are shown)

A term-level context Γ is well-formed under a type-level context Δ when it is either empty or extended by a term variable binding $x : A$ whose type A is well-kinded under Δ.

Type constructors and their kinding rules (Figs. 1 and 2). We extend the type constructor syntax by three constructs, and extend the kinding rules accordingly.

$\lambda i^{A}.F$ is the type-level abstraction over an index (or, index abstraction). Index abstractions introduce indexed arrow kinds by the kinding rule (λi). Note, the use of the new form of context extension, i^{A}, in the kinding rule (λi).

$F\, \{s\}$ is the type-level term-index application. In contrast to the ordinary type-level type-application $(F\, G)$ where the argument (G) is a type (of arbitrary kind). The argument of an term-index application $(F\, \{s\})$ is a term (s). We use the curly bracket notation around an index argument in a type to emphasize the transition from ordinary type to term, and to emphasize that s is a term-index, which is erasable. Index applications eliminate indexed arrow kinds by the kinding rule $(@i)$. Note, we type check the term-index (s) under the current type-level context paired with the empty term-level context $(\Delta; \cdot)$ since we do not want the term-index (s) to depend on any term-level bindings. Otherwise, we would admit value dependencies in types.

$\forall i^{A}.B$ is an index polymorphic type. The formation of indexed polymorphic types is governed by the kinding rule $\forall i$, which is very similar to the formation rule $(\forall)$ for ordinary polymorphic types.

In addition to the rules (λi), $(@i)$, and $(\forall i)$, we need a conversion rule $(Conv)$ at kind level. This is because the new extension to the kind syntax $A \to \kappa$ involves types. Since kind syntax involves types, we need more than simple structural equality over kinds (see Fig. 3). For instance, $A \to \kappa$ and $A' \to \kappa$ equivalent kinds when A' and A are equivalent types. Only the key equality rules are shown in

Fig. 3, and the other structural rules (one for each sorting/kinding/typing rule) and the congruence rules (symmetry, transitivity) are omitted.

Terms and their typing rules (Figs. 1 and 2). The term syntax is exactly the same as other Curry-style calclui. We write x for ordinary term variables introduced by term-level abstractions ($\lambda x.t$). We write i for index variables introduced by index abstractions ($\lambda i^A.F$) and by index polymorphic types ($\forall i^A.B$). As discussed earlier, the distinction between x and i is only for readability.

Since F_i has index polymorphic types ($\forall i^A.B$), we need typing rules for index polymorphism: $(\forall Ii)$ for index generalization and $(\forall Ei)$ for index instantiation. These rules are similar to the type generalization $(\forall I)$ and the type instantiation $(\forall I)$ rules, but involve indices, rather than types, and have additional side conditions compared to their type counterparts.

The additional side condition $i \notin \mathrm{FV}(t)$ in the $(\forall Ii)$ rule prevents terms from accessing the type-level index variables introduced by index polymorphism. Without this side condition, $\forall$-binder would no longer behave polymorphically, but instead would behave as a dependent function binder, which are usually denoted by Π in dependent type theories. Such side conditions on generalization rules for polymorphism are fairly standard in dependent type theories that distinguish between polymorphism (or, erasable arguments) and dependent functions (e.g., IPTS[17], ICC[16]).

The index instantiation rule $(\forall Ei)$ requires that the term-index s, which instantiates i, be well-typed in the current type-level context paired with the empty term-level context $(\Delta;\cdot)$ rather than the current term-level context, since we do not want indices to depend on term-level bindings.

In addition to the rules $(\forall Ii)$ and $(\forall Ei)$ for index polymorphism, we need an additional variable rule $(:i)$ to access index variables already in scope. In examples like $(\lambda i^A.F\{s\})$ and $(\forall i^A.F\{s\})$, the term (s) should be able to access the index variable (i) already in scope.

5 Metatheory

The expectation is that System F_i has all the nice properties of System F_ω, yet is more expressive (i.e., can state finer grained program properties) because of the addition of term-indexed types.

We show some basic well-formedness properties for the judgments of F_i in Sect. 5.1. We prove erasure properties of F_i, which capture the idea that indices are erasable since they are irrelevant for reduction in Sect. 5.2. We show strong normalization, logical consistence, and subject reduction for F_i by reasoning about well-known calculi related to F_i in Sect. 5.3.

5.1 Well-Formedness and Substitution Lemmas

We want to show that kinding and typing derivations give well-formed results under well-formed contexts. That is, kinding derivations $(\Delta \vdash F : \kappa)$ result in well-sorted kinds ($\vdash \kappa$) under well-formed type-level contexts ($\vdash \Delta$) (Proposition 1), and typing derivations ($\Delta;\Gamma \vdash t : A$) result in well-kinded types ($\Delta;\Gamma \vdash A : *$) under well-formed type and term-level contexts (Proposition 2).

Proposition 1. $\dfrac{\vdash \Delta \quad \Delta \vdash F : \kappa}{\vdash \kappa : \Box}$ **Proposition 2.** $\dfrac{\Delta \vdash \Gamma \quad \Delta;\Gamma \vdash t : A}{\Delta \vdash A : *}$

We can prove these well-formedness properties by induction over the judgment[4] and using the substitution lemma below.

Lemma 1 (substitution)

1. (type substitution) $\dfrac{\Delta, X^\kappa \vdash F : \kappa' \quad \Delta \vdash G : \kappa}{\Delta \vdash F[G/X] : \kappa'}$

2. (index substitution) $\dfrac{\Delta, i^A \vdash F : \kappa \quad \Delta; \cdot \vdash s : A}{\Delta \vdash F[s/i] : \kappa}$

3. (term substitution) $\dfrac{\Delta;\Gamma, x : A \vdash t : B \quad \Delta;\Gamma \vdash s : A}{\Delta;\Gamma \vdash t[s/x] : B}$

These substitution lemmas are fairly standard, comparable to substitution lemmas in other well-known systems such as F_ω or ICC.

5.2 Erasure Properties

We define a meta-operation of index erasure that projects F_i-types to F_ω-types.

Definition 1 (index erasure)

$\boxed{\kappa^\circ}$ $\quad *^\circ = * \qquad (\kappa_1 \to \kappa_2)^\circ = {\kappa_1}^\circ \to {\kappa_2}^\circ \qquad (A \to \kappa)^\circ = \kappa^\circ$

$\boxed{F^\circ}$ $\quad X^\circ = X \qquad (A \to B)^\circ = A^\circ \to B^\circ$

$(\lambda X^\kappa.F)^\circ = \lambda X^{\kappa^\circ}.F^\circ \qquad (\lambda i^A.F)^\circ = F^\circ$

$(F\ G)^\circ = F^\circ\ G^\circ \qquad (F\,\{s\})^\circ = F^\circ$

$(\forall X^\kappa.B)^\circ = \forall X^{\kappa^\circ}.B^\circ \qquad (\forall i^A.B)^\circ = B^\circ$

$\boxed{\Delta^\circ}$ $\quad \cdot^\circ = \cdot \qquad (\Delta, X^\kappa)^\circ = \Delta^\circ, X^{\kappa^\circ} \qquad (\Delta, i^A)^\circ = \Delta^\circ$

$\boxed{\Gamma^\circ}$ $\quad \cdot^\circ = \cdot \qquad (\Gamma, x : A)^\circ = \Gamma^\circ, x : A^\circ$

In addition, we define another meta-operation, which selects out all the index variable bindings from the type-level context. We use this in Theorem 6.

[4] The proof for Propositions 1 and 2 are mutually inductive. So, we prove these two propositions at the same time, using a combined judgment J, which is either a kinding judgment or a typing judgment (i.e., $J ::= \Delta \vdash F : \kappa \mid \Delta;\Gamma \vdash t : A$).

Definition 2 (index variable selection)

$$\boxed{\Delta^\bullet} \qquad \cdot^\bullet = \cdot \qquad (\Delta, X^\kappa)^\bullet = \Delta^\bullet \qquad (\Delta, i^A)^\bullet = \Delta^\bullet, i : A$$

Theorem 1 (index erasure on well-sorted kinds). $\dfrac{\vdash \kappa : \Box}{\vdash \kappa^\circ : \Box}$

Proof. By induction on the sort (κ). ∎

Remark 1. For any well-sorted kind κ in F_i, κ° is a well-sorted kind in F_ω.

Theorem 2 (index erasure on well-formed type-level contexts). $\dfrac{\vdash \Delta}{\vdash \Delta^\circ}$

Proof. By induction on the type-level context (Δ) and using Theorem 1. ∎

Remark 2. For any well-formed type-level context Δ in F_i, Δ° is a well-formed type-level context in F_ω.

Theorem 3 (index erasure on kind equality). $\dfrac{\vdash \kappa = \kappa' : \Box}{\vdash \kappa^\circ = \kappa'^\circ : \Box}$

Proof. By induction on the kind equality derivation $(\vdash \kappa = \kappa' : \Box)$. ∎

Remark 3. For any well-sorted kind equality $\vdash \kappa = \kappa' : \Box$ in F_i, κ° and κ'° are the syntactically same F_ω kinds. Note that no variables can appear in the erased kinds by definition of the erasure operation on kinds.

Theorem 4 (index erasure on well-kinded type constructors)

$$\frac{\vdash \Delta \qquad \Delta \vdash F : \kappa}{\Delta^\circ \vdash F^\circ : \kappa^\circ}$$

Proof. By induction on the kinding derivation $(\Delta \vdash F : \kappa)$. We use Theorem 2 in the (Var) case, Theorem 3 in the $(Conv)$ case, and Theorem 1 in the (λ) and $(\forall)$ cases. ∎

Remark 4. In the theorem above, F° is a well-kinded type constructor in F_ω.

Lemma 2. $(F[G/X])^\circ = F^\circ[G^\circ/X]$ **Lemma 3.** $(F[s/i])^\circ = F^\circ$

Theorem 5 (index erasure on type constructor equality)

$$\frac{\Delta \vdash F = F' : \kappa}{\Delta^\circ \vdash F^\circ = F'^\circ : \kappa^\circ}$$

Proof. By induction on the derivation of the type constructor equality judgment $(\Delta \vdash F = F' : \kappa)$. We also use Proposition 1 and Lemmas 2 and 3. ∎

Remark 5. When $\Delta \vdash F = F' : \kappa$ is a valid type constructor equality in F_i, $\Delta^\circ \vdash F^\circ = F'^\circ : \kappa^\circ$ is a valid type constructor equality in F_ω.

Theorem 6 (index erasure on well-formed term-level contexts prepended by index variable selection)

$$\frac{\Delta \vdash \Gamma}{\Delta^\circ \vdash (\Delta^\bullet, \Gamma)^\circ}$$

Proof. By induction on the term-level context (Γ) and using Theorem 4. ∎

Remark 6. We can also show that $\dfrac{\Delta \vdash \Gamma}{\Delta^\circ \vdash \Gamma^\circ}$ and prove Corollary 1 directly.

Theorem 7 (index erasure on well-typed terms). $\dfrac{\Delta \vdash \Gamma \qquad \Delta; \Gamma \vdash t : A}{\Delta^\circ; (\Delta^\bullet, \Gamma)^\circ \vdash t : A^\circ}$

Proof. By induction on the typing derivation ($\Delta; \Gamma \vdash t : A$). We also make use of Theorems 1, 4, 5, and 6. ∎

Remark 7. In the theorem above, t is a well typed term in F_ω as well as in F_i.

Corollary 1 (index erasure on index-free well-typed terms)

$$\frac{\Delta \vdash \Gamma \qquad \Delta; \Gamma \vdash t : A}{\Delta^\circ; \Gamma^\circ \vdash t : A^\circ} \ (\mathsf{dom}(\Delta) \cap \mathrm{FV}(t) = \emptyset)$$

5.3 Strong Normalization and Logical Consistency

Strong normalization is a corollary of the erasure property since we know that System F_ω is strongly normalizing. Index erasure also implies logical consistency. By index erasure, we know that any well-typed term in F_i is a well-typed term in F_ω with its erased type. That is, there are no extra well-typed terms in F_i that are not well-typed in F_ω. By the saturated sets model (as in [1]), we know that the void type ($\forall X^*.X$) in F_ω is uninhabited. Therefore, the void type ($\forall X^*.X$) in F_i is uninhabited since it erases to the same void type in F_ω. Alternatively, logical consistency of F_i can be drawn from ICC. System F_i is a restriction of the *restricted implicit calculus* [15] or ICC$^-$ [4], which are restrictions of ICC [16] known to be logically consistent.

6 Encodings of Term-Indexed Datatypes

Recall that our motivation was a foundational calculus that can encode term-indexed datatypes. In Sect. 2, we gave Church encodings of `List` (a regular datatype), `Powl` (a type-indexed datatype), and `Vec` (a term-indexed datatype). In this section, we discuss a more complex datatype [6] involving nested term-indices, and several encoding schemes that we have seen used in practice – first, encoding indexed datatypes using equality constraints [8, 18] and second, encoding datatypes in the Mendler-style [2, 3].

Nested term-indices: System F_i is able to express datatypes with *nested term-indices* – term-indices which are themselves term-indexed datatypes. Consider the resource-state tracking environment [6] in Nax-like syntax below:

```
data Env  : ({st} -> *) -> {Vec st {n}} -> * where
  Extend  : res {x} -> Env res {xs} -> Env res {VCons x xs}
  Empty   : Env res {VNil}
```

Note that `Env` has a term-index of type `Vec`, which is again indexed by `Nat`. For simplicity,[5] assume that `n` is some fixed constant (e.g., `S(S(S Z))`, i.e., 3). Then, an `Env` tracks 3 independent resources (`res`), each which could be in a different state (`st`). For example, 3 files in different states – one open for reading, the next open for writing, and the third closed. We can encode `Env` in F_i as follows:

$$\mathtt{Env} \triangleq \lambda Y^{\,\mathtt{st}\to *}.\, \lambda i^{(\mathtt{Vec\;st\;n})}.\, \forall X^{(\mathtt{Vec\;st}\,\{\mathtt{n}\})\to *}.$$
$$(\forall j^{\mathtt{st}}.\, \forall k^{(\mathtt{Vec\;st\;n})}.\, Y\{j\} \to X\{k\} \to X\{\mathtt{VCons}\; j\; k\}) \to X\{\mathtt{VNil}\} \to X\{i\}$$

The term encodings for `Extend` and `Empty` are exactly the same as the term encodings for `Cons` and `Nil` of the `List` datatype in Sect. 2.

Encoding indexed datatypes using equality constraints: Systematic encodings of GADTs [8, 18], which are used in practical implementations, typically involve equality constraints and existential quantification. Here, we want to emphasize that such encoding schemes are expressible within System F_i, since it is possible to define equalities and existentials over both types and term-indices in F_i.

It is well known that Leibniz equality over type constructors can be defined within System F_ω as $(\overset{\kappa}{=}) \triangleq \lambda X_1^\kappa.\lambda X_2^\kappa.\forall X^{\kappa\to *}.X X_1 \to X X_2$. Similarly, Leibniz equality over term-indices is defined as $(\overset{A}{=}) \triangleq \lambda i^A.\,\lambda j^A.\,\forall X^{A\to *}.\,X\{i\} \to X\{j\}$ in System F_i. Then, we can encode `Vec` as the sum of its two data constructor types:

$$\mathtt{Vec} \triangleq \lambda A^*.\,\lambda i^{\mathtt{Nat}}.\,\forall X^{\mathtt{Nat}\to *}.\,(\exists j^{\mathtt{Nat}}.\,(\mathtt{S}\; j \overset{\mathtt{Nat}}{=} i) \times A \times X\{j\}) + (\mathtt{Z} \overset{\mathtt{Nat}}{=} i)$$

where $+$ and $\times$ are the usual impredicative encoding of sums and products. We can encode the existential quantification over indices ($\exists$ used in the encoding of `Vec` above) as $\exists i^A.B \triangleq \forall X^*.(\forall i^A.B \to X) \to X$, which is similar to the usual encoding of the existential quantification over types in System F or F_ω.

Compared to the simple Church encoded versions in Sect. 2, the encodings using equality constraints work particularly well with encodings of functions that constrain their domain types by restricting their indices. For instance, the function $\mathtt{safeTail} : \mathtt{Vec}\; a\; \{\mathtt{S}\; n\} \to \mathtt{Vec}\; a\; \{n\}$, which can only be applied to non-empty length indexed lists due the index of the domain type $(\mathtt{S}\; n)$.

[5] Nax supports rank-1 kind-level polymorphism. It would be virtually useless if nested term-indices were only limited to constants rather than polymorphic variables. We strongly believe rank-1 kind polymorphism does not introduce inconsistency, since rank-1 polymorphic systems are essentially equivalent to simply-typed systems by inlining the polymorphic definition with the instantiated arguments in each instantiation site.

The Mendler-style encoding: Recursive type theories that extend higher-order polymorphic lambda calculi typically come with a built-in recursive type operator $\mu_\kappa : (\kappa \to \kappa) \to \kappa$ for each kind κ, which yields recursive types $(\mu_\kappa\, F : \kappa)$ when applied to type constructors of appropriate kind $(F : \kappa \to \kappa)$. For instance, $\texttt{List} \triangleq \lambda Y^*. \mu_*(\lambda X^*.Y \times X + \mathbb{1})$ where $\mathbb{1}$ is the unit type. One benefit of factoring out the recursion at type-level (e.g., μ_*) from the base structure (e.g., $\lambda X^*.Y \times X + \mathbb{1}$) of recursive types is that such factorized (or, two-level) representations are more amenable to express generic recursion schemes (e.g., catamorphism) that work over different recursive datatypes. Interestingly, there exists an encoding scheme, namely the Mendler style, which can embed μ_κ within Systems like F_ω or F_i. The Mendler-style encoding can keep the theoretical basis small, while enjoying the benefits of factoring out the recursion at type-level.

7 Related Work

System F_i is most closely related to Curry-style System F_ω [2, 12] and the Implicit Calculus of Constructions (ICC) [16]. All terms typable in a Curry-style System F_ω are typable (with the same type) in System F_i and all terms typable in F_i are typable (with the same type[6]) in ICC.

As mentioned in Sect. 5.3, we can derive strong normalization of F_i from System F_ω, and derive logical consistency of F_i from certain restrictions of ICC [4, 15]. In fact, ICC is more than just an extension of System F_i with dependent types and stratified universes, since ICC includes η-reduction and η-equivalence. We do not foresee any problems adding η-reduction and η-equivalence to System F_i. Although System F_i accepts fewer terms than ICC, it enjoys simpler erasure properties (Theorem 7 and Corollary 1) just by looking at the syntax of kinds and types, which ICC cannot enjoy due to its support for full dependent types. In System F_i, term-indices appearing in types (e.g., s in $F\{s\}$) are always erasable. Mishra-Linger and Sheard [17] generalized the ICC framework to one which describes erasure on arbitrary Church-style calculi (EPTS) and Curry-style calculi (IPTS), but only consider β-equivalence for type conversion.

In the practical setting of programming language implementation, Yorgey et al. [19], inspired by McBride [14], recently designed an extension to Haskell's GADTs by allowing datatypes to be used as kinds. For instance, `Bool` is promoted to a kind (i.e., $\texttt{Bool} : \Box$) and its data constructors `True` and `False` are promoted to types. They extended System F_C (the Glasgow Haskell Compiler's intermediate core language) to support *datatype promotion* and named it System $F_C^{\uparrow}$. The key difference between $F_C^{\uparrow}$ and F_i is in their kind syntax:

$$\begin{array}{ll} F_C^{\uparrow}\ \textbf{kinds} & \kappa ::= * \mid \kappa \to \kappa \mid F\boldsymbol{\kappa} \mid \mathcal{X} \mid \forall \mathcal{X}.\kappa \mid \cdots \\ \mathsf{F}_i\ \textbf{kinds} & \kappa ::= * \mid \kappa \to \kappa \mid A \to \kappa \end{array}$$

In $F_C^{\uparrow}$, all type constructors (F) are promotable to the kind level and become kinds when fully applied to other kinds ($F\boldsymbol{\kappa}$). On the other hand, in F_i, a type

[6] The $*$ kind in F_ω and F_i corresponds to Set in ICC.

can only appear as the domain of an index arrow kind ($A \to \kappa$). The ramifications of this difference is that $F_C^{\uparrow}$ can express type-level data structures but not nested term-indices, while F_i supports the converse. Intuitively, a type constructor like $\texttt{List} : * \to *$ is promoted to a kind constructor $\texttt{List} : \Box \to \Box$, which enables type-level data structures such as $[\texttt{Nat}, \texttt{Bool}, \texttt{Nat} \to \texttt{Bool}] : \texttt{List}\,*$. Type-level data structures motivate type-level computations over promoted data. This is made possible by type families[7]. The promotion of polymorphic types naturally motivates kind polymorphism ($\forall \mathcal{X}.\kappa$). Kind polymorphism of arbitrary rank is known to break strong normalization and cause logical inconsistency [13]. In a *programming language*, inconsistency is not an issue. However, when studying logically consistent systems, we need a more conservative approach, as in F_i.

8 Summary and Ongoing Work

System F_i is a strongly-normalizing, logically-consistent, higher-order polymorphic lambda calculus that was designed to support the definition of datatypes indexed by both terms and types. In terms of expressivity, System F_i sits between System F_ω and ICC. We designed System F_i as a tool to reason about programming languages with term-indexed datatypes. System F_i can express a large class of term-indexed datatypes, including datatypes with nested term-indices.

One limitation of System F_i is that it cannot express type-level data structures such as lists that contain type elements. We hope to overcome this limitation by extending F_i with first-class type representations [9], which reflect types as term-level data (a sort of a fully reflective version of `TypeRep` from Sect. 1).

Our goal is to build a unified programming and reasoning system, which supports (1) an expressive class of datatypes including nested term-indexed datatypes and negative datatypes, (2) logically consistent reasoning about program properties, and (3) Hindley–Milner-style type inference. Towards this goal, we are developing the programming language Nax based on System F_i. Nax is given semantics in terms of System F_i. That is, all the primitive language constructs of Nax that are not present in F_i have translations into System F_i. Such constructs include Mendler-style eliminators, recursive type operators, and pattern matching.

Some language features we want to include in Nax go beyond F_i. One of them is a recursion scheme that guarantee normalization due to paradigmatic use of indices in datatypes. For instance, some recursive computations always reduce a natural number term-index in every recursive call. Although such computations obviously terminate, we cannot express them in System F_i, since term-indices in them are erasable – F_i only accepts terms that are already type-correct in F_ω. We plan to explore extensions to System F_i that enable such computations while maintaining logical consistency.

[7] A GHC extension to define type-level functions in Haskell.

References

1. Abel, A., Matthes, R.: Fixed points of type constructors and primitive recursion. In: Marcinkowski, J., Tarlecki, A. (eds.) CSL 2004. LNCS, vol. 3210, pp. 190–204. Springer, Heidelberg (2004)
2. Abel, A., Matthes, R., Uustalu, T.: Iteration and coiteration schemes for higher-order and nested datatypes. TCS 333(1-2), 3–66 (2005)
3. Ahn, K.Y., Sheard, T.: A hierarchy of Mendler-style recursion combinators: Taming inductive datatypes with negative occurrences. In: ICFP 2011, pp. 234–246. ACM (2011)
4. Barras, B., Bernardo, B.: The implicit calculus of constructions as a programming language with dependent types. In: Amadio, R.M. (ed.) FOSSACS 2008. LNCS, vol. 4962, pp. 365–379. Springer, Heidelberg (2008)
5. Böhm, C., Berarducci, A.: Automatic synthesis of typed lambda-programs on term algebras. TCS 39, 135–154 (1985)
6. Brady, E., Hammond, K.: Correct-by-construction concurrency: Using dependent types to verify implementations of effectful resource usage protocols. Fundam. Inform. 102(2), 145–176 (2010)
7. Coquand, T., Huet, G.: The calculus of constructions. Rapport de Recherche 530, INRIA, Rocquencourt, France (May 1986)
8. Crary, K., Weirich, S., Morrisett, G.: Intensional polymorphism in type-erasure semantics. In: ICFP 1998, pp. 301–312. ACM (1998)
9. Dagand, P.E., McBride, C.: Transporting functions across ornaments. In: ICFP 1998, ICFP 2012, pp. 103–114. ACM (2012)
10. Garrigue, J., Normand, J.L.: Adding GADTs to OCaml: the direct approach. In: ML 2011. ACM (2011)
11. Geuvers, H.: Induction is not derivable in second order dependent type theory. In: Abramsky, S. (ed.) TLCA 2001. LNCS, vol. 2044, pp. 166–181. Springer, Heidelberg (2001)
12. Giannini, P., Honsell, F., Rocca, S.R.D.: Type inference: Some results, some problems. Fundam. Inform. 19(1/2), 87–125 (1993)
13. Girard, J.-Y.: Interprétation Fonctionnelle et Élimination des Coupures de l'Arithmétique d'Ordre Supérieur. Thèse de doctorat d'état, Université Paris VII (June 1972)
14. McBride, C.: Homepage of the Strathclyde Haskell Enhancement (SHE) (2009), `http://personal.cis.strath.ac.uk/conor/pub/she/`
15. Miquel, A.: A model for impredicative type systems, universes, intersection types and subtyping. In: LICS, pp. 18–29. IEEE Computer Society (2000)
16. Miquel, A.: The implicit calculus of constructions. In: Abramsky, S. (ed.) TLCA 2001. LNCS, vol. 2044, pp. 344–359. Springer, Heidelberg (2001)
17. Mishra-Linger, N., Sheard, T.: Erasure and polymorphism in pure type systems. In: Amadio, R.M. (ed.) FOSSACS 2008. LNCS, vol. 4962, pp. 350–364. Springer, Heidelberg (2008)
18. Sheard, T., Pašalić, E.: Meta-programming with built-in type equality. In: LFM 2004, pp. 106–124 (2004)
19. Yorgey, B.A., Weirich, S., Cretin, J., Jones, S.L.P., Vytiniotis, D., Magalhães, J.P.: Giving Haskell a promotion. In: TLDI, pp. 53–66. ACM (2012)

Non-determinism, Non-termination and the Strong Normalization of System T

Federico Aschieri[1] and Margherita Zorzi[2]

[1] Équipe Plume, LIP (UMR 5668), École Normale Supérieure de Lyon, France
[2] Università degli Studi di Verona, Italy

Abstract. We consider a de'Liguoro-Piperno-style extension of the pure lambda calculus with a non-deterministic choice operator as well as a non-deterministic iterator construct, with the aim of studying its normalization properties. We provide a simple characterization of non-strongly normalizable terms by means of the so called "zoom-in" perpetual reduction strategy. We then show that this characterization implies the strong normalization of the simply typed version of the calculus. As straightforward corollary of these results we obtain a new proof of strong normalization of Gödel's System T by a simple translation of this latter system into the former.

1 Introduction

The idea of defining the concept of redundancy or *detour* in an arithmetical proof [17] and the result that shows the possibility of eliminating all the detours in any proof, are real cornerstones of modern logic. Such results, known under the name of *normalization* or strong normalization, are interesting for a great deal of reasons. For example:

- Via the Curry-Howard correspondence, they can be translated as proofs of the termination of programs written in typed system like Gödel's T [6], Spector's B [19] or Girard's F [6]. Indeed, this is now the standard way of presenting normalization results.
- They are tools for proving consistency of logical systems and thus give rise, in the classical case, to Tarski models (see for example [12]).
- Many intuitionistic ([10]) and classical realizabilities ([12]), as well as functional interpretations [19], are built on ideas coming from normalization techniques.

Unfortunately, proving normalization properties of strong logical systems is difficult and when one succeeds, the resulting proof is often of little combinatorial information. This is of course due to the famous Gödel incompleteness theorems, which force normalization proofs to employ powerful mathematical methods.

In the case of the strong normalization of Gödel's System T, the most flexible and elegant proof is due to Tait (see [6]), which uses the abstract concept of reducibility. In our opinion, there are at least two reasons why the proof is not

M. Hasegawa (Ed.): TLCA 2013, LNCS 7941, pp. 31–47, 2013.

very intuitive from the combinatorial point of view. First, the predicates of reducibility are defined by formulas with arbitrarily many nested quantifiers. This is a strength, because, to put it as Girard [6], "the deep reason why reducibility works where combinatorial intuition fails, is its logical complexity" . However, this complexity also hampers a concrete understanding of the normalization process and, in fact, condemns combinatorial intuition to failure. Secondly, reducibility is in reality an instance of much more general techniques that can be used for proving a variety of results (for example, weak Church-Rosser property) in an elegant way. We are of course referring to logical relations and realizability. This is evident in Krivine's work [11], where realizability is carried out as a generalization of the Tait-Girard methods. Thus reducibility appears not to be tailored specifically for normalization problems (this observation can also be addressed to Sanchis' technique [18], which allows to reason in a well-founded way about terms of System T, and can be exploited to prove strong normalization).

Among other known normalization techniques one finds the one using infinite terms of Tait [22], more interesting combinatorially, but not suitable to prove strong normalization, the one of Gandy [8], the one of Joachimiski-Matthes [14], similar in spirit to that of Sanchis, and the one using ordinal analysis of Howard [7].

In this paper, we return to the problem of the strong normalization of Gödel's T, with the aim of better understanding its combinatorial structure. That is, we want to provide a *concrete* normalization proof instead of an *abstract* one. In particular, we show how strong normalization can be derived by just examining the terms produced by a simple reduction strategy. For this purpose, we start from some combinatorial ideas, due essentially to Van Daalen [20] and Levy [13] (but also present in Nederpelt [16]) and extended later by David and Nour to various systems of simple types [4]. These ideas inspired Melliès [15] to define a perpetual reduction strategy, so called zoom-in (discovered independently by Plaisted, Sorensen and Gramlich, see [9] for more details), which will be the heart of our method. In [1] (see also [2]), the zoom-in strategy has been employed to characterize non-strongly normalizable lambda terms, and derive as corollary the strong normalization of the simply typed lambda calculus and the intersection types. These latter results were obtained also by Melliès and David. The novelty in [1] consisted in the explicit statement of a characterization theorem. If with $t_1 \ldots t_2$ we denote any term of the form $(((u)u_1)\ldots)u_n$, where $u = t_1$ and $u_n = t_2$, then it is proved that:

Theorem 1 (Characterization of non-strongly normalizable terms). *Let u be an non-strongly normalizable lambda term. Then there exists an infinite reduction $u_1, u_2, \ldots, u_n, \ldots$ and an infinite sequence of terms $t_1, t_2, \ldots, t_n, \ldots$ such that $u_1 = u$ and for every i, u_i contains a subterm of the form $t_i \ldots t_{i+1}$.*

(for a more detailed formulation, see section §2). Notice how the strong normalization easily follows from the Characterization Theorem: non-strongly normalizable Church-typed lambda terms cannot exist, otherwise the type of each t_i would strictly contain the type of t_{i+1}. Remark also how each one of these terms t_i dynamically passes from the status of argument to the status of function applied to some other arguments: this is the crucial property of the reduction.

One natural question was then whether the Characterization Theorem could be extended to a pure lambda calculus with pairs and constants, containing at least booleans, numerals, the if-then-else if and iterator It constructs. In such a way, one would also obtain as corollary the strong normalization of its typed version, i.e. System T. Unfortunately, the Characterization Theorem does not extend so easily. The reason is that one passes from a pure, functional world – the lambda calculus – to an impure world in which booleans and numerals are treated as basic objects, but also retain a sort of functional behavior.

For example, one has a rule if $\texttt{False}\, u\, v \mapsto v$. But what would be the difference with a hypothetical reduction $\texttt{False}\, u\, v \mapsto v$, in which False would behave as the encoding of false in lambda calculus $\lambda x \lambda y.\, y$? Syntactically, if is treated just as a placeholder, being the boolean False the one which comes makes the real job. Similarly, one has a rule $\mathsf{It}\, u\, v\, \overline{2} \mapsto (v)(v)u$. But what would be the difference with a hypothetical reduction $\overline{2}uv \mapsto (v)(v)u$, in which $\overline{2}$ would behave as the Church-numeral two $\lambda f \lambda x.\, (f)(f)x$?

As a consequence of the use of objects as "hidden" functionals, one loses the Characterization Theorem: when one of the t_i above is, say, False or $\overline{2}$, we cannot expect it to pass from argument to head position in any meaningful way. The solution to this issue is radical: remove reduction rules involving booleans and numerals and simulate them with actual functionals. The idea is to use *non-determinism.* As in de' Liguoro and Piperno [5], we add to lambda-calculus a non-deterministic choice operator $\mathsf{if}^\star$, with rules $\mathsf{if}^\star\, u\, v \mapsto u$ and $\mathsf{if}^\star\, u\, v \mapsto v$, in order to simulate all possible if reductions. We also add a non-deterministic iterator operator $\mathsf{It}^\star$, with rules of the form $\mathsf{It}^\star\, u\, v \mapsto (v)\dots(v)u$ (one for each possible number of occurrences of v), in order to simulate all possible It reductions. We obtain as a result a non-deterministic lambda calculus $\Lambda^\star$ which enjoys the Characterization Theorem; its typed version $\mathsf{T}^\star$ will thus have the strong normalization property. We shall then prove strong normalization of System T by translating it into $\mathsf{T}^\star$ – almost trivially. It will be enough to substitute the normal versions of if and It with their non-deterministic counterparts $\mathsf{if}^\star$ and $\mathsf{It}^\star$.

Plan of the Paper. In Section §2 we introduce the non-deterministic lambda calculus $\Lambda^\star$ and prove the Characterization Theorem of its non-strongly normalizable terms. In Section §3, as a corollary, we prove the strong normalization of the non-deterministic typed system $\mathsf{T}^\star$. Section §4 is finally devoted to the proof of the strong normalization of Gödel's System T, by translation into $\mathsf{T}^\star$.

2 The Non-deterministic Lambda Calculus $\Lambda^\star$ with Pairs and Constants

In this section we define and study the non-deterministic lambda calculus $\Lambda^\star$, whose typed version will serve in section §4 to interpret Gödel's System T. In particular, we are going to give a syntactical characterization of the non-strongly normalizable terms of $\Lambda^\star$.

The non-deterministic lambda calculus $\Lambda^\star$ is formally described in Figure 1. Its deterministic part is a standard lambda calculus (for which we refer to [11]) augmented with pairs, projections, and some arbitrary set of constants $\mathsf{c}_0, \mathsf{c}_1, \ldots$ without any associated reduction rule. In this latter set, one will typically put 0, S, `True`, `False`, but no assumption will be made in this section. The non-deterministic part of $\Lambda^\star$ comprises as constants the non-deterministic choice operator $\mathsf{if}^\star$, as in de' Liguoro-Piperno [5], Dal Lago-Zorzi [3], and the non-deterministic iterator $\mathsf{It}^\star$. For $\mathsf{It}^\star\, u\, v$ one has denumerably many possible reductions:

$$\mathsf{It}^\star\, u\, v \mapsto u, \quad \mathsf{It}^\star\, u\, v \mapsto (v)u, \quad \mathsf{It}^\star\, u\, v \mapsto (v)(v)u, \quad \mathsf{It}^\star\, u\, v \mapsto (v)(v)(v)u\ \ldots$$

We point out that, as remarked in [5], $\mathsf{It}^\star$ can already be defined by $\mathsf{if}^\star$. But that is no longer possible in a typed setting, and so we had to leave $\mathsf{It}^\star$ in the syntax.

We now recall some very basic facts and definitions. We retain the Krivine parenthesis convention for pure lambda calculus and extend it to $\Lambda^\star$. The term $(t)u$ will be written as tu and $[u]\pi_i$ as $u\pi_i$, if there is no ambiguity. Thus every lambda term t can be uniquely written in the form $\lambda x_1 \ldots \lambda x_m.\ vt_1 \ldots t_n$, where $m, n \geq 0$, for every i, t_i is either a term or the symbol π_0 or π_1, and v is a variable or a constant or pair $\langle t, u\rangle$ or one of the following redexes: $(\lambda x.u)t$, $\mathsf{if}^\star\, t\, u$, $\mathsf{It}^\star\, t\, u$, $[\langle t, u\rangle]\pi_i$. If v is a redex, v is called the head redex of t. A term is said to be an application if it is of the form tu, an abstraction if it is of the form λxu. If t' is a subterm of t we will write $t \trianglerighteq t'$, reading t *contains* t'. Finally:

Definition 1 (Strongly Normalizable Terms). *We write $t \mapsto t'$ iff t' is obtained from t by contracting a redex in t according to the reduction rules in Figure 1. A sequence (finite or infinite) of terms $t_1, t_2, \ldots, t_n, \ldots$ is said to be a reduction of t, if $t = t_1$, and for all i, $t_i \mapsto t_{i+1}$. A term t of $\Lambda^\star$ is strongly normalizable if there is no infinite reduction of t. We denote with* SN *the set of strongly normalizable terms of $\Lambda^\star$.*

The reduction tree of a strongly normalizable lambda term is well-founded. It is well-known that it is possible to assign to each node of a well-founded tree an ordinal number, that it decreases passing from a node to any of its sons. We will call the *ordinal size* of a lambda term $t \in \mathsf{SN}$ the ordinal number assigned to the root of its reduction tree and we denote it by $h(t)$; thus, if $t \mapsto u$, then $h(t) > h(u)$. To fix ideas, one may define $h(t) := \mathsf{sup}\{h(u) + 1 \mid t \mapsto u\}$.

2.1 The Zoom-in Reduction

In order to really understand the phenomenon of non-termination in lambda calculus it is crucial to isolate the mechanisms that are really essential to produce it. For example, in the term $(\lambda y.\, y)(\lambda x.\, xx)\lambda x.\, xx$ (beware Krivine's notation!), the part that generates an infinite reduction is $(\lambda x.\, xx)\lambda x.\, xx$; the term $\lambda y.\, y$ is only a disturbing context and should be ignored. This is because the smallest non-strongly normalizing subterm of our term is $(\lambda x.\, xx)\lambda x.\, xx$. We thus arrive at the notion of elementary term: a non-strongly normalizable term that cannot be decomposed into smaller non-strongly normalizable terms.

Definition 2 (Elementary Terms). *A term tu is said to be* elementary *if $t \in \mathsf{SN}$, $u \in \mathsf{SN}$ and $tu \notin \mathsf{SN}$.*

Constants $\mathsf{c} ::= \mathsf{It}^\star \mid \mathsf{if}^\star \mid \mathsf{c}_0 \mid \mathsf{c}_1 \ldots$

Terms $t, u ::= x \mid \lambda x.t \mid (t)u \mid \langle t, u\rangle \mid [t]\pi_0 \mid [t]\pi_1 \mid \mathsf{c}$

Reduction Rules

$$(\lambda x.u)t \mapsto u[t/x] \qquad [\langle u_0, u_1\rangle]\pi_i \mapsto u_i, \text{ for i=0,1}$$

$$\mathsf{if}^\star u\, v \mapsto u \qquad \mathsf{if}^\star u\, v \mapsto v$$

$$\mathsf{It}^\star uv \mapsto \overbrace{(v)\ldots(v)}^{n\ times}\, u, \qquad \text{for each natural number } n$$

Fig. 1. Non-Deterministic Lambda Calculus $\Lambda^\star$

We observe that an elementary term cannot be of the form $xt_1 \ldots t_n$, since $t_1, \ldots, t_n \in \mathsf{SN}$, and hence $xt_1 \ldots\ t_n \in \mathsf{SN}$. Similarly, it cannot be neither of the form $\mathsf{c}_i t_1 \ldots t_n$ nor $(\langle t, u\rangle)t_1 \ldots t_n$ nor $[\lambda xt]\pi_i t_1 \ldots\ t_n$. Therefore, every elementary lambda term is either of the form $(\lambda xu)tt_1 \ldots t_n$ or $\mathsf{if}^\star t\, u\, t_1 \ldots t_n$ or $\mathsf{It}^\star t\, u\, t_1 \ldots t_n$ or $[\langle t, u\rangle]\pi_i t_1 \ldots t_n$ (and clearly $u, t, t_1, \ldots t_n \in \mathsf{SN}$).

Proposition 1. *Suppose $v \notin \mathsf{SN}$. Then v has an elementary subterm.*

Proof. By induction on v.

- If $v = x$ or $v = \mathsf{c}$, it is trivially true.
- If $v = ut$, and $u \in \mathsf{SN}$ and $t \in \mathsf{SN}$, v is elementary; if instead $u \notin \mathsf{SN}$ or $t \notin \mathsf{SN}$, by induction hypothesis u or t contains an elementary subterm, and hence v.
- If $v = \lambda xu$ or $v = u\pi_i$, then $u \notin \mathsf{SN}$, and by induction hypothesis u contains an elementary subterm, and thus also v.
- If $v = \langle t, u\rangle$, then $t \notin \mathsf{SN}$ or $u \notin \mathsf{SN}$, and by induction hypothesis u contains an elementary subterm, and thus also v.

The next proposition tells that it is always possible to contract the head redex of an elementary term in such a way to preserve its property of being non-strongly normalizable.

Proposition 2 (Saturation). *Suppose that v is elementary. Then:*

1. *If $v = (\lambda xu)tt_1 \ldots t_n \notin \mathsf{SN}$, then $u[t/x]t_1 \ldots t_n \notin \mathsf{SN}$.*
2. *If $v = [\langle u_0, u_1\rangle]\pi_i t_1 \ldots t_n \notin \mathsf{SN}$, then $u_i t_1 \ldots t_n \notin \mathsf{SN}$.*
3. *If $v = \mathsf{if}^\star t\, u\, t_1 \ldots t_n \notin \mathsf{SN}$, then $ut_1 \ldots t_n \notin \mathsf{SN}$ or $tt_1 \ldots t_n \notin \mathsf{SN}$*
4. *If $v = \mathsf{It}^\star t\, u\, t_1 \ldots t_n \notin \mathsf{SN}$, then for some $m \in \mathbb{N}$, $(\overbrace{(u)\ldots(u)}^{m\ times}\, t)t_1 \ldots t_n \notin \mathsf{SN}$.*

Proof.

1. By lexicographic induction on the $(n+2)$-tuple $(h(u), h(t), h(t_1), \ldots, h(t_n))$. Since by hypothesis $(\lambda xu)tt_1 \ldots t_n \notin \mathsf{SN}$, there exists a $w \notin \mathsf{SN}$ such that

$(\lambda xu)tt_1 \dots t_n \mapsto w$. There are two cases. First case: w is either $(\lambda xu')tt_1 \dots t_n$ or $(\lambda xu)t't_1 \dots t_n$ or $(\lambda xu)tt_1 \dots t'_i \dots t_n$ with $u \mapsto u'$, $t \mapsto t'$ and $t_i \mapsto t'_i$ $(i = 1, \dots n)$ respectively. We have $h(u') < h(u), h(t') < h(t), h(t'_i) < h(t_i)$. Then, by induction hypothesis, $u[t/x]t_1 \dots t_n \mapsto u'[t/x]t_1 \dots t_n \notin \mathsf{SN}$, $u[t/x]t_1 \dots t_n \mapsto u[t'/x]t_1 \dots t_n \notin \mathsf{SN}$ and
$u[t/x]t_1 \dots t_i \dots t_n \mapsto u[t/x]t_1 \dots t'_i \dots t_n \notin \mathsf{SN}$ $(i = 1 \dots n)$. Second case: $w = u[t/x]t_1 \dots t_n$. We conclude $u[t/x]t_1 \dots t_n \notin \mathsf{SN}$ by hypothesis on w.
2. The other cases are similar.

Let $v \notin \mathsf{SN}$ and s be an elementary subterm of v. Then $s = (\lambda xu)tt_1 \dots t_n$ or $s = \mathsf{if}^\star t\, u\, t_1 \dots t_n$ or $s = \mathsf{It}^\star t\, u\, t_1 \dots t_n$ or $s = [\langle t, u\rangle]\pi_i t_1 \dots t_n$. By Proposition 2, there exists an $s' \notin \mathsf{SN}$ such that s' is obtained from s by the contraction of its head redex. In particular, either $s' = u[t/x]t_1 \dots t_n$, $s' = ut_1 \dots t_n$ or $s' = tt_1 \dots t_n$ or $s' = ((u) \dots (u)t)t_1 \dots t_n$. This provides the justification for the next definition and proposition.

Definition 3 (Zoom-in Reduction). *Let $t \notin \mathsf{SN}$ and s be an elementary subterm of t. We write $t \overset{\mathsf{z}}{\mapsto} u$ if u has been obtained from t by replacing s with an $s' \notin \mathsf{SN}$ such that s' results from s by a contraction of the head redex of s. A sequence (finite or infinite) of terms $t_1, t_2, \dots, t_n, \dots$ is said to be a* zoom-in reduction *of t if $t = t_1$, and for all i, $t_i \overset{\mathsf{z}}{\mapsto} t_{i+1}$; if $i \le j$, we write $t_i \overset{\mathsf{z*}}{\longmapsto} t_j$.*

Proposition 3. *Suppose $t \notin \mathsf{SN}$. There is an infinite zoom-in reduction of t.*

The zoom-in reduction strategy was studied in Melliès's PhD Thesis [15]. It is a *perpetual* reduction (see [21]), in the sense it preserves non-strong normalization. The idea is to contract each time a redex which is *essential* in order to produce an infinite reduction. In this way, one concentrates on a minimal amount of resources sufficient to generate non-termination. For example, the reduction $(\lambda y.\, y)(\lambda x.\, xx)\lambda x.\, xx \mapsto (\lambda x.\, xx)\lambda x.\, xx$ is smartly avoided by the relation $\overset{\mathsf{z}}{\mapsto}$, because the reduction of the first redex is not strictly necessary. Instead, one has $(\lambda y.\, y)(\lambda x.\, xx)\lambda x.\, xx \overset{\mathsf{z}}{\mapsto} (\lambda y.\, y)(\lambda x.\, xx)\lambda x.\, xx$ by contraction of the second redex.

We now study what happens when the zoom-in reduction strategy is applied to elementary terms. The goal is to prove an Inversion Property (Proposition 6). That is, starting from an elementary term ut, we want to show that t will necessarily be used in head position as an active function in the future of the zoom-in reduction of ut. In this sense, there will be an *inversion* of the roles of argument and function. We break the result in two steps.

The first observation is that the zoom-in reduction of ut will contract redexes inside u as long as the term is "blocked", i.e. u does not transform into a function.

Proposition 4. *Let ut be elementary. Then one of the following cases occurs:*

1. *There exists a term $(\lambda xv)t$ such that $ut \overset{\mathsf{z*}}{\longmapsto} (\lambda xv)t$.*
2. *There exists a term $\mathsf{It}^\star v\, t$ such that $ut \overset{\mathsf{z*}}{\longmapsto} \mathsf{It}^\star v\, t$*

Proof. By induction on $h(u)$. There are two cases:

- The head redex of ut is in u. Then $ut \overset{\mathsf{z}}{\mapsto} u't$, with $u \mapsto u'$ and $h(u') < h(u)$. By induction hypothesis, $u't \overset{\mathsf{z}*}{\mapsto} (\lambda xv)t$ or $u't \overset{\mathsf{z}*}{\mapsto} \mathsf{It}^\star v\, t$, and we are done.
- The head redex of ut is ut itself. If $ut = (\lambda xv)t$ or $ut = \mathsf{It}^\star v\, t$, we have the thesis. Moreover, those are the only possible cases, for neither $ut = \mathsf{if}^\star v\, t$ nor $ut = [\langle v, v' \rangle]\pi_i$ can hold; otherwise by Proposition 2, $v \notin \mathsf{SN}$ or $v' \notin \mathsf{SN}$ or $t \notin \mathsf{SN}$, but since ut is elementary, $v, v', t \in \mathsf{SN}$.

The second observation is that in a zoom-in reduction of a term $u[t/x] \notin \mathsf{SN}$, with $u, t \in \mathsf{SN}$, t will necessarily be used at some point in head position, because at some point one will run out of redexes in u.

Proposition 5. *Suppose $u, t \in \mathsf{SN}$ and $u[t/x] \notin \mathsf{SN}$. Then there exists v such that $u[t/x] \overset{\mathsf{z}*}{\mapsto} v$ and v has an elementary subterm of the form $tt_1 \ldots t_n$ $(n > 0)$.*

Proof. By induction on $h(u)$. Assume $u[t/x] \overset{\mathsf{z}}{\mapsto} w$; let s be the elementary subterm of $u[t/x]$ whose head redex is contracted in order to obtain w. We have the following possibilities:

- A redex inside u has been contracted, obtaining $u'[t/x]$, with $u \mapsto u'$. Then, $h(u') < h(u)$ and the proposition immediately follows by induction hypothesis.
- A redex inside t has been contracted. Since $t \in \mathsf{SN}$, s is not a subterm of t; moreover, since the head redex of s must have been contracted, $s = tt_1 \ldots t_n$.
- A redex which is neither in t nor in u has been contracted. Then, t is a lambda abstraction or a pair or $\mathsf{if}^\star$ or $\mathsf{It}^\star$, u has a subterm of the form $xu_1 \ldots u_n$ and $s = (xu_1 \ldots u_n)[t/x] = tt_1 \ldots t_n$. Of course, $n > 0$, since $t \in \mathsf{SN}$ and $s \notin \mathsf{SN}$.

We are now able to prove the Inversion Property, the most crucial result.

Proposition 6 (Inversion Property). *Let ut be elementary. Then there exists w such that $ut \overset{\mathsf{z}*}{\mapsto} w$ and w has a subterm of the form $tt_1 \ldots t_n$ $(n > 0)$.*

Proof. By Propositions 4 and 5, one of the following cases occur:

1. $ut \overset{\mathsf{z}*}{\mapsto} (\lambda x.v)t \mapsto v[t/x] \overset{\mathsf{z}*}{\mapsto} w$, with w containing an elementary subterm of the form $tt_1 \ldots t_n$ $(n > 0)$.
2. $ut \overset{\mathsf{z}*}{\mapsto} \mathsf{It}^\star v\, t \mapsto (t) \ldots (t)v = (x) \ldots (x)v[t/x] \overset{\mathsf{z}*}{\mapsto} w$ (for some x not free in v), with w containing an elementary subterm of the form $tt_1 \ldots t_n$ $(n > 0)$.

By iteration of the Inversion Property, we finally obtain our characterization of non-strongly normalizable terms.

Theorem 2 (Characterization of non-strongly normalizable terms). *Let $u \notin \mathsf{SN}$. Then there exists an infinite sequence of terms $u_1, u_2, \ldots, u_n, \ldots$ such that $u_1 = u$, for all i, $u_i \overset{\mathsf{z}*}{\mapsto} u_{i+1}$ and:*

$$u_1 \trianglerighteq t_1 \ldots t_2, \quad u_2 \trianglerighteq t_2 \ldots t_3, \quad u_3 \trianglerighteq t_3 \ldots t_4, \quad \ldots, \quad u_n \trianglerighteq t_n \ldots t_{n+1} \quad \ldots$$

where for all i, $t_i \ldots t_{i+1}$ is an elementary term.

Proof. We set $u_1 = u$. Supposing u_n to have been defined, and that $u_n \trianglerighteq t_n \ldots t_{n+1}$ elementary. By Proposition 6, we can set u_{n+1} as the term obtained from u_n by substituting $t_n \ldots t_{n+1}$ with a v' such that $t_n \ldots t_{n+1} \overset{\mathsf{z}*}{\mapsto} v'$ and v' contains an elementary subterm of the form $t_{n+1} \ldots t_{n+2}$.

3 The System $\mathsf{T}^\star$ and Its Strong Normalization

As well as one can consider a simply typed version of the ordinary lambda calculus with pairs, we now introduce a simply typed version of the non-deterministic lambda calculus $\Lambda^\star$. We call it System $\mathsf{T}^\star$, since it will be interpreted as a non-deterministic version of Gödel's System T. $\mathsf{T}^\star$ is formally described in Figure 2. The basic objects of $\mathsf{T}^\star$ are numerals and booleans, its basic computational constructs are primitive iterator at all types, if-then-else and pairs; $\overline{n}$ is the usual encoding $\mathsf{S}\ldots\mathsf{S}0$ of the natural number n. The strong normalization of $\mathsf{T}^\star$ can be readily proved from the Characterization Theorem 2.

Types

$$\sigma, \tau ::= \mathtt{N} \mid \mathtt{Bool} \mid \sigma \to \tau \mid \sigma \times \tau$$

Constants

$$\mathsf{c} ::= \mathsf{It}^\star_\tau \mid \mathsf{if}^\star_\tau \mid 0 \mid \mathsf{S} \mid \mathtt{True} \mid \mathtt{False}$$

Terms

$$t, u ::= \mathsf{c} \mid x^\tau \mid (t)u \mid \lambda x^\tau u \mid \langle t, u\rangle \mid [t]\pi_0 \mid [t]\pi_1$$

Typing Rules for Variables and Constants

$$x^\tau : \tau$$

$$0 : \mathtt{N}, \mathsf{S} : \mathtt{N} \to \mathtt{N}$$

$$\mathtt{True} : \mathtt{Bool}, \mathtt{False} : \mathtt{Bool}$$

$$\mathsf{if}^\star_\tau : \tau \to \tau \to \tau$$

$$\mathsf{It}^\star_\tau : \tau \to (\tau \to \tau) \to \tau$$

Typing Rules for Composed Terms

$$\frac{t : \sigma \to \tau \qquad u : \sigma}{tu : \tau} \qquad \frac{u : \tau}{\lambda x^\sigma u : \sigma \to \tau}$$

$$\frac{u : \sigma \qquad t : \tau}{\langle u, t\rangle : \sigma \times \tau} \qquad \frac{u : \tau_0 \times \tau_1}{\pi_i u : \tau_i}\ i \in \{0, 1\}$$

Reduction Rules The same reduction rules of $\Lambda^\star$, restricted to the terms of $\mathsf{T}^\star$.

Fig. 2. The system $\mathsf{T}^\star$

Theorem 3 (Strong Normalization Theorem for $\mathsf{T}^\star$). *Every term w of $\mathsf{T}^\star$ is strongly normalizable.*

Proof. Suppose for the sake of contradiction that $w \notin \mathsf{SN}$. By the Characterization Theorem 2 (which can clearly be applied also to the terms of $\mathsf{T}^\star$), we obtain the existence of an infinite sequence of typed elementary terms $t_1 \ldots t_2$, $t_2 \ldots t_3$, $\ldots, t_n \ldots t_{n+1} \ldots$. which yields a contradiction, since for every i, the type of t_i is strictly greater than the type of t_{i+1}.

4 The System T and Its Strong Normalization

In this section we will prove the strong normalization theorem for System T. Syntax and typing rules of T are formally described in Figure 3.

Types

$$\sigma, \tau ::= \mathtt{N} \mid \mathtt{Bool} \mid \sigma \to \tau \mid \sigma \times \tau$$

Constants

$$c ::= \mathsf{It}_\tau \mid \mathsf{if}_\tau \mid 0 \mid \mathsf{S} \mid \mathtt{True} \mid \mathtt{False}$$

Terms

$$t, u ::= \ c \mid x^\tau \mid (t)u \mid \lambda x^\tau u \mid \langle t, u\rangle \mid [u]\pi_0 \mid [u]\pi_1$$

Typing Rules for Variables and Constants

$$x^\tau : \tau$$
$$0 : \mathtt{N}, \mathsf{S} : \ \mathtt{N} \to \mathtt{N}$$
$$\mathtt{True} : \mathtt{Bool}, \mathtt{False} : \ \mathtt{Bool}$$
$$\mathsf{if}_\tau : \mathtt{Bool} \to \tau \to \tau \to \tau$$
$$\mathsf{It}_\tau : \tau \to (\tau \to \tau) \to \mathtt{N} \to \tau$$

Typing Rules for Composed Terms

$$\frac{t : \sigma \to \tau \qquad u : \sigma}{(t)u : \tau} \qquad \frac{u : \tau}{\lambda x^\sigma u : \sigma \to \tau}$$

$$\frac{u : \sigma \qquad t : \tau}{\langle u, t\rangle : \sigma \times \tau} \qquad \frac{u : \tau_0 \times \tau_1}{[u]\pi_i : \tau_i}\ i \in \{0, 1\}$$

Fig. 3. Syntax and Typing Rules for Gödel's system T

Strong normalization follows as a corollary of Theorem 3. We define a simple translation mapping terms of System T into terms of System $\mathsf{T}^\star$:

Definition 4 (Translation of T into $\mathsf{T}^\star$). *We define a translation* $\_^* : \mathsf{T} \to \mathsf{T}^\star$*, leaving types unchanged. In the case of constants of the form* $\mathsf{if}_\tau, \mathsf{It}_\tau$*, we set:*

$$(\mathsf{if}_\tau)^* := \lambda b^{\mathtt{Bool}}.\, \mathsf{if}^\star_\tau \qquad\qquad (\mathsf{It}_\tau)^* := \lambda x^\tau \lambda y^{\tau\to\tau} \lambda z^{\mathtt{N}}.\, \mathsf{It}^\star{}_\tau xy$$

For all other terms t of Gödel's System T*, we set t^* as the term of* $\mathsf{T}^\star$ *obtained from t by replacing all its constants* if_τ *with* $(\mathsf{if}_\tau)^*$ *and all its constants* It_τ *with* $(\mathsf{It}_\tau)^*$.

In the following, we will proceed by endowing T with two distinct reduction strategies, respectively dubbed as $\mapsto_{\mathsf{v}}$ and $\mapsto$. Informally, $\mapsto_{\mathsf{v}}$ forces a call-by-value discipline on the datatype N. The second one, $\mapsto$, is the usual strategy T is endowed with. We will prove the strong normalization property in both cases. Whereas the goal is straightforward for $\mapsto_{\mathsf{v}}$, in the second case a bit of work is required.

4.1 Strong Normalization for System T with the strategy $\mapsto_{\mathsf{v}}$

The reduction strategy $\mapsto_{\mathsf{v}}$ is formally defined in Figure 4. Strong normalization theorem for T with $\mapsto_{\mathsf{v}}$ easily follows from Theorem 3. As a matter of fact, each computational step in T (with $\mapsto_{\mathsf{v}}$ reductions' set) can be plainly simulated in $\mathsf{T}^\star$ by a non-deterministic guess. In particular, each reduction step between T terms corresponds to *at least* a step between their translations:

Proposition 7 (Preservation of the Reduction Relation). *Let v be any term of* T*. Then* $v \mapsto_{\mathsf{v}} w \implies v^* \mapsto^+ w^*$

Reduction strategy $\mapsto_{\mathsf{v}}$

$$(\lambda x^{\tau} u)t \mapsto_{\mathsf{v}} u[t/x^{\tau}]$$

$$[\langle u_0, u_1\rangle]\pi_i \mapsto_{\mathsf{v}} u_i, \text{ for i=0,1}$$

$$\mathsf{It}_{\tau} uv\overline{n} \mapsto_{\mathsf{v}} \overbrace{(v)\dots(v)}^{n\ times}u$$

$$\mathsf{if}_{\tau}\ \mathtt{True}\, u\, v \mapsto_{\mathsf{v}} u \qquad \mathsf{if}_{\tau}\ \mathtt{False}\, u\, v \mapsto_{\mathsf{v}} v$$

Fig. 4. Reduction strategy $\mapsto_{\mathsf{v}}$ for T

Proof. It is sufficient to prove the proposition when v is a redex r. We have several possibilities:

1. $r = (\lambda x^{\tau} u)t \mapsto_{\mathsf{v}} u[t/x^{\tau}]$. We verify indeed that
$$((\lambda x^{\tau} u)t)^* = (\lambda x^{\tau} u^*)t^* \mapsto u^*[t^*/x^{\tau}] = u[t/x^{\tau}]^*$$
2. $r = \langle u_0, u_1\rangle \pi_i \mapsto_{\mathsf{v}} u_i$. We verify indeed that
$$(\langle u_0, u_1\rangle \pi_i)^* = \langle u_0^*, u_1^*\rangle \pi_i \mapsto u_i^*$$
3. $r = \mathsf{if}\, \mathtt{True}\, t\, u \mapsto_{\mathsf{v}} t$ or $r = \mathsf{if}\, \mathtt{False}\, t\, u \mapsto_{\mathsf{v}} u$. We verify indeed – by choosing the appropriate reduction rule for $\mathsf{if}^{\star}$ – that
$$(\mathsf{if}\, \mathtt{True}\, t\, u)^* = (\mathsf{if})^*\, \mathtt{True}\, t^*\, u^* \mapsto \mathsf{if}^{\star} t^* u^* \mapsto t^*$$
$$(\mathsf{if}\, \mathtt{False}\, t\, u)^* = (\mathsf{if})^*\, \mathtt{False}\, t^*\, u^* \mapsto \mathsf{if}^{\star} t^* u^* \mapsto u^*$$
4. $r = \mathsf{It}\, u\, t\, \overline{n} \mapsto_{\mathsf{v}} \overbrace{(t)\dots(t)}^{n\ times}\, u$. We verify indeed – by choosing the appropriate reduction rule for $\mathsf{It}^{\star}$ – that
$$(\mathsf{It}\, u\, t\, \overline{n})^* = (\mathsf{It})^* u^* t^* \overline{n} \mapsto^* \mathsf{It}^{\star} u^* t^* \mapsto (t^*)\dots(t^*)u^*$$

Theorem 4 (Strong Normalization for System T with $\mapsto_{\mathsf{v}}$ strategy). *Any term t of System T is strongly normalizable with respect to the relation $\mapsto_{\mathsf{v}}$.*

Proof. By Proposition 7, any infinite reduction $t = t_1, t_2, \dots, t_n, \dots$ in System T gives rise to an infinite reduction $t^* = t_1^*, t_2^*, \dots, t_n^*, \dots$ in System $\mathsf{T}^{\star}$. By the strong normalization Theorem 3 for $\mathsf{T}^{\star}$, infinite reductions of the latter kind cannot occur; thus neither of the former.

We have just proved the strong normalization theorem for T with the call-by-value restriction on the datatype $\mathtt{N}$. In any "practical" application (such as realizability, functional interpretation, program extraction from logical proofs), this evaluation discipline is perfectly suitable. From the constructive point of view, the call-by-value evaluation on natural numbers is even *desirable.* In fact, what essentially distinguishes the constructive reading of the iteration from the classical one is that the first requires complete knowledge of the number of times a functional will be iterated *before the actual execution* of the iteration. Call-by-value performs exactly this task: in a term $\mathsf{It}\,u\,v\,t$, it first completely evaluates t to a numeral, so providing a precise account about the number of times the function v will be called. Even if that is constructively satisfying, for the sake of completeness we will prove strong normalization with respect to the most general reduction strategy. This is the aim of the following section.

4.2 Strong Normalization of System T with the Strategy $\mapsto$

The reduction strategy $\mapsto$ is formally defined in Figure 5. Notice that the only difference with respect to the call-by-value strategy $\mapsto_{\mathsf{v}}$ is that the term t in the reduction rule for It is not necessarily a numeral. We define SN_{T} to be the set of strongly normalizable terms of T with respect to the strategy $\mapsto$ and E_{T} to be the set of elementary terms of T with respect to the strategy $\mapsto$. We observe that it is still true that each term of T not in SN_{T} contains a term in E_{T}.

Reduction Strategy $\mapsto$

$$(\lambda x^{\tau} u)t \mapsto u[t/x^{\tau}]$$

$$[\langle u_0, u_1\rangle]\pi_i \mapsto u_i, \text{ for i=0,1}$$

$$\mathsf{It}_{\tau} uv0 \mapsto u \qquad \mathsf{It}_{\tau} uv(\mathsf{S}t) \mapsto v(\mathsf{It}_{\tau} uvt)$$

$$\mathsf{if}_{\tau}\mathtt{True}\, u\, v \mapsto u \qquad \mathsf{if}_{\tau}\mathtt{False}\, u\, v \mapsto v$$

Fig. 5. Reduction Strategy $\mapsto$ for System T

One may be tempted to proceed as in the previous section, by directly simulating $\mapsto$-reduction steps in T with reduction steps in $\mathsf{T}^{\star}$. Unfortunately, this is not possible. On the $\mathsf{T}^{\star}$ side, in order to interpret $\mathsf{It}\,u\,v\,t$, one has to "guess" the value of t by means of $\mathsf{It}^{\star}$. But it can very well happen that t is open, for example, so without value. To solve this issue, we are going to define an "almost" reduction relation $\mathscr{P}$ which can instead be simulated in $\mathsf{T}^{\star}$. In fact, $\mathscr{P}$ turns out to be a version of $\overset{\mathsf{z}}{\mapsto}$ adapted to System T, which can be proved perpetual (Proposition 10). As a first step, we need to widen the class of numerals:

Definition 5 (Generalized Numerals). *A* generalized numeral *is a term of* T *of the form* $\mathsf{S}\ldots\mathsf{S}t$, *with* $t \in \mathsf{NF}$, $t \neq \mathsf{S}u$; GN *is the set of generalized numerals. If* $\mathsf{S}\ldots\mathsf{S}t$ *is a generalized numeral and* v *occurs in the head of* $(v)\ldots(v)u$ *as many times as* S *occurs in the prefix of* $\mathsf{SS}\ldots\mathsf{S}t$, *then* $(v)\ldots(v)u$ *is said to be the* expansion *of* $\mathsf{It}\,u\,v\,(\mathsf{S}\ldots\mathsf{S}t)$.

We remark that one could have equivalently defined GN as the set of type-$\mathtt{N}$ terms; this latter definition however does not generalized to untyped lambda calculus, while our results probably do, with some adaptation.

As a second step, we need to define a relation $\mathscr{P}$.

Definition 6 (Perpetual Relation $\mathscr{P}$). *Let* $t \notin \mathsf{SN}_{\mathsf{T}}$ *and* $s \in \mathsf{E}_{\mathsf{T}}$ *be a subterm of* t. *We write* $t \mathrel{\mathscr{P}} u$ *if* u *has been obtained from* t *by replacing* s *with an* s' *such that:*

- $s = (\lambda x^\tau u)tt_1 \dots t_n \implies s' = u[t/x]t_1 \dots t_n$;
- $s = [\langle u_0, u_1\rangle]\pi_i t_1 \dots t_n \implies s' = u_i t_1 \dots t_n$;
- $s = \text{if}\,\texttt{True}\, t\, u\, t_1 \dots t_n \implies s' = tt_1 \dots t_n$;
- $s = \text{if}\,\texttt{False}\, t\, u\, t_1 \dots t_n \implies s' = ut_1 \dots t_n$;
- $s = \mathsf{It}\, u\, v\, t\, t_1 \dots t_n$ *and* $t \mapsto^* t' \in \mathsf{GN} \implies s' = ((v)(v)\dots(v)u)t_1 \dots t_n$, *where* $(v)\dots(v)u$ *is the expansion of* $\mathsf{It}\, u\, v\, t'$.

The idea behind $\mathscr{P}$ is to make it behave like a call-by-value strategy on N, even when it should not be possible, by considering a term in GN as a "numeral". In order to show that $\mathscr{P}$ is perpetual, we need some technical but quite simple results.

The following lemma states that the set of non-strongly normalizable terms is closed w.r.t. the substitution of subterms in SN_T with their normal forms.

Lemma 1. *Assume* $t_1, \dots, t_n : \texttt{N}$ *and* $t_1, \dots, t_n \in \mathsf{SN}_\mathsf{T}$. *Let* $s_1, \dots, s_n : \texttt{N}$ *be such that, for all* $i = 1 \dots n$, s_i *is the normal form of* t_i. *Then, given any term* u *of* T:

$$u[t_1/x_1, \dots, t_n/x_n] \notin \mathsf{SN}_\mathsf{T} \implies u[s_1/x_1, \dots, s_n/x_n] \notin \mathsf{SN}_\mathsf{T}$$

Proof. It suffices to prove that there exist terms u', $t_1', \dots, t_m' \in \mathsf{SN}_\mathsf{T}$ and $s_1', \dots, s_n'$ such that for $i = 1, \dots, m$, and

$$u'[t_1'/x_1, \dots, t_m'/x_m] \notin \mathsf{SN}_\mathsf{T}$$

and

$$u[s_1/x_1, \dots, s_n/x_n] \mapsto^+ u'[s_1'/x_1, \dots, s_m'/x_m]$$

where again each s_i' is the normal form of t_i'. Since the end terms of the two lines above satisfy the hypothesis of the proposition, one may iterate this construction infinitely many times and obtains an infinite reduction of $u[s_1/x_1, \dots, s_n/x_n]$.

In order to show that, let us consider an infinite reduction of $u[t_1/x_1, \dots, t_n/x_n]$. Since $t_1, \dots, t_n \in \mathsf{SN}_\mathsf{T}$, only finitely many reduction steps can be performed inside them. So the infinite reduction has a first segment of the shape:

$$u[t_1/x_1, \dots, t_n/x_n] \mapsto^* u[t_1'/x_1, \dots, t_n'/x_n] \mapsto w \notin \mathsf{SN}_\mathsf{T}$$

with $t_i \mapsto^* t_i'$. We have now two possibilities, depending on the kind of redex that has been contracted in order to obtain w:

1. $w = u'[t_1'/x_1, \dots, t_n'/x_n]$, with $u \mapsto u'$. Then also $u[s_1/x_1, \dots, s_n/x_n] \mapsto u'[s_1/x_1, \dots, s_n/x_n]$ and we are done.
2. w has been obtained from $u[t_1'/x_1, \dots, t_n'/x_n]$ by reduction of a redex created by the substitution t_i'/x_i. In this case, since $t_1', \dots, t_n' : \texttt{N}$, the only possible redex of that kind has the form $(\mathsf{It}uvx_i)[t_1'/x_1, \dots t_i'/x_i \dots t_n'/x_n]$, with $\mathsf{It}uvx_i$ subterm of u and $t_i' = \mathsf{S}t_{n+1}$. Then v is obtained by replacing

$$\mathsf{It}uvx_i[t_1'/x_1, \dots t_i'/x_i \dots t_n'/x_n] = \mathsf{It}u'v'\mathsf{S}(t_{n+1})$$

with

$$(v')\mathsf{It}u'v't_{n+1} = (v)\mathsf{It}uvx_{n+1}[t'_1/x_1, \ldots t'_i/x_i \ldots t'_n/x_n \; t_{n+1}/x_{n+1}]$$

where x_{n+1} is a fresh variable. If we define $u' := u[(\mathsf{It}uvx_i) := (v)\mathsf{It}uvx_{n+1}]$ (i.e. u' is obtained from u by replacing $\mathsf{It}uvx_i$ with $(v)\mathsf{It}uvx_{n+1}$) we then have

$$v = u'[t'_1/x_1 \ldots t'_n/x_n \; t_{n+1}/x_{n+1}]$$

Since s_i is the normal form of $t'_i = \mathsf{S}t_{n+1}$, we have $s_i = \mathsf{S}s_{n+1}$, where s_{n+1} is the normal form of t_{n+1}. As before,

$$\mathsf{It}uvx_i[s_1/x_1, \ldots s_i/x_i \ldots s_n/x_n] \mapsto (v)\mathsf{It}uvx_{n+1}[s_1/x_1, \ldots s_i/x_i \ldots s_n/x_n \; s_{n+1}/x_{n+1}]$$

which implies $u[s_1/x_1 \ldots s_n/x_n] \mapsto u'[s_1/x_1, \ldots s_i/x_i \ldots s_n/x_n \; s_{n+1}/x_{n+1}]$ and we are done.

By means of Lemma 1 it is possible to prove:

Proposition 8. *If* $(\mathsf{It}uvt)t_1 \ldots t_n \in \mathsf{E_T}$ *and* $t \mapsto^* t' \in \mathsf{GN}$, *then* $(\mathsf{It}uvt')$ $t_1 \ldots t_n \in \mathsf{E_T}$.

Proof. By Lemma 1, applied to the terms $\mathsf{It}\,u\,v\,x\,t_1 \ldots t_n[t/x]$ and $\mathsf{It}\,u\,v\,x\,t_1 \ldots t_n[t'/x]$ (x fresh).

Lemma 2 is similar to Lemma 1: the set of non-strongly normalizable terms can be proved to be closed w.r.t. the substitution of subterms with their expansions.

Lemma 2. *Let* $t_1, \ldots, t_n, s_1, \ldots, s_n$ *be a sequence of terms such that for* $i = 1 \ldots n$, s_i *is the expansion of* t_i *and all the proper subterms of* t_i *are in* $\mathsf{SN_T}$. *Then given any term* u *of* T,

$$u[t_1/x_1, \ldots, t_n/x_n] \notin \mathsf{SN_T} \implies u[s_1/x_1, \ldots, s_n/x_n] \notin \mathsf{SN_T}$$

Proof. It suffices to prove that there exist terms u' and $t'_1, \ldots, t'_m, s'_1, \ldots, s'_m$ such that for $i = 1, \ldots, m$, s'_i is the expansion of t'_i, all the strict subterms of t'_i are in $\mathsf{SN_T}$ and

$$u'[t'_1/x_1, \ldots, t'_m/x_m] \notin \mathsf{SN_T} \quad \text{and}$$
$$u[s_1/x_1, \ldots, s_n/x_n] \mapsto^+ u'[s'_1/x_1, \ldots, s'_m/x_m]$$

Since the end terms of the two lines above satisfy the hypothesis of the proposition, one may iterate this construction infinite times and obtains an infinite reduction of $u[s_1/x_1, \ldots, s_n/x_n]$.

In order to show that, let us consider an infinite reduction of $u[t_1/x_1, \ldots, t_n/x_n]$. By definition 5, $t_i = \mathsf{It}\,u_i\,v_i\,n_i$, for some u_i, v_i and generalized numeral n_i. Since $u_i, v_i \in \mathsf{SN_T}$, only finitely many reduction steps can be performed inside them. So the infinite reduction has a first segment of the shape:

$$u[t_1/x_1, \ldots, t_n/x_n] \mapsto^* u[t'_1/x_1, \ldots, t'_n/x_n] \mapsto v \notin \mathsf{SN_T}$$

with $t'_i = \mathsf{It}\,u'_i\,v'_i\,n_i$ and $u_i \mapsto^* u'_i, v_i \mapsto^* v'_i$. We have now two possibilities, depending on the kind of redex that has been contracted in order to obtain v (we notice that it must be already in u or in some t'_i):

1. $v = u'[t'_1/x_1, \ldots, t'_n/x_n]$, with $u \mapsto u'$. Let now, for $i = 1, \ldots, n$, s'_i be the expansion of t'_i. Then
$$s_i = (v_i) \ldots (v_i)u_i \mapsto^* (v'_i) \ldots (v'_i)u'_i = s'_i$$
Therefore $u[s_1/x_1, \ldots, s_n/x_n] \mapsto^+ u'[s'_1/x_1, \ldots, s'_n/x_n]$ and we are done.
2. v has been obtained from $u[t'_1/x_1, \ldots, t'_n/x_n]$ by replacing one of the occurrences of $t'_i = \mathsf{It}\, u'_i\, v'_i\, n_i$ with $(v'_i)\mathsf{It}\, u'_i\, v'_i\, m_i$ (assuming that $n_i = \mathsf{S}m_i$). Let $t'_{n+1} := \mathsf{It}\, u'_i\, v'_i\, m_i$. Then there exists a term u' (obtained from u by replacing a suitable occurrence of x_i with $(v'_i)x_{n+1}$, where x_{n+1} fresh) such that
$$v = u'[t'_1/x_1, \ldots, t'_n/x_n\; t'_{n+1}/x_{n+1}]$$
Let now, for $i = 1, \ldots, n+1$, s'_i be the expansion of t'_i. We want to show that
$$u[s_1/x_1, \ldots, s_n/x_n] \mapsto^+ u'[s'_1/x_1, \ldots, s'_n/x_n\; s'_{n+1}/x_{n+1}]$$
As before, $u[s_1/x_1, \ldots, s_n/x_n] \mapsto^* u[s'_1/x_1, \ldots, s'_n/x_n]$. Moreover, since s'_i is the expansion of $\mathsf{It}\, u'_i\, v'_i\, \mathsf{S}m_i$ and s'_{n+1} is the expansion of $\mathsf{It}\, u'_i\, v'_i\, m_i$, we have $s'_i = (v'_i)s'_{n+1}$. Therefore
$$x_i[s'_i/x_i] = s'_i = (v'_i)s'_{n+1} = (v'_i)x_{n+1}[s'_{n+1}/x_{n+1}]$$
and thus
$$u[s'_1/x_1, \ldots, s'_n/x_n] = u'[s'_1/x_1, \ldots, s'_n/x_n\; s'_{n+1}/x_{n+1}]$$
which concludes the proof.

The set $\mathsf{E_T}$ is closed w.r.t. the expansion of a head It redex of an elementary term:

Proposition 9. *Suppose that s' is the expansion of s. Then*
$$st_1 \ldots t_n \in \mathsf{E_T} \implies s't_1 \ldots t_n \in \mathsf{E_T}$$

Proof. By Lemma 2, applied to $xt_1 \ldots t_n[s/x]$ and $xt_1 \ldots t_n[s'/x]$ (x fresh).

Finally, the perpetuality of $\mathscr{P}$ follows from Propositions 8 and 9.

Proposition 10 (Perpetuality of $\mathscr{P}$). *If $t \notin \mathsf{SN_T}$ and $t \mathrel{\mathscr{P}} u$, then $u \notin \mathsf{SN_T}$.*

Proof. Assume u is obtained from t by replacing an elementary subterm s of u with s'; we show that $s' \notin \mathsf{SN_T}$. The only case not covered by a straightforward adaptation of Proposition 2 is the one in which $s = \mathsf{It}\, u\, v\, t\, t_1 \ldots t_n$ and $t \mapsto^* t' \in \mathsf{GN} \implies s' = ((v)(v) \ldots (v)u)t_1 \ldots t_n$, where $(v) \ldots (v)u$ is the expansion of $\mathsf{It}uvt'$. Now, by Proposition 8, we obtain that $\mathsf{It}uvt'$ is elementary; by Proposition 9, we obtain that $((v)(v) \ldots (v)u)t_1 \ldots t_n$ is elementary too.

The perpetual relation $\mathscr{P}$ is simulated in $\mathsf{T}^\star$ by means of the translation $\_^*$.

Proposition 11 (Simulation of the Perpetual relation in $\mathsf{T}^\star$). *Let v be any term of T. Then $v \mathrel{\mathscr{P}} w \implies v^* \mapsto^+ w^*$.*

Proof. The proof is the same as that of proposition 7.

We are now able to prove the Strong Normalization Theorem for T:

Theorem 5 (Strong Normalization for System T). *Every term t of Gödel's System* T *is strongly normalizable with respect to the relation* $\mapsto$.

Proof. Suppose for the sake of contradiction that $t \notin \mathsf{SN}_{\mathsf{T}}$. By Proposition 10, there is an infinite sequence of terms $t = t_1, t_2, \ldots, t_n, \ldots$ in System T such that for all i, $t_i \mathscr{P} t_{i+1}$. By Proposition 11 that gives rise to an infinite reduction $t^* = t_1^*, t_2^*, \ldots, t_n^*, \ldots$ in System $\mathsf{T}^\star$. By the strong normalization Theorem 3 for $\mathsf{T}^\star$, infinite reductions of the latter kind cannot occur: contradiction.

5 Conclusions and Related Works

Most of the proofs in this paper are intuitionistic. We remark however that our proof of the Characterization Theorem 2 is classical, since the excluded middle is used in a crucial way to prove Proposition 1. But this is not an issue: it is nowadays well-known how to interpret constructively classical proofs, especially when so limited a use of classical reasoning is made. One may thus obtain, by using classical realizabilities [12] or functional interpretations [19], non-trivial programs providing arbitrarily long approximations of the sequence of terms proved to exists in the Characterization Theorem. The same considerations apply to the proofs of the strong normalization theorems: it is possible to extract *directly* from them normalization algorithms (giving a nice case study in the field of program-extraction from classical proofs).

Our proofs of strong normalizations bear similarities with others. In [22], the iterator It_τ is translated as the infinite term

$$\lambda x^\tau \lambda f^{\tau\to\tau} \lambda n^{\mathtt{N}}.\ \langle x, (f)x, (f)(f)x, (f)(f)(f)x, \ldots \rangle\, n$$

and a weak normalization theorem is proven with respect to the new infinite calculus. On our side, the use of the non-deterministic operator $\mathsf{It}^\star$ clearly allows to simulate that infinite term. On a first thought, the move may not seem a big deal, but, surprisingly, the gain is considerable. First, one radically simplifies Tait's calculus by avoiding infinite terms. Secondly, the Characterization Theorem for $\Lambda^\star$ and $\mathsf{T}^\star$ *does not hold* for Tait's infinite calculus, since this latter does not enjoy its main corollary, strong normalization (an infinite term may contain infinite redexes). Last, with our technique we obtain strong normalization for T.

Our work has also some aspects in common with the technique of Joachimski-Matthes [14], which provides an adaptation of the technique in [18] that works for the lambda formulation of System T. For example, our use of generalized numerals is similar to the evaluation function of [14] used to inject Ω in SN. Indeed, we consider our work to be a refinement and an extension to an untyped setting of the methods of [18,14]. In fact, we claim to be also able to prove the strong normalization theorem for System $\mathsf{T}^\star$ *directly*, in a Van Daalen style (see also [4]). In other words, one can simplify both our proof for T (call-by-value) and the one in [14] by avoiding to reason on a inductively defined set of "SN" terms and instead use a triple induction. This is possible since the non-deterministic reduction relation of T* allows to express in a natural way a heavy inductive load, which is performed in [18,14] by defining a set of "regular" terms and the set "SN" by an omega-rule. Indeed, we believe that the idea of using non-determinism to simplify the study of strong normalization can be applied in other situations as well: we shall show that in future papers. Moreover, our technique makes explicit as a perpetual reduction the "reduction" hidden in the family of proofs in [18,20,14]. This enables not only to prove normalization, but also to increase the qualitative understanding of non-termination in lambda calculus with explicit recursion and to explain why it is avoided in the typed version. As for [1], we consider our extension of the Characterization Theorem from lambda calculus to T* as a genuine advancement: for quite a while, such a generalization seemed hopeless for a system which can simulate in a so direct way System T.

References

1. Aschieri, F.: Una caratterizzazione dei lambda termini non fortemente normalizzabili. Master Degree Thesis, Università degli Studi di Verona (2007)
2. Biasi, C., Aschieri, F.: A Term Assignment for Polarized Bi-Intuitionistic Logic and its Strong Normalization. Fundamenta Informaticae 84(02) (2008)
3. Dal Lago, U., Zorzi, M.: Probabilistic Operational Semantics for the Lambda Calculus. RAIRO-ITA 46(03), 413–450 (2012), doi:10.1051/ita/2012012
4. David, R., Nour, K.: A short proof of the strong normalization of the simply typed lambda-mu-calculus. Schedae Informaticae 12, 27–34 (2003)
5. de' Liguoro, U., Piperno, A.: Non-Deterministic Extensions of Untyped Lambda-Calculus. Information and Computation 122, 149–177 (1995)
6. Girard, J.-Y., Lafont, Y., Taylor, P.: Proofs and Types. Cambridge University Press (1989)
7. Howard, W.A.: Ordinal analysis of terms of finite type. The Journal of Symbolic Logic 45(3), 493–504 (1980)
8. Gandy, R.O.: Proofs of Strong Normalization. Essays on Combinatoriay Logic, Lambda Calculus and Formalism, pp. 457–477. Academic Press, London (1980)
9. Khasidashvili, Z., Ogawa, M.: Perpetualilty and Uniform Normalization. In: Hanus, M., Heering, J., Meinke, K. (eds.) ALP 1997 and HOA 1997. LNCS, vol. 1298, pp. 240–255. Springer, Heidelberg (1997)
10. Kreisel, G.: Interpretation of Analysis by Means of Constructive Functionals of Finite Types. In: Constructivity in Mathematics, pp. 101–128. North-Holland (1959)
11. Krivine, J.-L.: Lambda-calcul types et modèles, Masson, Paris. Studies in Logic and Foundations of Mathematics, pp. 1–176 (1990)
12. Krivine, J.-L.: Classical Realizability. In: Interactive Models of Computation and Program Behavior. Panoramas et Synthèses, vol. 27, pp. 197–229. Société Mathématique de France (2009)
13. Levy, J.-J.: Reductions correctes et optimales dans le lambda-calcul. PhD Thesis, Université Paris 7 (1978)

14. Joachimski, F., Matthes, R.: Short proofs of normalization for the simply-typed lambda-calculus, permutative conversions and Gödel's T. Archive of Mathematical Logic 42(1), 49–87 (2003)
15. Melliès, P.-A.: Description Abstraite des Systèmes de Réécriture. PhD Thesis, Université Paris 7 (1996)
16. Nederpelt, R.-P.: Strong Normalization in Typed Lambda Calculus with Lambda Structured Types. PhD Thesis, Eindhoven University of Technology (1973)
17. Prawitz, D.: Ideas and Results in Proof Theory. In: Proceedings of the Second Scandinavian Logic Symposium (1971)
18. Sanchis, L.-E.: Functionals Defined by Recursion. Notre Dame Journal of Formal Logic VIII(3), 161–174 (1967)
19. Spector, C.: Provably recursive functionals of analysis: a consistency proof of analysis by an extension of principles in current intuitionistic mathematics. In: Proceedings of Symposia in Pure Mathematics, vol. 5, pp. 1–27. AMS (1962)
20. van Daalen, D.: The language theory of Automath. PhD Thesis, Eindhoven University of Technology (1977)
21. van Raamsdonk, F., Severi, P., Sørensen, M.H., Rensen, M.H., Xi, H.: Perpetual Reductions in Lambda-Calculus. Information and Computation 149, 173–225 (1999)
22. Tait, W.: Infinitely Long Terms of Transfinite Type. Formal Systems and Recursive Functions 40, 176–185 (1965)

Proof-Relevant Logical Relations for Name Generation

Nick Benton[1], Martin Hofmann[2], and Vivek Nigam[3]

[1] Microsoft Research, Cambridge
[2] LMU, Munich
[3] UFPB, João Pessoa
nick@microsoft.com, hofmann@ifi.lmu.de, vivek.nigam@gmail.com

Abstract. Pitts and Stark's ν-calculus is a paradigmatic total language for studying the problem of contextual equivalence in higher-order languages with name generation. Models for the ν-calculus that validate basic equivalences concerning names may be constructed using functor categories or nominal sets, with a dynamic allocation monad used to model computations that may allocate fresh names. If recursion is added to the language and one attempts to adapt the models from (nominal) sets to (nominal) domains, however, the direct-style construction of the allocation monad no longer works. This issue has previously been addressed by using a monad that combines dynamic allocation with continuations, at some cost to abstraction.

This paper presents a direct-style model of a ν-calculus-like language with recursion using the novel framework of *proof-relevant logical relations*, in which logical relations also contain objects (or proofs) demonstrating the equivalence of (the semantic counterparts of) programs. Apart from providing a fresh solution to an old problem, this work provides an accessible setting in which to introduce the use of proof-relevant logical relations, free of the additional complexities associated with their use for more sophisticated languages.

1 Introduction

Reasoning about contextual equivalence in higher-order languages that feature dynamic allocation of names, references, objects or keys is challenging. Pitts and Stark's ν-calculus boils the problem down to its purest form, being a total, simply-typed lambda calculus with just names and booleans as base types, an operation `new` that generates fresh names, and equality testing on names. The full equational theory of the ν-calculus is surprisingly complex and has been studied both operationally and denotationally, using logical relations [16,11], environmental bisimulations [6] and nominal game semantics [1,17].

Even before one considers 'exotic' equivalences, there are two basic equivalences that hold for essentially all forms of generativity:

$$(\texttt{let}\ x \Leftarrow \texttt{new}\ \texttt{in}\ e) = e, \text{ provided } x \text{ is not free in } e. \qquad \text{(Drop)}$$
$$(\texttt{let}\ x \Leftarrow \texttt{new}\ \texttt{in}\ \texttt{let}\ y \Leftarrow \texttt{new}\ \texttt{in}\ e) = (\texttt{let}\ y \Leftarrow \texttt{new}\ \texttt{in}\ \texttt{let}\ x \Leftarrow \texttt{new}\ \texttt{in}\ e) \text{ (Swap).}$$

The (Drop) equivalence says that removing the generation of unused names preserves behaviour; this is sometimes called the 'garbage collection' rule. The (Swap) equivalence says that the order in which names are generated is immaterial. These two equations also appear as structural congruences for name restriction in the π-calculus.

M. Hasegawa (Ed.): TLCA 2013, LNCS 7941, pp. 48–60, 2013.

Denotational models for the ν-calculus validating (Drop) and (Swap) may be constructed using (pullback-preserving) functors in $Set^{\mathbf{W}}$, where $\mathbf{W}$ is the category of sets and injections [16], or in FM-sets [10]. These models use a dynamic allocation monad to interpret possibly-allocating computations. One might expect that moving to $Cpo^{\mathbf{W}}$ or FM-cpos would allow such models to adapt straightforwardly to a language with recursion, and indeed Shinwell, Pitts and Gabbay originally proposed [15] a dynamic allocation monad over FM-cpos. However, it turned out that the underlying FM-cppo of such monad does not have least upper bounds for all finitely-supported chains. A counter-example is given in Shinwell's thesis [13, page 86]. To avoid the problem, Shinwell and Pitts subsequently [14] moved to an *indirect-style* model, using a *continuation monad* [11]: $(-)^{\top\top} \stackrel{def}{=} (- \to 1_\bot) \to 1_\bot$ to interpret computations. In particular, one shows that two programs are equivalent by proving that they co-terminate in any context. The CPS approach was also adopted by Benton and Leperchey [7] for modelling a language with references.

In the context of our on-going research on the semantics of effect-based program transformations [5], we have been developing *proof-relevant* logical relations [3]. These interpret types not merely as partial equivalence relations, as is commonly done, but as a proof-relevant generalization thereof: *setoids*. A setoid is like a category all of whose morphisms are isomorphisms (a groupoid) with the difference that no equations between these morphisms are imposed. The objects of a setoid establish that values inhabit semantic types, whilst its morphisms are understood as explicit proofs of semantic equivalence. This paper shows how we can use proof-relevant logical relations to give a direct-style model of a language with name generation and recursion, validating (Drop) and (Swap). Apart from providing a fresh approach to an old problem, our aim in doing this is to provide a comparatively accessible presentation of proof-relevant logical relations in a simple setting, free of the extra complexities associated with specialising them to abstract regions and effects [3].

Section 2 sketches the language with which we will be working, and a naive 'raw' domain-theoretic semantics for it. This semantics does not validate interesting equivalences, but is adequate. By constructing a realizability relation between it and the more abstract semantics we subsequently introduce, we will be able to show adequacy of the more abstract semantics. In Section 3 we introduce our category of setoids; these are predomains where there is a (possibly-empty) set of 'proofs' witnessing the equality of each pair of elements. We then describe pullback-preserving functors from the category of worlds $\mathbf{W}$ into the category of setoids. Such functors will interpret types of our language in the more abstract semantics, with morphisms between them interpreting terms. The interesting construction here is that of a dynamic allocation monad over the category of pullback-preserving functors. Section 4 shows how the abstract semantics is defined and related to the more concrete one. Section 5 then shows how the semantics may be used to establish equivalences involving name generation.

2 Syntax and Semantics

We work with an entirely conventional CBV language, featuring recursive functions and base types that include names, equipped with equality testing and fresh name generation (here + is just a representative operation on integers):

$$\tau := \mathtt{int} \mid \mathtt{bool} \mid \mathtt{name} \mid \tau \to \tau'$$
$$v := x \mid b \mid i \mid \mathtt{rec}\, f\, x = e$$
$$e := v \mid v + v' \mid v = v' \mid \mathtt{new} \mid \mathtt{let}\, x \Leftarrow e\, \mathtt{in}\, e' \mid v\, v'$$
$$\mathtt{if}\, v\, \mathtt{then}\, e\, \mathtt{else}\, e'$$
$$\Gamma := x_1 : \tau_1, \ldots, x_n : \tau_n$$

There are typing judgements for values, $\Gamma \vdash v : \tau$, and computations, $\Gamma \vdash e : \tau$, defined as usual. In particular, $\Gamma \vdash \mathtt{new} : \mathtt{name}$. We define a simple-minded concrete denotational semantics $\lceil\!\lceil \cdot \rceil\!\rceil$ for this language using predomains and continuous maps. For types we take

$$\lceil\!\lceil \mathtt{int} \rceil\!\rceil = \mathbb{Z} \qquad \lceil\!\lceil \mathtt{bool} \rceil\!\rceil = \mathbb{B} \qquad \lceil\!\lceil \mathtt{name} \rceil\!\rceil = \mathbb{N}$$
$$\lceil\!\lceil \tau \to \tau' \rceil\!\rceil = \lceil\!\lceil \tau \rceil\!\rceil \to (\mathbb{N} \to \mathbb{N} \times \lceil\!\lceil \tau' \rceil\!\rceil)_\bot$$
$$\lceil\!\lceil x_1 : \tau_1, \ldots, x_n : \tau_n \rceil\!\rceil = \lceil\!\lceil \tau_1 \rceil\!\rceil \times \cdots \times \lceil\!\lceil \tau_n \rceil\!\rceil$$

and there are then conventional clauses defining

$$\lceil\!\lceil \Gamma \vdash v : \tau \rceil\!\rceil : \lceil\!\lceil \Gamma \rceil\!\rceil \to \lceil\!\lceil \tau \rceil\!\rceil \qquad \text{and}$$
$$\lceil\!\lceil \Gamma \vdash e : \tau \rceil\!\rceil : \lceil\!\lceil \Gamma \rceil\!\rceil \to (\mathbb{N} \to \mathbb{N} \times \lceil\!\lceil \tau \rceil\!\rceil)_\bot$$

Note that this semantics just uses naturals to interpret names, and a state monad over names to interpret possibly-allocating computations. For allocation we take

$$\lceil\!\lceil \Gamma \vdash \mathtt{new} : \mathtt{name} \rceil\!\rceil(\eta) = [\lambda n.(n+1, n)]$$

returning the next free name and incrementing the name supply. This semantics validates no interesting equivalences involving names, but is adequate for the obvious operational semantics. Our more abstract semantics, $[\![\cdot]\!]$, will be related to $\lceil\!\lceil \cdot \rceil\!\rceil$ in order to establish *its* adequacy.

3 Proof-Relevant Logical Relations

We define the *category of setoids* as the exact completion of the category of predomains, see [9,8]. We give here an elementary description using the language of dependent types. A *setoid* A consists of a predomain $|A|$ and for any two $x, y \in |A|$ a set $A(x, y)$ of "proofs" (that x and y are equal). The set of triples $\{(x, y, p) \mid p \in A(x, y)\}$ must itself be a predomain and the first and second projections must be continuous. Furthermore, there are continuous functions $r_A : \Pi x \in |A|.A(x, x)$ and $s_A : \Pi x, y \in |A|.A(x, y) \to A(y, x)$ and $t_A : \Pi x, y, z.A(x, y) \times A(y, z) \to A(x, z)$, witnessing reflexivity, symmetry and transitivity; note that no equations between these are imposed.

We should explain what continuity of a dependent function like $t(-,-)$ is: if $(x_i)_i$ and $(y_i)_i$ and $(z_i)_i$ are ascending chains in A with suprema x, y, z and $p_i \in A(x_i, y_i)$ and $q_i \in A(y_i, z_i)$ are proofs such that $(x_i, y_i, p_i)_i$ and $(y_i, z_i, q_i)_i$ are ascending chains, too, with suprema (x, y, p) and (y, z, q) then $(x_i, z_i, t(p_i, q_i))$ is an ascending chain of proofs (by monotonicity of $t(-,-)$) and its supremum is $(x, z, t(p, q))$. Formally, such dependent functions can be reduced to non-dependent ones using pullbacks, that is t would be a function defined on the pullback of the second and first projections from $\{(x, y, p) \mid p \in A(x, y)\}$ to $|A|$, but we find the dependent notation to be much more readable. If $p \in A(x, y)$ we may write $p : x \sim y$ or simply $x \sim y$. We also omit $|-|$ wherever appropriate. We remark that "setoids" also appear in constructive mathematics and formal proof, see *e.g.*, [2], but the proof-relevant nature of equality proofs is not exploited there and everything is based on sets (types) rather than predomains. A morphism from setoid A to setoid B is an equivalence class of pairs $f = (f_0, f_1)$ of continuous functions where $f_0 : |A| \to |B|$ and $f_1 : \Pi x, y \in |A|.A(x, y) \to B(f_0(x), f_0(y))$. Two such pairs $f, g : A \to B$ are *identified* if there exists a continuous function $\mu : \Pi a \in |A|.B(f(a), g(a))$.

Proposition 1. *The category of setoids is cartesian closed; moreover, if D is a setoid such that $|D|$ has a least element $\bot$ and there is also a least proof $\bot \in D(\bot, \bot)$ then there is a morphism of setoids $Y : [D \to D] \to D$ satisfying the usual fixpoint equations.*

Definition 1. *A setoid D is discrete if for all $x, y \in D$ we have $|D(x, y)| \leq 1$ and $|D(x, y)| = 1 \iff x = y$.*

Thus, in a discrete setoid proof-relevant equality and actual equality coincide and moreover any two equality proofs are actually equal (proof irrelevance).

3.1 Pullback Squares

Pullback squares are a central notion in our framework. As it will become clear later, they are the "proof-relevant" component of logical relations. Recall that a morphism u in a category is a monomorphism if $ux = ux'$ implies $x = x'$ for all morphisms x, x'. A commuting square $xu = x'u'$ of morphisms is a *pullback* if whenever $xv = x'v'$ there is unique t such that $v = ut$ and $v' = u't$. This can be visualized as follows:

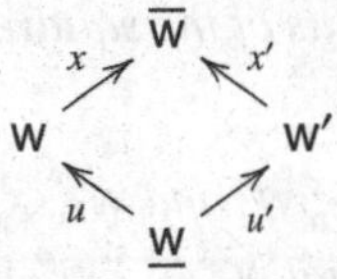

We write ${}^{x}_{u}\Diamond^{x'}_{u'}$ or $\mathsf{w}\,{}^{x}_{u}\Diamond^{x'}_{u'}\,\mathsf{w}'$ (when $\mathsf{w}^{(\prime)} = \mathrm{dom}(x^{(\prime)})$) for such a pullback square. We call the common codomain of x and x' the *apex* of the pullback, written $\overline{\mathsf{w}}$, while the common domain of u, u' is the *low point* of the square, written $\underline{\mathsf{w}}$. A pullback square $xu = x'u'$ is *minimal* if whenever $fx = gx$ and $fx' = gx'$ then $f = g$, in other words, x and x' are *jointly epic*. A pair of morphisms u, u' with common domain is a span, a pair of morphisms x, x' with common codomain is a co-span. A category has pullbacks if every co-span can be completed to a pullback square.

In our more general treatment of proof-relevant logical relations for reasoning about stateful computation [3], we treat worlds axiomatically, defining a category of worlds to be a category with pullbacks in which every span can be completed to a minimal pullback square, and all morphisms are monomorphisms. That report gives various useful examples, including ones built from PERs on heaps. For the simpler setting of this paper, however, we fix on one particular instance:

Definition 2 (Category of worlds). *The* category of worlds $\mathbf{W}$ *has finite sets of natural numbers as objects and injective functions for morphisms.*

An object w of $\mathbf{W}$ is a set of generated/allocated names, with injective maps corresponding to renamings and extensions with newly generated names.

Given $f : X \to Z$ and $g : Y \to Z$ forming a co-span in $\mathbf{W}$, we form their pullback as $X \xleftarrow{f^{-1}} fX \cap gY \xrightarrow{g^{-1}} Y$. This is minimal when $fX \cup gY = Z$. Conversely, given a span $Y \xleftarrow{f} X \xrightarrow{g} Z$, we can complete to a minimal pullback by

$$(Y \setminus fX) \uplus fX \xrightarrow{[in_1, in_3 \circ f^{-1}]} (Y \setminus fX) + (Z \setminus gX) + X \xleftarrow{[in_2, in_3 \circ g^{-1}]} (Z \setminus gX) \uplus gX$$

where $[-,-]$ is case analysis on the disjoint union $Y = (Y \setminus fX) \uplus fX$. Thus a minimal pullback square in $\mathbf{W}$ is of the form:

$$\begin{array}{ccc} & X_1' \cup X_2' & \\ {}^{x}\nearrow & & \nwarrow^{x'} \\ X_1 \cong X_1' & & X_2 \cong X_2' \\ {}_{u}\nwarrow & & \nearrow_{u'} \\ & X_1' \cap X_2' & \end{array}$$

Such a minimal pullback corresponds to a *partial bijection* between X_1 and X_2, as used in other work on logical relations for generativity [12,4]. We write $u : x \hookrightarrow y$ to mean that u is a subset inclusion and note that if we have a span u, u' then we can choose x, x' so that ${}^{x}_{u}\diamond^{x'}_{u'}$ is a minimal pullback and x' is an inclusion, too. To do that, we simply replace the apex of any minimal pullback completion with an isomorphic one. The analogous property holds for completion of co-spans to pullbacks.

Definition 3. *Two pullbacks* $\mathsf{w}\,{}^{x}_{u}\diamond^{x'}_{u'}\,\mathsf{w}'$ *and* $\mathsf{w}\,{}^{y}_{v}\diamond^{y'}_{v'}\,\mathsf{w}'$ *are isomorphic if there is an isomorphism f between the two low points of the squares so that $vf = u$ and $v'f = u'$, thus also $uf^{-1} = v$ and $u'f^{-1} = v'$.*

Lemma 1. *If* $\mathsf{w}, \mathsf{w}', \mathsf{w}'' \in \mathbf{W}$, *if* $\mathsf{w}\,{}^{x}_{u}\diamond^{x'}_{u'}\,\mathsf{w}'$ *and* $\mathsf{w}'\,{}^{y}_{v}\diamond^{y'}_{v'}\,\mathsf{w}''$ *are pullback squares as indicated then there exist z, z', t, t' such that* $\mathsf{w}\,{}^{zx}_{ut}\diamond^{z'y'}_{v't'}\,\mathsf{w}''$ *is also a pullback.*

Proof. Choose z, z', t, t' in such a way that ${}^{z}_{x'}\diamond^{z'}_{y}$ and ${}^{u'}_{t}\diamond^{v}_{t'}$ are pullbacks. The verifications are then an easy diagram chase.

We write $r(\mathsf{w})$ for $\mathsf{w}\,{}^{1}_{1}\diamond^{1}_{1}\,\mathsf{w}$ and $s({}^{x}_{u}\diamond^{x'}_{u'}) = {}^{x'}_{u'}\diamond^{x}_{u}$ and $t({}^{x}_{u}\diamond^{x'}_{u'}, {}^{y}_{v}\diamond^{y'}_{v'}) = {}^{zx}_{z'y'}\diamond^{ut}_{v't'}$ where z, z', t, t' are given by Lemma 1 (which requires choice).

Lemma 2. *A pullback square* ${}^{x}_{u}\diamond^{x'}_{u'}$ *in* $\mathbf{W}$ *is isomorphic to* $t({}^{x}_{1}\diamond^{1}_{x}, {}^{1}_{x'}\diamond^{x'}_{1})$.

3.2 Setoid-Valued Functors

A functor A (actually a pseudo functor) from the category of worlds $\mathbf{W}$ to the category of setoids comprises as usual for each $\mathsf{w} \in \mathbf{W}$ a setoid $A\mathsf{w}$ and for each $u : \mathsf{w} \to \mathsf{w}'$ a morphism of setoids $Au : A\mathsf{w} \to A\mathsf{w}'$ preserving identities and composition; for an identity morphism id, a continuous function of type $\Pi a.A\mathsf{w}(a, (Aid)\,a)$; and for two morphisms $u : \mathsf{w} \to \mathsf{w}_1$ and $v : \mathsf{w}_1 \to \mathsf{w}_2$ a continuous function of type $\Pi a.A\mathsf{w}_2(Av(Au\,a), A(vu)\,a)$.

If $u : \mathsf{w} \to \mathsf{w}'$ and $a \in A\mathsf{w}$ we may write $u.a$ or even ua for $Au(a)$ and likewise for proofs in $A\mathsf{w}$. Note that there is a proof of equality of $(uv).a$ and $u.(v.a)$.

In the sequel, we shall abbreviate these setoid-valued (pseudo-)functors as s.v.f.

Intuitively, s.v.f. will become the denotations of value types and computations. Thus, an element of $A\mathsf{w}$ represents values involving the names in w. If $u : \mathsf{w} \to \mathsf{w}_1$ then $A\mathsf{w} \ni a \mapsto u.a \in A\mathsf{w}_1$ represents renaming and possible weakening by names not "actually" occurring in a. Note that due to the restriction to injective functions identification of names ("contraction") is precluded. This is in line with Stark's use of set-valued functors on the category **W** to model fresh names.

Definition 4. *We call a functor A pullback-preserving (s.v.f.) if for every pullback square $\mathsf{w}{}^{x}_{u}\!\Diamond^{x'}_{u'}\mathsf{w}'$ with apex $\overline{\mathsf{w}}$ and low point $\underline{\mathsf{w}}$ the diagram $A\mathsf{w}{}^{Ax}_{Au}\!\Diamond^{Ax'}_{Au'}A\mathsf{w}'$ is a pullback in Std. This means that there is a continuous function of type*

$$\Pi a \in A\mathsf{w}.\Pi a' \in A\mathsf{w}'.A\overline{\mathsf{w}}(x.a, x'.a') \to \Sigma \underline{a} \in A\underline{\mathsf{w}}.A\mathsf{w}(u.\underline{a}, a) \times A\mathsf{w}'(u'.\underline{a}, a')$$

Thus, if two values $a \in A\mathsf{w}$ and $a' \in A\mathsf{w}'$ are equal in a common world $\overline{\mathsf{w}}$ then this can only be the case because there is a value in the "intersection world" $\underline{\mathsf{w}}$ from which both a, a' arise.

All the s.v.f. that we define in this paper will turn out to be pullback-preserving. However, for the results described in this paper pullback preservation is not needed. Thus, we will not use it any further, but note that there is always the option to require that property should the need arise subsequently.

Lemma 3. *If A is a s.v.f., $u : \mathsf{w} \to \mathsf{w}'$ and $a, a' \in A\mathsf{w}$, there is a continuous function $A\mathsf{w}'(u.a, u.a') \to A\mathsf{w}(a, a')$. Moreover, the "common ancestor" $\underline{a}$ of a and a' is unique up to $\sim$.*

Note that the ordering on worlds and world morphisms is discrete so that continuity only refers to the $A\mathsf{w}'(u.a, u.a')$ argument.

Definition 5 (Morphism of functors). *If A, B are s.v.f., a morphism from A to B is a pair $e = (e_0, e_1)$ of continuous functions where $e_0 : \Pi\mathsf{w}.A\mathsf{w} \to B\mathsf{w}$ and $e_1 : \Pi\mathsf{w}.\Pi\mathsf{w}'.\Pi x : \mathsf{w} \to \mathsf{w}'.\Pi a \in A\mathsf{w}.\Pi a' \in A\mathsf{w}'.A\mathsf{w}'(x.a, a') \to B\mathsf{w}'(x.e_0(a), e_0(a'))$. A proof that morphisms e, e' are equal is given by a continuous function $\mu : \Pi\mathsf{w}.\Pi a \in A\mathsf{w}.B\mathsf{w}(e(a), e'(a))$.*

These morphisms compose in the obvious way and so the s.v.f. and morphisms between them form a category.

3.3 Instances of Setoid-Valued Functors

We now describe some concrete functors that will allow us to interpret types of the ν-calculus as s.v.f. The simplest one endows any predomain with the structure of a s.v.f. where the equality is proof-irrelevant and coincides with standard equality. The second one generalises the function space of setoids and is used to interpret function types. The third one is used to model dynamic allocation and is the only one that introduces proper proof-relevance.

Constant Functor. Let D be a predomain. Then the s.v.f. over this domain, written also as D, has D itself as underlying set (irrespective of w), i.e., $D\mathsf{w} = D$ and $D\mathsf{w}(d, d')$ is given by a singleton set, say, $\{\star\}$ if $d = d'$ and is empty otherwise.

Names. The s.v.f. N of names is given by $N\mathsf{w} = \mathsf{w}$ where w on the right hand side stands for the discrete setoid over the discrete cpo of locations in w. Thus, e.g. $N\{1, 2, 3\} = \{1, 2, 3\}$.

Product. Let A and B be s.v.f. The product $A \times B$ is the s.v.f. given as follows. We have $(A \times B)\mathsf{w} = A\mathsf{w} \times B\mathsf{w}$ (product predomain) and $(A \times B)\mathsf{w}((a, b), (a', b')) = A\mathsf{w}(a, a') \times B\mathsf{w}(b, b')$. This defines a cartesian product on the category of s.v.f. More generally, we can define indexed products $\prod_{i \in I} A_i$ of a family $(A_i)_i$ of s.v.f.

Function Space. Let A and B be s.v.f. The function space $A \Rightarrow B$ is the s.v.f. given as follows. We have $(\mathsf{f}_0, \mathsf{f}_1) \in (A \Rightarrow B)\mathsf{w}$ when f_0 has type $\Pi \mathsf{w}_1 . \Pi u : \mathsf{w} \to \mathsf{w}_1 . A\mathsf{w}_1 \to B\mathsf{w}_1$, that is, it takes a morphism $u : \mathsf{w} \to \mathsf{w}_1$ and an object in $A\mathsf{w}_1$ and returns an object in $B\mathsf{w}_1$. The second component, f_1, which takes care of proofs is a bit more complicated, having type:

$$\begin{aligned}&\Pi \mathsf{w}_1 . \Pi \mathsf{w}_2 . \Pi u : \mathsf{w} \to \mathsf{w}_1 . \Pi v : \mathsf{w}_1 \to \mathsf{w}_2 . \Pi a \in A\mathsf{w}_1 . \Pi a' \in A\mathsf{w}_2 .\\ &\quad A\mathsf{w}_2(v.a, a') \to B\mathsf{w}_2(v.\mathsf{f}_0(u, a), \mathsf{f}_0(vu, a'))\end{aligned}$$

Intuitively, the definition above encompasses two desired properties. The first one is when v is instantiated as the identity yielding a function of mapping proofs in $A\mathsf{w}_1$ to proofs in $B\mathsf{w}_1$:

$$\Pi \mathsf{w}_1 . \Pi u : \mathsf{w} \to \mathsf{w}_1 . \Pi a \in A\mathsf{w}_1 . \Pi a' \in A\mathsf{w}_1 . A\mathsf{w}_1(a, a') \to B\mathsf{w}_1(\mathsf{f}_0(u, a), \mathsf{f}_0(u, a'))$$

We note that, since A is only a pseudo functor we must compose with a proof that $id.f_0(u, a)$ equals $f_0(u, a)$.

The second desired property is that the proof in $B\mathsf{w}_2$ can be achieve either by obtaining an object, $\mathsf{f}_0(vu, a')$, the directly from $A\mathsf{w}_2$, or by first obtaining an object $\mathsf{f}_0(u, a)$ in $B\mathsf{w}_1$ and then taking it to $B\mathsf{w}_2$ by using v.

Definition 6. *A s.v.f. A is discrete if* $A\mathsf{w}$ *is a discrete setoid for every world* w.

The constructions presented so far only yield discrete s.v.f., *i.e.*, proof relevance is merely propagated but never actually created. This is not so for the next operator on s.v.f. which is to model dynamic allocation.

Dynamic Allocation Monad. Finally, the third instantiation is the dynamic allocation monad T. For natural number n let us write $[n]$ for the set $\{1, \dots, n\}$.

Let A be a s.v.f., then the elements of $(TA)\mathsf{w}$ are again pairs $(\mathsf{c}_0, \mathsf{c}_1)$ where c_0 is of type

$$\Pi n \in \{n \mid [n] \supseteq \mathsf{w}\} . (\Sigma \mathsf{w}_1 . I(\mathsf{w}, \mathsf{w}_1) \times A\mathsf{w}_1 \times \{n_1 \mid [n_1] \supseteq \mathsf{w}_1\})_\bot$$

where $I(\mathsf{w}, \mathsf{w}_1)$ is the set of inclusions $u : \mathsf{w} \hookrightarrow \mathsf{w}_1$ and such that either $\mathsf{c}_0(n) = \bot$ for all n such that $[n] \supseteq \mathsf{w}$ or else $\mathsf{c}_0(n) \neq \bot$ for all such n. The naturals n and n_1 represent concrete allocator states, whilst w and w_1 are smaller sets of names on which values actually depend.

The second component c_1 assigns to any two n, n' with $[n] \supseteq \mathsf{w}, [n'] \supseteq \mathsf{w}$ where $\mathsf{c}_0(n) = (\mathsf{w}_1, u, \mathsf{v}, n_1)$ and $\mathsf{c}_0(n') = (\mathsf{w}'_1, u', \mathsf{v}', n'_1)$ a co-span x, x' such that $xu = x'u'$ and a proof $p \in A\overline{\mathsf{w}}(x.\mathsf{v}, x'.\mathsf{v}')$ with $\overline{\mathsf{w}}$ the apex of the co-span.

The ordering on $(TA)\mathsf{w}$ is given by $(\mathsf{c}_0, \mathsf{c}_1) \sqsubseteq (\mathsf{c}'_0, \mathsf{c}'_1)$ just when $\mathsf{c}_0 \sqsubseteq \mathsf{c}'_0$ in the natural componentwise fashion (the second components are ignored).

A proof in $TA\mathsf{w}((\mathsf{c}_0, \mathsf{c}_1), (\mathsf{c}'_0, \mathsf{c}'_1))$ is defined analogously. For any n such that $[n] \supseteq \mathsf{w}$ it must be that $\mathsf{c}_0(n) = \bot \iff \mathsf{c}'_0(n) = \bot$ (otherwise there is no proof) and if $\mathsf{c}_0(n) = (\mathsf{w}_1, u, \mathsf{v}, n_1)$ and $\mathsf{c}'_0(n) = (\mathsf{w}'_1, u', \mathsf{v}', n'_1)$ then the proof must assign a co-span x, x' such that $xu = x'u'$ and a proof $p \in A\overline{\mathsf{w}}(x.\mathsf{v}, x'.\mathsf{v}')$ with $\overline{\mathsf{w}}$ the apex of the co-span. If $\mathsf{c}_0(n) = \mathsf{c}'_0(n) = \bot$ then the proof is trivial (need not return anything).

To make TA a pseudo-functor, we also have to give its action on morphisms. Assume that $u : \mathsf{w} \to \mathsf{q}$ is a morphism in $\mathbf{W}$. We want to construct a morphism $(TA)u : (TA)\mathsf{w} \to (TA)\mathsf{q}$ in *Std*, so assume $(\mathsf{c}_0, \mathsf{c}_1) \in (TA)\mathsf{w}$ and $[m] \supseteq \mathsf{q}$. We let n be the largest element of w, an arbitrary choice ensuring $[n] \supseteq \mathsf{w}$. If $\mathsf{c}_0(n) = \bot$, then define $\mathsf{d}_0(m) = \bot$ too. Otherwise $\mathsf{c}_0(n) = (\mathsf{w}_1, i : \mathsf{w} \hookrightarrow \mathsf{w}_1, \mathsf{v}, n_1)$, and we define $\mathsf{d}_0(m) = (\mathsf{q}_1, i', u_1.\mathsf{v}, m_1)$ where $\mathsf{q}_1, i' : \mathsf{q} \hookrightarrow \mathsf{q}_1$ and $u_1 : \mathsf{w}_1 \to \mathsf{q}_1$ are chosen to make $\mathsf{q}\,{}^{i'}_{u}\!\diamond^{u_1}_{i}\,\mathsf{w}_1$ a minimal pullback, and m_1 is (again arbitrarily) the largest element of q_1. We then take $(TA)(u)(\mathsf{c}_0, \mathsf{c}_1)$ to be $(\mathsf{d}_0, \mathsf{d}_1)$, where d_1 just has to return identity co-spans. This specifies how the functor TA transports objects from w to q using the morphism u.

The following diagram illustrates how an equality proof in $TA((\mathsf{c}_0, \mathsf{c}_1), (\mathsf{c}'_0, \mathsf{c}'_1))$ is transported to an equality proof in $TA(u(\mathsf{c}_0, \mathsf{c}_1), u(\mathsf{c}'_0, \mathsf{c}'_1))$.

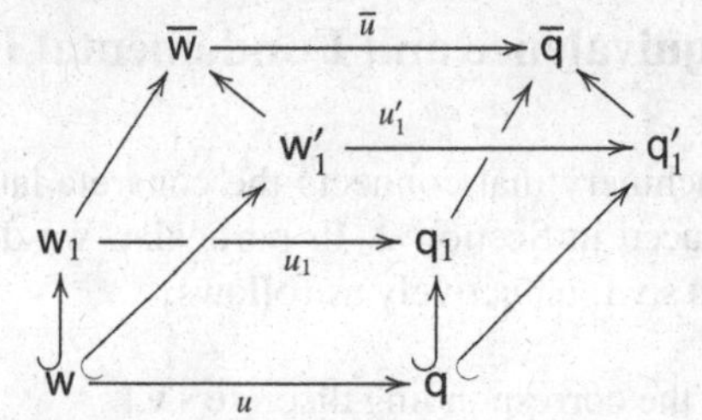

Here $\mathsf{w}_1 \diamond \mathsf{w}'_1$ and $\mathsf{q}_1 \diamond \mathsf{q}'_1$ are pullback squares. It is easy to check how the morphisms u_1, u'_1 and $\overline{u}$ are constructed. Then we can take the values a and a' in $A\mathsf{w}_1$ and $A\mathsf{w}'_1$ and the proof p in $A\overline{\mathsf{w}}$ to the pullback square $\mathsf{q}_1 \diamond \mathsf{q}'_1$, by using u_1, u'_1 and $\overline{u}$, *i.e.*, $u_1.a \in A\mathsf{q}_1$, $u'_1.a' \in A\mathsf{q}'_1$ and $\overline{u}.p \in A\overline{\mathsf{q}}$.

The following is direct from the definitions.

Proposition 2. *T is a monad on the category of s.v.f.; the unit sends $\mathsf{v} \in A\mathsf{w}$ to $(\mathsf{w}, id_{\mathsf{w}}, \mathsf{v}, n) \in (TA)\mathsf{w}$ and the multiplication sends $(\mathsf{w}_1, u, (\mathsf{w}_2, v, \mathsf{v}, n_2), n_1) \in (TTA)\mathsf{w}$ to $(\mathsf{w}_2, vu, \mathsf{v}, n_2) \in TA\mathsf{w}$. If $\mu : A \to B$ then $T\mu : TA \to TB$ at world w sends $(\mathsf{w}_1, u, \mathsf{v}, n_1) \in TA\mathsf{w}$ to $(\mathsf{w}_1, u, \mu u(\mathsf{v}), n_1) \in TB\mathsf{w}$.*

Comparison with FM domains. It is well-known that Gabbay-Pitts FM-sets [10] are equivalent to pullback-preserving functors from our category of worlds $\mathbf{W}$ to the category of sets. Likewise, Pitts and Shinwell's FM-domains are equivalent to pullback preserving functors from **W** to the category of domains, thus corresponding exactly to the pullback-preserving discrete s.v.f.

As mentioned in the introduction, Mark Shinwell discusses a flawed attempt at defining a name allocation monad on the category of FM-domains which when transported along the equivalence between FM-domains and discrete s.v.f. would look as follows: Given a discrete s.v.f. A and world w define $SA\mathsf{w}$ as the set of triples $(\mathsf{w}_1, u, \mathsf{v})$ where $u : \mathsf{w} \hookrightarrow \mathsf{w}_1$ and $\mathsf{v} \in A\mathsf{w}_1$ modulo the equivalence relation generated by the identification of $(\mathsf{w}_1, u, \mathsf{v})$ with $(\mathsf{w}'_1, u', \mathsf{v}')$ if there exists a co-span v, v' such that $vu = v'u'$ and $v.\mathsf{v} = v'.\mathsf{v}'$.

As for the ordering, the only reasonable choice is to decree that on representatives $(\mathsf{w}_1, u, \mathsf{v}) \leq (\mathsf{w}'_1, u', \mathsf{v}')$ if $v.\mathsf{v} \leq v'.\mathsf{v}'$ for some co-span v, v' with $vu = v'u'$. However, while this defines a partial order it is not clear why it should have suprema of ascending chains and indeed, Shinwell's thesis [13] contains a concrete counterexample.

We also remark that this construction *does* work if we work with sets rather than predomains and thus do not need orderings or suprema. However, the exact completion of the category sets being equivalent to the category of sets itself is not very surprising.

The previous solution to this conundrum was to move to a continuation-passing style semantics or, equivalently, to use $\top\top$-closure. Intuitively, rather than quantifying existentially over sets of freshly allocated names, one quantifies universally over continuations, which has better order-theoretic properties. Using continuations, however, makes the derivation of concrete equivalences much more difficult and in some cases we still do not know whether it is possible at all.

4 Observational Equivalence and Fundamental Lemma

We now construct the machinery that connects the concrete language with the denotational machinery introduced in Section 2. In particular, we define the semantics of types, written using $[\![\cdot]\!]$, as s.v.f. inductively as follows:

- For basic types $[\![\tau]\!]$ is the corresponding discrete s.v.f..
- $[\![\tau \to \tau']\!]$ is defined as the function space $[\![\tau]\!] \to T[\![\tau]\!]$, where T is the dynamic allocation monad.
- For typing context Γ we define $[\![\Gamma]\!]$ as the indexed product of s.v.f. $\prod_{x \in \mathrm{dom}(\Gamma)} [\![\Gamma(x)]\!]$.

To each term in context $\Gamma \vdash e : \tau$ we can associate a morphism $[\![e]\!]$ from $[\![\Gamma]\!]$ to $T[\![\tau]\!]$ by interpreting the syntax in the category of s.v.f. using cartesian closure and the fact that T is a monad. We omit the straightforward but perhaps slightly tedious definition and only give the clause for "new" here:

$$[\![\texttt{new}\,]\!]\mathsf{w}(n) = (\mathsf{w} \cup \{n+1\}, u, n+1, n+1)$$

Here $u : \mathsf{w} \hookrightarrow \mathsf{w} \cup \{n+1\}$ is the inclusion. Note that since $[n] \supseteq \mathsf{w}$ we have $n+1 \notin \mathsf{w}$.

Our aim is now to relate these morphisms to the computational interpretation $\lceil\!\lceil e \rceil\!\rceil$.

Definition 7. *For each type τ and world w we define a relation $\Vdash^{\tau}_{\mathsf{w}} \subseteq [\![\tau]\!] \times [\![\tau]\!]\mathsf{w}$:*

$$b \Vdash^{\mathtt{bool}}_{\mathsf{w}} \mathsf{b} \iff b = \mathsf{b}$$
$$i \Vdash^{\mathtt{int}}_{\mathsf{w}} \mathsf{i} \iff i = \mathsf{i}$$
$$l \Vdash^{\mathtt{name}}_{\mathsf{w}} k \iff l = k$$
$$f \Vdash^{\tau \to \tau'}_{\mathsf{w}} g \iff \forall \mathsf{w}_1. \forall u : \mathsf{w} \hookrightarrow \mathsf{w}_1. \forall v\, \mathsf{v}. v \Vdash^{\tau}_{\mathsf{w}_1} \mathsf{v} \Rightarrow f(v) \Vdash^{T\tau'}_{\mathsf{w}_1} g_0(u, \mathsf{v})$$

$$c \Vdash^{T\tau}_{\mathsf{w}} \mathsf{c} \iff \begin{array}{l} \forall n. \mathsf{w} \subseteq [n] \Rightarrow (\mathsf{c}(n) = \bot \Leftrightarrow c(n) = \bot) \wedge \\ (\mathsf{c}(n) = (\mathsf{w}_1, u : \mathsf{w} \hookrightarrow \mathsf{w}_1, \mathsf{v}, n_1) \wedge c(n) = (n_1', v) \Rightarrow n_1 = n_1' \wedge v \Vdash^{\tau}_{\mathsf{w}_1} \mathsf{v})). \end{array}$$

The realizability relation for the allocation monad thus specifies that the abstract computation c is related to the concrete computation c at world w if they co-terminate, and if they do terminate then the resulting values are also related.

The following is a direct induction on types.

Lemma 4. *If $u : \mathsf{w} \hookrightarrow \mathsf{w}_1$ is an inclusion as indicated and $v \Vdash^{\tau}_{\mathsf{w}} \mathsf{v}$ then $v \Vdash^{\tau}_{\mathsf{w}_1} u.\mathsf{v}$, too.*

We extend $\Vdash$ to typing contexts by putting

$$\eta \Vdash^{\Gamma}_{\mathsf{w}} \gamma \iff \forall x \in \mathrm{dom}(\Gamma). \eta(x) \Vdash^{\Gamma(x)}_{\mathsf{w}} \gamma(x)$$

for $\eta \in [\![\Gamma]\!]$ and $\gamma \in [\![\Gamma]\!]$.

Theorem 1 (Fundamental lemma). *If $\Gamma \vdash e : \tau$ then whenever $\eta \Vdash^{\Gamma}_{\mathsf{w}} \gamma$ then $[\![e]\!]\eta \Vdash^{T\tau}_{\mathsf{w}}$ $[\![e]\!](\gamma)$.*

Proof. By induction on typing rules.

The most interesting case is for the `let` case: Assume that $\Gamma \vdash \mathtt{let}\, x {\Leftarrow} e_1\, \mathtt{in}\, e_2 : \tau_2$, where $\Gamma \vdash e_1 : \tau_1$ and $\Gamma, x : \tau_1 \vdash e_2 : \tau_2$. Moreover, assume that $\eta \Vdash^{\Gamma}_{\mathsf{w}} \gamma$, where w is an initial world and that (H1) $[\![e_1]\!]\eta \Vdash^{T\tau_1}_{\mathsf{w}} [\![e_1]\!](\gamma)$ and (H2) $[\![e_2]\!](\eta, x) \Vdash^{T\tau_2}_{\mathsf{w}_1} [\![e_2]\!](\gamma, [\![x]\!])$ for all $x \Vdash^{\tau_1}_{\mathsf{w}} [\![x]\!]$ and world extension w_1, that is, a world for which there is an inclusion $u : \mathsf{w} \hookrightarrow \mathsf{w}_1$. We define $[\![\mathtt{let}\, x {\Leftarrow} e_1\, \mathtt{in}\, e_2]\!](\mathsf{w})(\gamma)(n)$ for some n where $\mathsf{w} \subseteq [n]$ as follows: If $[\![e_1]\!]\mathsf{w}(\gamma)(n) = (\mathsf{w}_1, u_1 : \mathsf{w} \hookrightarrow \mathsf{w}_1, \mathsf{v}_1, n_1)$ and that $[\![e_2]\!]\mathsf{w}_1(\gamma, \mathsf{v}_1)(n_1) = (\mathsf{w}_2, u_2 : \mathsf{w}_1 \hookrightarrow \mathsf{w}_2, \mathsf{v}_2, n_2)$. Then

$$[\![\mathtt{let}\, x {\Leftarrow} e_1\, \mathtt{in}\, e_2]\!](\mathsf{w})(\gamma)(n) = (\mathsf{w}_2, u_2 u_1 : \mathsf{w} \hookrightarrow \mathsf{w}_2, \mathsf{v}_2, n_2).$$

Otherwise $[\![\mathtt{let}\, x {\Leftarrow} e_1\, \mathtt{in}\, e_2]\!](\mathsf{w})(\gamma)(n) = \bot$ if $[\![e_1]\!]\mathsf{w}(\gamma)(n) = \bot$ or if $[\![e_1]\!]\mathsf{w}(\gamma)(n) = (\mathsf{w}_1, u_1 : \mathsf{w} \hookrightarrow \mathsf{w}_1, \mathsf{v}_1, n_1)$, but $[\![e_2]\!]\mathsf{w}_1(\gamma, \mathsf{v}_1)(n_1) = \bot$.

We only show the case where $[\![\mathtt{let}\, x {\Leftarrow} e_1\, \mathtt{in}\, e_2]\!](\mathsf{w})(\gamma)(n)$ is different from $\bot$. The other cases are straighforward. Assume that $[\![e_1]\!]\eta(n) = (v_1, n_1')$. From (H1), we have $n_1' = n_1$ and that $v_1 \Vdash^{\tau_1}_{\mathsf{w}_1} \mathsf{v}_1$. Thus from Lemma 4, we have $\eta, v_1 \Vdash^{\tau_1}_{\mathsf{w}_1} \gamma, \mathsf{v}_1$. Now, assume that $[\![e_2]\!](\eta, v_1)(n_1) = (v_2, n_2')$. Thus from (H2), we have that $n_2 = n_2'$ and that $v_2 \Vdash^{\tau_2}_{\mathsf{w}_2}$ v_2. This finishes the proof, as it is enough to conclude that $[\![\mathtt{let}\, e_1 {\Leftarrow} e_2\, \mathtt{in}\,]\!]\eta \Vdash^{T\tau_2}_{\mathsf{w}}$ $[\![\mathtt{let}\, e_1 {\Leftarrow} e_2\, \mathtt{in}\,]\!]\gamma$.

It is now possible to validate a number of equational rules on the level of the setoid semantics $[\![-]\!]$ including transitivity, $\beta\eta$, fixpoint unrolling, and congruence rules. We omit the definition of such an equational theory here and refer to [3] for details on how this could be set up. As we now show equality on the level of the setoid semantics entails observational equivalence on the level of the raw denotational semantics.

4.1 Observational Equivalence

Definition 8. *Let τ be a type. We define an* observation of type τ *as a closed term* $\vdash o : \tau \to \texttt{bool}$. *Two values* $v, v' \in \llbracket\tau\rrbracket$ *are* observationally equivalent at type τ *if for all observations* o *of type* τ *one has that* $\llbracket o\rrbracket(v)(0)$ *is defined iff* $\llbracket o\rrbracket(v')(0)$ *is defined and when* $\llbracket o\rrbracket(v)(0) = (n_1, v_1)$ *and* $\llbracket o\rrbracket(v')(0) = (n'_1, v'_1)$ *then* $v_1 = v'_1$.

We now show how the proof-relevant semantics can be used to deduce observational equivalences.

Theorem 2 (Observational equivalence). *If* τ *is a type and* $v \Vdash^{\tau}_{\emptyset} e$ *and* $v' \Vdash^{\tau}_{\emptyset} e'$ *with* $e \sim e'$ *in* $\llbracket\tau\rrbracket\emptyset$ *then* v *and* v' *are observationally equivalent at type* τ.

Proof. Let o be an observation at type τ. By the Fundamental Lemma (Theorem 1) we have $\llbracket o\rrbracket \Vdash^{\tau\to\texttt{bool}}_{\emptyset} \llbracket o\rrbracket$.

Now, since $e \sim e'$ we also have $\llbracket o\rrbracket(e) \sim \llbracket o\rrbracket(e')$ and, of course, $\llbracket o\rrbracket(v) \Vdash^{T\texttt{bool}}_{\emptyset} \llbracket o\rrbracket(e)$ and $\llbracket o\rrbracket(v') \Vdash^{T\texttt{bool}}_{\emptyset} \llbracket o\rrbracket(e')$.

From $\llbracket o\rrbracket(e) \sim \llbracket o\rrbracket(e')$ we conclude that either $\llbracket o\rrbracket(e)(0)$ and $\llbracket o\rrbracket(e')(0)$ both diverge in which case the same is true for $\llbracket o\rrbracket(v)(0)$ and $\llbracket o\rrbracket(v')(0)$ by definition of $\Vdash^{T\tau}$. Secondly, if $\llbracket o\rrbracket(e)(0)) = (\_, \_, b, \_)$ and $\llbracket o\rrbracket(e')(0)) = (\_, \_, b', \_)$ for booleans b, b' then, by definition of $\sim$ at $\llbracket T\tau\rrbracket$ we get $b = b'$ and, again by definition of $\Vdash^{T\tau}$ this then implies that $\llbracket o\rrbracket(v)(0) = (\_, b)$ and $\llbracket o\rrbracket(v)(0) = (\_, b')$ with $b = b'$, hence the claim.

5 Direct-Style Proofs

We now have enough machinery to provide a direct-style proofs for equivalences involving name generation.

Drop equation. We start with the following equation, which allows to eliminate a dummy allocation:

$$c = (\texttt{let}\, x \Leftarrow \texttt{new in}\, e) = e, \text{ provided } x \text{ is not free in } e = c'.$$

Assume an initial world w and suppose that $c' \Vdash^{TA}_{\mathsf{w}} \mathsf{c}'$, where c' is an abstract computation related to c' at world w. We provide a semantic computation c, such that $c \Vdash^{TA}_{\mathsf{w}} \mathsf{c}$, that is, it is related to the computation that performs a dummy allocation, and we also provide a proof $\mathsf{c} \sim \mathsf{c}'$. From Theorem 2, this means that the two computations are observationally equivalent. Let $\mathsf{c} = (\mathsf{w}, id : \mathsf{w} \hookrightarrow \mathsf{w}, c', n)$, which does not advance the world. We can show that it is related to the expression c, with the dummy allocation, *i.e.*, $c \Vdash^{\Gamma\vdash TA}_{\mathsf{w}} \mathsf{c}$ by opening its definition, stated in Definition 7:

$$\begin{aligned}&\forall n.\mathsf{w} \subseteq [n] \Rightarrow (\mathsf{c} = \bot \Leftrightarrow c(n) = \bot) \wedge \\ &\quad (\mathsf{c} = (\mathsf{w}, id : \mathsf{w} \hookrightarrow \mathsf{w}, \mathsf{c}', n) \wedge c(n) = ([n_1], c') \Rightarrow (\mathsf{w} \subseteq [n_1] \wedge c' \Vdash^{A}_{\mathsf{w}} \mathsf{c}')).\end{aligned}$$

where the value c' resulting is exactly the function without the dummy allocation, thus $\mathsf{c} \sim \mathsf{c}'$ with the identity pullback square. The key observation is that heaps $[n]$ are allowed to contain more locations that those in w, containing the locations that one actually needs.

Notice as well that if we were to annotate monads with the corresponding effects of the function, such as read, write or allocation effects, as done in [4], from the proof above the first allocation in c with the dummy allocation would not need to flag an allocation effect. That is, that step could be considered pure.

Swap equation. Let us now consider the following equivalence where the order in which the names are generated is switched:

$$c = (\texttt{let}\, x \Leftarrow \texttt{new in let}\, y \Leftarrow \texttt{new in}\, e) = (\texttt{let}\, y \Leftarrow \texttt{new in let}\, x \Leftarrow \texttt{new in}\, e) = c'.$$

For showing that these programs are equivalent, we will need to consider world advancements. Assume that we start from an initial world w. Assume the abstract computations $\mathsf{c}_1 = (\mathsf{w} \cup \{\mathsf{l}_1\}, u_1 : \mathsf{w} \hookrightarrow \mathsf{w} \cup \{\mathsf{l}_1\}, c_2, n_1)$ and $\mathsf{c}_1' = (\mathsf{w} \cup \{\mathsf{l}'\}, u_1' : \mathsf{w} \hookrightarrow \mathsf{w} \cup \{\mathsf{l}'\}, c_2', n_1')$, where l and l' are the first proper concrete locations allocated. Moreover, let $c_2 = (\mathsf{w} \cup \{\mathsf{l}_1, \mathsf{l}_2\}, u_2 : \mathsf{w} \cup \{\mathsf{l}\} \hookrightarrow \mathsf{w} \cup \{\mathsf{l}_1, \mathsf{l}_2\}, c, n_2)$ and $c_2' = (\mathsf{w} \cup \{\mathsf{l}_1', \mathsf{l}_2'\}, u_2' : \mathsf{w} \cup \{\mathsf{l}'\} \hookrightarrow \mathsf{w} \cup \{\mathsf{l}_1', \mathsf{l}_2'\}, c', n_2)$, where the second location is allocated. The proof is now the pullback square $\mathsf{w} \cup \{\mathsf{l}_1, \mathsf{l}_2\} \,{}_{u_2 u_1}^{\;\;id}\!\!\diamond_{u_2' u_1'}^{x'}\, \mathsf{w} \cup \{\mathsf{l}_1', \mathsf{l}_2'\}$, with $\overline{\mathsf{w}} = \mathsf{w} \cup \{\mathsf{l}_1, \mathsf{l}_2\}$ and where x' fixes everything except that it maps l_2' to l_1 and l_1' to l_2, *i.e.*, it permutes the allocation order. In this way we get that $id.\mathsf{c} \sim x'.\mathsf{c}'$.

6 Discussion

We have introduced proof-relevant logical relations and shown how they may be used to model and reason about simple equivalences in a higher-order language with recursion and name generation. A key innovation compared with previous functor category models is the use of functors valued in setoids (which are here also built on predomains), rather than plain sets. One payoff is that we can work with a direct style model rather than one based on continuations (which, in the absence of control operators in the language, is less abstract).

The technical machinery used here is not *entirely* trivial, and the reader might be forgiven for thinking it slightly excessive for such a simple language and rudimentary equations. However, our aim has not been to present impressive new equivalences, but rather to present an accessible account of how the idea of proof relevant logical relations works in a simple setting. The companion report [3] gives significantly more advanced examples of applying the construction to reason about equivalences justified by abstract semantic notions of effects and separation, but the way in which setoids are used is there somewhat obscured by the details of, for example, much more sophisticated categories of worlds and a generalization of s.v.f.s for modelling computation types. Our hope is that this account will bring the idea to a wider audience, make the more advanced applications more accessible, and inspire others to investigate the construction in their own work.

Thanks to Andrew Kennedy for numerous discussions, and to an anonymous referee for suggesting that we write up the details of how proof-relevance applies to pure name generation.

References

1. Abramsky, S., Ghica, D.R., Murawski, A.S., Ong, C.-H.L., Stark, I.D.B.: Nominal games and full abstraction for the nu-calculus. In: Proc. 19th Annual IEEE Symposium on Logic in Computer Science (LICS 2004). IEEE Computer Society (2004)
2. Barthe, G., Capretta, V., Pons, O.: Setoids in type theory. J. Funct. Program. 13(2), 261–293 (2003)
3. Benton, N., Hofmann, M., Nigam, V.: Abstract effects and proof-relevant logical relations. CoRR abs/1212.5692 (2012)
4. Benton, N., Kennedy, A., Beringer, L., Hofmann, M.: Relational semantics for effect-based program transformations with dynamic allocation. In: Proc. Ninth International ACM SIGPLAN Symposium on Principles and Practice of Declarative Programming (PPDP 2007). ACM (2007)
5. Benton, N., Kennedy, A., Hofmann, M., Beringer, L.: Reading, writing and relations: Towards extensional semantics for effect analyses. In: Kobayashi, N. (ed.) APLAS 2006. LNCS, vol. 4279, pp. 114–130. Springer, Heidelberg (2006)
6. Benton, N., Koutavas, V.: A mechanized bisimulation for the nu-calculus. In: Higher-Order and Symbolic Computation (to appear, 2013)
7. Benton, N., Leperchey, B.: Relational reasoning in a nominal semantics for storage. In: Urzyczyn, P. (ed.) TLCA 2005. LNCS, vol. 3461, pp. 86–101. Springer, Heidelberg (2005)
8. Birkedal, L., Carboni, A., Rosolini, G., Scott, D.S.: Type theory via exact categories. In: Proc. 13th Annual IEEE Symposium on Logic in Computer Science (LICS 1998). IEEE Computer Society Press (1998)
9. Carboni, A., Freyd, P.J., Scedrov, A.: A categorical approach to realizability and polymorphic types. In: Main, M.G., Mislove, M.W., Melton, A.C., Schmidt, D. (eds.) MFPS 1987. LNCS, vol. 298, pp. 23–42. Springer, Heidelberg (1988)
10. Gabbay, M., Pitts, A.M.: A new approach to abstract syntax with variable binding. Formal Asp. Comput. 13(3-5), 341–363 (2002)
11. Pitts, A., Stark, I.: Operational reasoning for functions with local state. In: Higher Order Operational Techniques in Semantics, pp. 227–273. Cambridge University Press (1998)
12. Pitts, A.M., Stark, I.D.B.: Observable properties of higher-order functions that dynamically create local names, or what's new? In: Borzyszkowski, A.M., Sokolowski, S. (eds.) MFCS 1993. LNCS, vol. 711, pp. 122–141. Springer, Heidelberg (1993)
13. Shinwell, M.R.: The Fresh Approach: functional programming with names and binders. PhD thesis, University of Cambridge (2004)
14. Shinwell, M.R., Pitts, A.M.: On a monadic semantics for freshness. Theor. Comput. Sci. 342(1), 28–55 (2005)
15. Shinwell, M.R., Pitts, A.M., Gabbay, M.J.: FreshML: Programming with binders made simple. In: Proc. Eighth ACM SIGPLAN International Conference on Functional Programming (ICFP 2003). ACM (2003)
16. Stark, I.D.B.: Names and Higher-Order Functions. PhD thesis, University of Cambridge, Cambridge, UK, Also published as Technical Report 363, University of Cambridge Computer Laboratory (December 1994)
17. Tzevelekos, N.: Program equivalence in a simple language with state. Computer Languages, Systems and Structures 38(2) (2012)

Games with Sequential Backtracking and Complete Game Semantics for Subclassical Logics

Stefano Berardi[1] and Makoto Tatsuta[2]

[1] Torino University, C.so Svizzera 185, 10149 Torino, Italy
stefano@di.unito.it
[2] National Institute of Informatics, 2-1-2 Hitotsubashi, Tokyo 101-8430, Japan
tatsuta@nii.ac.jp

Abstract. This paper introduces a game semantics for Arithmetic with various sub-classical logics that have implication as a primitive connective. This semantics clarifies the infinitary sequent calculus that the authors proposed for intuitionistic arithmetic with Excluded Middle for Sigma-0-1-formulas, a formal system motivated by proof mining and by the study of monotonic learning, for which no game semantics is known. This paper proposes games with Sequential Backtracking, and proves that they provide a sound and complete semantics for the logical system and other various subclassical logics. In order for that, this paper also defines a one-sided version of the logical system, whose proofs have a tree isomorphism with respect to the winning strategies of the game semantics.

1 Introduction

We briefly describe motivations and state-of-the-art game semantics for various sub-classical logics, and for the λ-calculus, possibly extended with exceptions and continuations, and then we motivate games with sequential backtracking as an extension of the existing game semantics.

There are game models of functional languages due to Hyland and Ong [16], in which the main computational rule described by the semantics is the β-rule. In Logic, a similar model for intuitionistic arithmetic with primitive implication (but originally, without cut) was provided by Lorenzen [22]. There are game models of classical arithmetic without primitive implication, and with cut, due to Coquand [13], [14]. Herbelin [20] pointed out that these models are implicitly models of continuation-based programming languages. These game semantics have a wide range of applications. Hyland and Ong used them to prove the correctness of program transformation. Curien defined an abstract machine for a functional language simulating a debate between two players [12]. Coquand showed how algorithms learning by trial-and-error their result may be extracted from classical proofs using his game semantics, and Hayashi suggested to use games for *proof animation* [19], a method for checking whether an algorithm meets a given specification.

M. Hasegawa (Ed.): TLCA 2013, LNCS 7941, pp. 61–76, 2013.

What is missing is an attempt of merging the approaches of Hyland-Ong and Coquand, representing as particular cases of the same game model: intuitionism, classical arithmetic (with and without implication), and functional programming languages with and without continuations. There is also a special case of continuations we would like to model: exceptions, which we think as continuations $\lambda k.E$ in which the variable k is not used (as continuation erasing their environment).

The following is not yet done either: a game model for classical logic with primitive implication and two-sided sequent calculus is missing. Indeed, Coquand models classical arithmetic using only generalized disjunctions and conjunctions: implication is defined in a classical way, as a particular kind of disjunction, and his sequent calculus has only one side. In particular, there is a subclassical logic we would like to represent: $\mathtt{EM}_1$-*logic*, or Excluded Middle over Σ^0_1-statements, which is of interest both for proof mining [1], and for its relationship with monotonic learning algorithms ([2], [3], [4], [5]).

For what concerns programming languages, we would like to model both the β-rule and the continuations, because a part of programming may be more conveniently expressed by function definition, but exceptions and continuations are a desirable feature too. The two game semantics we would like to merge have complementary features. In Hyland-Ong, a move may be retracted and changed only if it is a move over a type occurring negatively: from the viewpoint of a programmer, only if the move is from a type representing an *input* of the first player. In Coquand, instead, a move may be retracted and changed only if it is a move from a disjunctive statement: from the viewpoint of a programmer, only if the move is from a type representing an *output* of the first player. As in [10], we merge the two approaches adding a "negative" and "positive" marker to all position of a play, representing, respectively, "input" and "output", or "left" and "right" sides of a sequent, and similar to the "question" and "answer" marks used by Hyland-Ong. Besides, we allow to retract moves at least from disjunctive statements and output types, as in Coquand, and we allow to restart any number of "conjunctive" sub-plays (sub-plays in which the opponent moves first) as in Hyland-Ong. By combining these two features, we introduce games with sequential backtracking, which are suitable to model intuitionistic arithmetic, $\mathtt{EM}_1$-arithmetic, and classical arithmetic, as particular cases. This work is preliminary. In the future, we hope to use games with sequential backtracking to model functional languages, functional languages plus exceptions, and functional languages plus continuation as particular cases of the same notion of games.

In this paper we study IPA^- [23], a system for intuitionistic arithmetic with implication and Excluded Middle for Σ^0_1-formulas, motivated by proof mining and by the study of monotonic learning. We introduce a game semantics for IPA^- as a particular case of our game semantics for classical arithmetic with primitive implication. We have two main results.

1. *(Soundness and Completeness)* Our game semantics is sound and complete for IPA^-, in particular it satisfies cut elimination.

2. *(A proof/strategy isomorphism for* $\mathtt{EM}_1$*-arithmetic)* There is a one-sided version ILW_1 of IPA^-, such that the cut-free proofs of ILW_1 are tree-isomorphic to the recursive winning strategy of our game semantics for $\mathtt{EM}_1$.

These results continue a long research line. Lorenzen's game semantics is proved sound and complete for Infinitary Intuitionistic Arithmetic [22], while Coquand's game semantics is proved sound and complete for Infinitary Classical Arithmetic by Herbelin [20]. Herbelin proved a tree isomorphism result for classical arithmetic: there is a tree isomorphism between cut-free infinitary recursive proof-trees of arithmetic and recursive winning strategies for the first player in Coquand's games. Like Curry-Howard isomorphism, this is an isomorphism between proofs of a formal system and some set of programs, and allows to "run" proofs as programs. Herbelin's result was adapted to arithmetic with $\mathtt{EM}_1$-logic and *without* implication by Berardi and Yamagata [9]. In this paper we consider the non-trivial problem of adding primitive implication, and obtain a game semantics for IPA^-, that is, arithmetic with $\mathtt{EM}_1$-logic and implication.

Together with Soundness and Completeness we can prove the admissibility of cut rule for our game semantics of IPA^-. In a forthcoming paper, we will strengthen the cut-elimination result we prove in this paper to a more game theoretical result: *any debate between two well-founded strategies terminates*, extending the result of Coquand [13], [14].

The main contribution of our game semantics is a new notion of backtracking, *sequential backtracking*, in which a player may come back to *any* previous position of the play. Backtracking may either be considered as an independent move, where it is considered as a duplication of the position we backtrack to, or it may be merged with the next move, according to the will of the player who does the backtracking.

This is the plan of the paper. In §2 we define "sequential backtracking", which is the core of our game semantics. In §3 we briefly recall the definition of arithmetical formulas and judgements and the axiom schema $\mathtt{EM}_1$. In §4 we define a game semantics for arithmetic with intuitionistic logic, $\mathtt{EM}_1$-logic, classical logic, as particular cases of games with sequential backtracking. In §5 we introduce ILW_1 and IPA^-, two equivalent sequent calculi for arithmetic with primitive implication, ω-rule and $\mathtt{EM}_1$-logic. In §6 we prove soundness, completeness, cut elimination of our semantics w.r.t. ILW_1 and IPA^-. We define a proof/strategy isomorphism between cut-free proofs of ILW_1 and winning strategies of our game semantics for $\mathtt{EM}_1$. In 7 we sketch some future work, including a definition of dialogue which should provide interpretation of the cut-elimination procedure.

2 A Game Semantics with Sequential Backtracking

We define the notion of sequential backtracking, generalizing Coquand's and Hyland and Ong's backtracking. We assume having two players: $\mathtt{E}$ (Eloise), $\mathtt{A}$ (Abelard). The names are taken from the history of logic, but the symbol $\mathtt{E}$ is intended to remember the connective $\exists$, and in general disjunctive formulas, while the symbol $\mathtt{A}$ is intended to remember the connective $\forall$, and in general conjunctive formulas. E is often called the *first player* or the *Player*, and A is called the *second player* or the *Opponent*. In Logic E defends the truth of a statement and A attacks it. In the semantics of programming languages E represents a computation strategy and A represents the environment. The initial definition of games is folklore (see for instance [21] or [6]):

Definition 1 (Games). *A game is any list* $\mathcal{G} = \langle G, \mathtt{turn}_G, W_{\mathtt{E},G}, W_{\mathtt{A},G} \rangle$*, where:*

1. *G is a tree, which is a set of nodes with father/child relation denoted by* $p <_{1,G} q$ *for a father p and a child q.*
2. $\mathtt{turn}_G : G \to \{\mathtt{E}, \mathtt{A}\}$ *is a map*
3. $(W_{\mathtt{E},G}, W_{\mathtt{A},G})$ *is a partition of the set of infinite branches of G.*

Let $g \in G$*. We use the following game terminology:*

1. *The root of G is the initial position of* $\mathcal{G}$
2. *any node of the tree is a position of* $\mathcal{G}$
3. *any edge between two nodes of G is a move of* $\mathcal{G}$
4. *any branch of G is a play of the game*
5. $\mathtt{turn}_G(g) \in \{\mathtt{E}, \mathtt{A}\}$ *is the player with the move obligation from g.*
6. *G is a well-founded tree if and only if G has no infinite branches.*

The initial position of a play is the root of the tree G. At each step the player with the move obligation moves to some child of the current node, and this child becomes the next current node. Eventually, either the player with the move obligation drops out and loses (usually because no move is available from the current position), or the play continues forever, defining some infinite branch α of the tree G. In the second case, if $\alpha \in W_{\mathtt{E},G}$ then E wins, and if $\alpha \in W_{\mathtt{A},G}$ then A wins.

We usually skip the subscript "G" and we denote the game $\mathcal{G}$ just with the tree G. To any game G we may add "backtracking" in the sense of Coquand, defining some extension $\mathsf{Coq}(G)$ of the game G. $\mathsf{Coq}(G)$ is a game biased in favor of the player E. At turn $i-1$, E may either move from the current position p_{i-1}, or may "backtrack" to any previous position p_j of the play, provided E moved from p_j. Then E retracts the move p_{j+1} that E did from it, and moves again, selecting some child p_i of p_j. This idea is formalized as follows. A position of $\mathsf{Coq}(G)$ is a pair $\langle \alpha, f \rangle$ of a sequence $\alpha = \langle p_0, \ldots, p_n \rangle$ of nodes, and a map $f : [1, n] \to [0, n-1]$. If $j = f(i)$, then p_j is the node to which the player who moved at turn $i-1$ backtracks. In particular, we always require that $f(i) \leq i-1$. When $j = f(i) = i-1$ we say that the move from p_j is an ordinary move of $\mathsf{Coq}(G)$, and when $j < i-1$ we say that the move from p_j is a backtracking move of $\mathsf{Coq}(G)$. When A has the move obligation from $p_{f(i)}$, then A moves, and we ask that the move is an ordinary move, that is: $j = f(i) = i-1$. We always ask that $p_{f(i)} <_1 p_i$: p_i is a move from $p_{f(i)}$ according to the *original* rules of G.

We do not formalize the game $\mathsf{Coq}(G)$, because in this paper we propose an extension $\mathsf{Seq}(G)$ of G with backtracking which is even more liberal than Coquand's. We call $\mathsf{Seq}(G)$ "G extended with *sequential backtracking*", a new kind of backtracking in which the entire sequence of moves is available for backtracking to E, and E may backtrack even if A has the move obligation. In $\texttt{Seq}(G)$, if E moves at turn j, then $j = f(i)$ may be *any* index less or equal than $i-1$: when E has the move obligation from p_j, if $j = i-1$ we say we have an ordinary move, and if $j < i-1$ we say we have a backtracking move. If A has the move obligation from p_j, then E may move $p_i = p_j$, duplicating p_j. In this case, E asks A to move from the copy p_i of p_j, that is, to repeat the move A did from p_j. When A moves from p_j and p_i is a child of p_j we say this is an ordinary move of A, and we ask that $j = i-1$. We now formalize the definition of $\texttt{Seq}(G)$. For the sake of simplicity we consider only the case G a well-founded tree: we assume that G has no infinite branches, that is, $W_{\texttt{E}} = W_{\texttt{A}} = \emptyset$. If α, α' are any two lists, we will write $\alpha <_1 \alpha'$ if $\alpha = \langle p_0, \dots, p_{l-1}\rangle$ and $\alpha' = \langle p_0, \dots, p_l\rangle$ for some $p_0, \dots, p_l$. We write $\le$ and $\ge$ for the prefix and the extension order on lists respectively: $\le$ is the reflexive and transitive closure of the relation $<_1$ on lists and $\ge$ is its dual.

Definition 2 (The game $\texttt{Seq}(G)$ with Sequential Backtracking). *Assume $\mathcal{G} = \langle G, \texttt{turn}, W_{\texttt{E}}, W_{\texttt{A}}\rangle$ is any game with $W_{\texttt{E}} = W_{\texttt{A}} = \emptyset$. We define the extension $\langle \texttt{Seq}(G), \texttt{turn}_{\texttt{Seq}(G)}, W_{\texttt{E},\texttt{Seq}(G)}, W_{\texttt{A},\texttt{Seq}(G)}\rangle$ of G with sequential backtracking.*

1. *$\texttt{Seq}(G)$ is the set of all $\langle \alpha, f\rangle$, such that $\alpha = \langle p_0, \dots, p_n\rangle$ is a list of nodes of G with $p_0 =$ the root of G, and $f : [1, n] \to [0, n-1]$ is such that, for all $i \in [1, n]$:*
   (1) $f(i) < i$
   (2) either $p_{f(i)} <_1 p_i$ or $p_{f(i)} = p_i$.
   (3) if $p_{f(i)} <_1 p_i$ and $\texttt{turn}(p_{f(i)}) = \texttt{A}$ then $f(i) = i - 1$.
2. *The father/child relation $<_1$ in the tree $\texttt{Seq}(G)$ is: $\langle \alpha, f\rangle <_1 \langle \alpha', f'\rangle$ if and only if $\alpha <_1 \alpha'$ and $f \subseteq f'$.*
3. *If $\alpha = \langle p_0, \dots, p_n\rangle$, then $\texttt{turn}_{\texttt{Seq}(G)}(\langle \alpha, f\rangle) = \texttt{turn}(p_n)$.*
4. *$W_{\texttt{E}} = \emptyset$, $W_{\texttt{A}} = \{$ all infinite plays $\}$.*

By definition, if a play in $\texttt{Seq}(G)$ is infinite, the winner is A. Why are infinite plays lost by E? Without backtracking, a play in $\texttt{Seq}(G)$ is a play in G and therefore finite, because we assumed that G is well-founded. An infinite play in $\texttt{Seq}(G)$ is the effect of an infinite backtracking by E. E, to avoid losing the play, may come back infinitely many times to the same position, or duplicate the same position infinitely many times, just to waste time. In this case E is penalized.

$\texttt{Seq}(G)$ is a game in a sense more general than the games considered by Def. 1. We may have *turn conflicts*, because E may backtrack even if the move obligation is for A. We define a map move which takes any position $\langle \alpha, f\rangle \in \texttt{Seq}(G)$ different from the initial position, and returns the player who moved to $\langle \alpha, f\rangle$. Assume $\alpha = \langle p_0, \dots, p_i\rangle$. When $p_{f(i)} <_1 p_i$ and $\texttt{turn}(p_{f(i)}) = \texttt{A}$ we set $\texttt{move}(\langle \alpha, f\rangle) = \texttt{A}$, otherwise $\texttt{move}(\langle \alpha, f\rangle) = \texttt{E}$. The definition of a play of $\texttt{Seq}(G)$ is like a play of G, except that even when A has the move obligation, E may backtrack. In this case we have a conflict, which is always solved in favor of E: we execute the move of E and the next position is updated accordingly.

We adapt the definition of a winning strategy S for a game G (see for instance [6]) to $\mathtt{Seq}(G)$. The only difference is in the case of conflicts: if both $\mathtt{E}$ and $\mathtt{A}$ may move from a given position, then S has two options: either S suggests a move to $\mathtt{E}$ or S provides a winning strategy for all possible moves of $\mathtt{A}$.

Definition 3 (Winning strategies for games with sequential backtracking). *Assume G is any game. A winning strategy of $\mathtt{E}$ on $\mathtt{Seq}(G)$ is any set $S \subseteq \mathtt{Seq}(G)$ of plays such that:*

1. *S is a well-founded tree w.r.t. $<_1$, with the root the unique play of length 1.*
2. *for all $\alpha \in S$: Either*
   (1) $\{\beta \in S | \beta >_1 \alpha\} = \{\beta_0\}$ for some β_0 such that $\mathtt{move}(\beta_0) = \mathtt{E}$, or
   (2) $\{\beta \in S | \beta >_1 \alpha\} = \{\beta \in \mathtt{Seq}(G) | \beta >_1 \alpha \wedge \mathtt{move}(\beta) = \mathtt{A}\}$ and $\mathtt{A}$ has the move obligation from α.

If S is any winning strategy for $\mathtt{E}$, then, no matter how $\mathtt{A}$ moves, $\mathtt{E}$ may move in such a way to maintain the play in S. In this case $\mathtt{E}$ always wins. Indeed, the play eventually terminates because S is well-founded, and when the play terminates, the last turn is $\mathtt{A}$ who drops out, because $\mathtt{E}$ has always some move to choose, suggested by S when $\mathtt{E}$ has the move obligation.

3 A Language for Arithmetic and Judgements

We will use sequential backtracking to define a game semantics for arithmetic and various subclassical logics. We define the language for arithmetic first.

Assume we have infinitely many term variables. Arithmetical terms are defined by the syntax: $t ::= k | x | f(t_1, \ldots, t_n)$, where k is any natural number, x any term variable, f is any n-ary primitive recursive function and $t_1, \ldots, t_n$ are arithmetical terms. A term is closed if there are no variables in it. Arithmetical formulas are defined by: $A ::= p(t_1, \ldots, t_n) | A \vee A | A \wedge A | A \to A | \exists x.A | \forall x.A$, where p is any primitive recursive predicate and $t_1, \ldots, t_n$ are arithmetical terms. We define an immediate subformula relation on *closed* formulas $A <_1 B$ by: $A, B <_1 A \vee B, A \wedge B, A \to B$ and: $A[t/x] <_1 \forall x.A, \exists x.A$ for all *closed* terms t. We call A a *negative* immediate subformula in the case $A <_1 A \to B$, and we call it a *positive* immediate subformula in all other cases.

Definition 4 (Judgements). *Let A be any closed arithmetical formula.*

1. *A judgement is any $J ::= \mathtt{t}.A | \mathtt{f}.A$.*
2. *A positive judgement is $\mathtt{t}.A$, and a negative judgement is $\mathtt{f}.A$.*
3. *A judgement has its truth value. If A is a true formula then $\mathtt{t}.A$ is a true judgement and $\mathtt{f}.A$ is a false judgement. If A is a false formula then $\mathtt{t}.A$ is a false judgement and $\mathtt{f}.A$ is a true judgement.*
4. *An atomic judgement is $\mathtt{t}.A$ or $\mathtt{f}.A$ where $A = p(t_1, \ldots, t_n)$.*
5. *A sequent is any ordered list $\Gamma = s_0.A_0, \ldots, s_{l-1}.A_{l-1}$ of signed formulas.*

6. *A disjunctive judgement is either some atomic false judgement* $s.p(t_1, \ldots, t_n)$*, or:*

$$\mathtt{t}.A \vee B,\ \mathtt{t}.\exists x A,\ \mathtt{t}.A \rightarrow B,\ \mathtt{f}.A \wedge B,\ \mathtt{f}.\forall x A.$$

7. *A conjunctive judgement is either some atomic true judgement* $s.p(t_1, \ldots, t_n)$*, or:*

$$\mathtt{f}.A \vee B,\ \mathtt{f}.\exists x A,\ \mathtt{f}.A \rightarrow B,\ \mathtt{t}.A \wedge B,\ \mathtt{t}.\forall x A.$$

8. *A sequent is positive (negative, disjunctive, conjunctive) if all its judgements are.*
9. $s.A <_1 t.B$ *if* $A <_1 B$ *and: either* A *is a positive immediate subformula of* B *and* $s = t$*, or* A *is a negative immediate subformula of* B *and* $s \neq t$*.*
10. $s.A \leq t.B$ *is the reflexive and transitive closure of* $<_1$*.*

We denote judgements with $s.A, t.B, \ldots$, a disjunctive judgement with $D, D', \ldots$, and a conjunctive judgement with $C, C', \ldots$. Atomic formulas are constructed from primitive recursive predicates, and therefore the truth value of atomic judgements is computable. A disjunctive judgement is equivalent to the disjunction of its immediate sub-judgements, a conjunctive judgement is equivalent to the conjunction of its immediate sub-judgements. In the case of atomic judgement, this holds because the disjunction of the empty set is false and the conjunction of the empty set is true. $\mathtt{EM}_1$-logic is obtained by adding to intuitionistic logic the axiom schema $\mathtt{EM}_1$.

Definition 5 (The axiom schemata $\mathtt{EM}_1$ and $\mathtt{EM}_2$).

1. $\mathtt{EM}_1 = \{\mathtt{t}.\forall x.(\exists y.p(x,y) \vee \forall y.p^{\perp}(x,y)) | p \text{ primitive recursive}\}$
2. $\mathtt{EM}_2 = \{\mathtt{t}.\forall x.(\exists y.\forall z.q(x,y,z) \vee \forall y.\exists z.q^{\perp}(x,y,z)) | q \text{ primitive recursive}\}$

where $p^{\perp}, q^{\perp}$ *denote the complement of* p*,* q *respectively.*

Let $s.A$ be any judgement. Our thesis is that if we choose $\mathtt{Tarski}(s.A)$ as the set of judgement chains from $s.A$ by $<_1$, then $\mathtt{Seq}(\mathtt{Tarski}(s.A))$ defines a sound and complete game semantics for classical arithmetic, and, if we add some restriction to backtracking, it defines that for various subclassical arithmetics. We believe that these semantics have the potentiality of adapting to pure functional languages possibly extended with exceptions, or with continuations.

4 A Game Semantics for $\mathtt{EM}_1$

For any judgement $s.A$ we define the Tarski game $\mathtt{Tarski}(s.A)$. Then we will show that there is some restriction $\mathtt{EM}_1(\mathtt{Tarski}(s.A))$ of $\mathtt{Seq}(\mathtt{Tarski}(s.A))$ which defines a sound and complete game semantics of intuitionistic arithmetic extended with $\mathtt{EM}_1$ and recursive ω-rule. In the next section we show that there is an infinitary semi-formal logical system ILW_1 whose proofs are tree-isomorphic to the winning strategies for $\mathtt{EM}_1(\mathtt{Tarski}(s.A))$: introducing this system is also useful for proving soundness and completeness of our $\mathtt{EM}_1$-game semantics.

Definition 6 (Tarski games). *Let $s.A$ be any judgement. The Tarski game $\langle \mathtt{Tarski}(s.A), \mathtt{turn}, W_{\mathtt{E}}, W_{\mathtt{A}} \rangle$ is defined as follows.*

1. $\mathtt{Tarski}(s.A)$ *is the tree* $\{\langle s_0.A_0, \ldots, s_n.A_n \rangle | s_0.A_0 = s.A \wedge \forall i \in [0, n-1]. s_i.A_i <_1 s_{i+1}.A_{i+1}\}$*, with root* $\langle s.A \rangle$*, and father/child relation the one-step extension* $<_1$ *on lists.*
2. $\mathtt{turn}(\langle s_0.A_0, \ldots, s_n.A_n \rangle) = \mathtt{turn}(s_n.A_n)$*, that is,* $= \mathtt{E}$ *if* $s_n.A_n$ *is disjunctive, and* $= \mathtt{A}$ *if* $s_n.A_n$ *is conjunctive.*
3. $W_{\mathtt{E}} = W_{\mathtt{A}} = \emptyset$.

For instance, E has the move obligation whenever $s_n.A_n$ is atomic false, A has the move obligation whenever $s_n.A_n$ is atomic true. The player with the move obligation from an atomic judgement cannot move and loses: E loses whenever $s_n.A_n$ is atomic false, A loses whenever $s_n.A_n$ is atomic true. For the sake of simplicity, we will denote a position $\langle s_0.A_0, \ldots, s_n.A_n \rangle$ of $\mathtt{Tarski}(s.A)$ simply with $s_n.A_n$. We formally define two possible restrictions to $\mathtt{Seq}(\mathtt{Tarski}(s.A))$. Each restriction is called with the kind of logic we want to model.

Definition 7 (Restrictions to backtracking). *Assume* $\langle \alpha, f \rangle \in \mathtt{Seq}(\mathtt{Tarski}(s.A))$ *is a position in a game with sequential backtracking, with* $\alpha = \langle s_0.A_0, \ldots, s_n.A_n \rangle$.

1. $\langle \alpha, f \rangle$ *satisfies the intuitionistic restriction if for all* $i \in [1, n]$ *we have: either* $s_{f(i)} = \mathtt{f}$*, or* $s_{f(i)+1} = \ldots = s_{i-1} = \mathtt{f}$.
2. $\langle \alpha, f \rangle$ *satisfies the* $\mathtt{EM}_1$*-restriction if for all* $i, j \in [1, n]$ *such that* $f(i) < f(j) < i < j$ *we have: either* $s_{f(i)} = \mathtt{f}$*, or* $s_{f(i)+1} = \ldots = s_{f(j)} = \mathtt{f}$.

$\mathtt{Int}(s.A)$*, and* $\mathtt{EM}_1(s.A)$ *are the set of positions of* $\mathtt{Seq}(\mathtt{Tarski}(s.A))$ *which satisfy: the intuitionistic restriction, and the* $\mathtt{EM}_1$*-restriction respectively.*

We explain the restrictions, then we include some examples of plays and winning strategies.

1. In the case of Intuitionistic Logic, if E backtracks from $s_{i-1}.A_{i-1}$ to $s_{f(i)}.A_{f(i)}$, and $s_{f(i)} = \mathtt{t}$, then we ask that all judgements between $s_{f(i)+1}.A_{f(i)+1}$ and $s_{i-1}.A_{i-1}$ included are negative. In other words: if E backtracks to a positive judgement, then this *is the last positive judgement.*
2. In the case of $\mathtt{EM}_1$-logic, we allow E to backtrack from $s_{i-1}.A_{i-1}$ to any positive judgement $s_{f(i)}.A_{f(i)}$, but then we ask that for all judgements between $s_{f(i)+1}.A_{f(i)+1}$ and $s_{i-1}.A_{i-1}$, those *after the first positive judgement* are no more available to E for backtracking. We express this request in the dual way: we ask that when E backtracks from some $s_j.A_j$ with $j > i$ to some $s_{f(j)}.A_{f(j)}$ with $f(i) < f(j) < i$, then all judgements from $s_{f(i)+1}.A_{f(i)+1}$ to $s_{f(j)}.A_{f(j)}$ included are *negative.*

E has a winning strategy in $\mathtt{Tarski}(\mathtt{t}.A)$ if and only if A is true, but the winning strategy is often not recursive and does not correspond to a program. The more we add backtracking, the more there are recursive winning strategies for E. Consider the judgements $\mathtt{t}.A = \mathtt{t}.\forall x.(\exists y.p(x,y) \vee (\exists y.p(x,y) \rightarrow \bot))$, for some Σ^0_1-complete $\exists y.p(x,y)$, and $\mathtt{t}.B = \forall x.(\exists y.\forall z.q(x,y,z) \vee \forall y.\exists z.q^\bot(x,y,z))$, for some Σ^0_2-complete $\exists y.\forall z.q(x,y,z)$. The formula A is equivalent to some instance of $\mathtt{EM}_1$, while B is some instance of $\mathtt{EM}_2$. E has winning strategy on Tarski(t.A) and on Tarski(t.B), but has no recursive winning strategy. E, thanks to a restricted possibility of backtracking, has a recursive winning strategy on $\mathtt{EM}_1$(t.A), yet no recursive winning strategy on $\mathtt{EM}_1$(t.B). E, thanks to an unrestricted possibility of backtracking, has a recursive winning strategy both on Seq(Tarski(t.A)) and on Seq(Tarski(t.B)): we informally explain why.

A *non-recursive* winning strategy for E on Tarski(t.A) is the following. A moves to $\mathtt{t}.(\exists y.p(a,y) \vee (\exists y.p(a,y) \rightarrow \bot))$ for some $a \in N$. Then if $\exists y.p(a,y)$ is true E moves $\mathtt{t}.\exists y.p(a,y)$, while if $\exists y.p(a,y)$ is false, then E moves $\mathtt{t}.\exists y.p(a,y) \rightarrow \bot$. In the first case, E moves $\mathtt{t}.p(a,b)$ for some $b \in N$ such that $p(a,b)$ is true, then wins, because A loses from $\mathtt{t}.p(a,b)$ with $p(a,b)$ true. In the second case, E moves to $\mathtt{f}.\exists y.p(a,y)$. Then A moves $\mathtt{f}.p(a,b)$, for some $b \in N$: $p(a,b)$ is false because $\exists y.p(a,y)$ is false. Also in this case A loses. This strategy is not recursive, because we cannot decide whether $\exists y.p(a,y)$ is true or false for a Σ^0_1-complete predicate $\exists y.p(a,y)$.

In the case of $\mathtt{EM}_1$(t.A), instead, E has a *recursive* winning strategy. A moves to the disjunctive judgement $\mathtt{t}.(\exists y.p(a,y) \vee (\exists y.p(a,y) \rightarrow \bot))$ for some $a \in N$. Then E moves to the disjunctive judgement $\mathtt{t}.\exists y.p(a,y) \rightarrow \bot$, then to the conjunctive judgement $\mathtt{f}.\exists y.p(a,y)$. A replies with $\mathtt{f}.p(a,b)$ for some $b \in N$. If $p(a,b)$ is false then A loses. If $p(a,b)$ is true then in a Tarski game E would lose, but in the game $\mathtt{EM}_1$(t.A) E *may backtrack* to the position t.A. Then E plays $\mathtt{t}.\exists y.p(a,y)$ this time. E plays $\mathtt{t}.p(a,b)$ from $\mathtt{t}.\exists y.p(a,y)$, and since $p(a,b)$ is true then E wins.

In order to have a *recursive* winning strategy in t.B, instead, E needs a more liberal notion of backtracking. To see that, assume that A moves from t.B to the disjunctive $\mathtt{t}.\exists y.\forall z.q(a,y,z) \vee \forall y.\exists z.q^\bot(a,y,z)$ for some $a \in N$. We denote it by C. E could move to the conjunctive $\mathtt{t}.\forall y.\exists z.q^\bot(a,y,z)$. Assume that A moves to the disjunctive $\mathtt{t}.\exists z.q^\bot(a,b,z)$. E does not know which value to choose for z, therefore she backtracks to t.C. According to the $\mathtt{EM}_1$ restriction, all positions after the first positive judgement after t.C, that is: $\mathtt{t}.\forall y.\exists z.q^\bot(a,y,z)$, $\mathtt{t}.\exists y.q^\bot(a,b,z)$ are no more available for backtracking, and this *is* a problem! Indeed, assume E moves to the disjunctive $\mathtt{t}.\exists y.\forall z.q(a,y,z)$, then to the conjunctive $\mathtt{t}.\forall z.q(a,b,z)$. A replies $\mathtt{t}.q(a,b,c)$ for some $c \in N$. If $q(a,b,c)$ is true then A loses, otherwise $q(a,b,c)$ is false. In this case $q^\bot(a,b,c)$ is true, and E should backtrack to $\mathtt{t}.\exists z.q^\bot(a,b,z)$, then move $\mathtt{t}.q^\bot(a,b,c)$ and win, because $q^\bot(a,b,c)$ is true. However, as we point out, *this backtracking is not available to* E *in* $\mathtt{EM}_1$(t.B). This backtracking is correct for Seq(Tarski(t.B)), in which there is no restriction to backtracking, and indeed E has a recursive winning strategy in Seq(Tarski(t.B)). Instead, by the result of [6], there is no recursive winning strategy for E on Tarski(t.A) and on $\mathtt{EM}_1$(t.B).

In §6 we prove that E has a recursive winning strategy on $\mathtt{EM}_1$(t.A) if and only if A is provable in IPA^-. We only quote another result not proved in this paper: the game semantics Seq(Tarski(t.A)) is complete with respect to the classical arithmetic plus recursive ω-rule, and since this latter is complete w.r.t. classical truth, we may conclude that $\mathtt{E}$ has a recursive winning strategy on $\mathtt{Seq}(\mathtt{Tarski}(\mathtt{t}.A))$ if and only if A is true.

5 ILW_1, a One-Sided Infinitary Sequent Calculus

We may express the formation rules for winning strategies of $\mathtt{EM}_1(s.A)$ through the inference rules for an infinitary semi-formal system ILW_1 with ω-rule: as a consequence, $\mathtt{EM}_1(s.A)$ is a sound and complete semantics for ILW_1. Later we will prove that ILW_1 is equivalent to $\mathtt{EM}_1$-arithmetic IPA^-: we will deduce that $\mathtt{EM}_1(s.A)$ is a sound and complete semantics for IPA^-.

Assume S is any winning strategy for $\mathtt{E}$ on $\mathtt{Seq}(\mathtt{Tarski}(s.A))$ and $\Theta \in S$: we analyze the possible branching of Θ in S and we read them as logical rules for a sequent calculus.

1. Assume S suggests to $\mathtt{E}$ to backtrack to some disjunctive judgement, say, to $\mathtt{t}.A_1 \vee A_2$. Then the position Θ is $\langle \Gamma, \mathtt{t}.A_1 \vee A_2, \Delta, f\rangle$ for some map f, and after backtracking to $\mathtt{t}.A_1 \vee A_2$, S suggests to $\mathtt{E}$ the move $\mathtt{t}.A_i$ for some $i \in \{1, 2\}$. The next position is $\langle \Gamma, \mathtt{t}.A_1 \vee A_2, \Delta, \mathtt{t}.A_i, f \cup \{k \mapsto l\}\rangle$ where k is the index of $\mathtt{t}.A_i$ and l is the index of $\mathtt{t}.A_1 \vee A_2$ in the sequent. If we read the tree S upside down, and we forget the map f, this case corresponds to the introduction rule for $\vee$: from $\Gamma, \mathtt{t}.A_1 \vee A_2, \Delta, \mathtt{t}.A_i$ for some i, deduce $\Gamma, \mathtt{t}.A_1 \vee A_2, \Delta$.
2. Assume S suggests to $\mathtt{E}$ to backtrack to some judgement J and to duplicate it. Then $\Theta = \langle \Gamma, J, \Delta, f\rangle$, and after backtracking to J, S suggests the move J again to $\mathtt{E}$. The next position is $\langle \Gamma, J, \Delta, J, f \cup \{k \mapsto l\}\rangle$. If we read the tree S upside down, this corresponds to the contraction rule: from Γ, J, Δ, J deduce Γ, J, Δ.
3. Assume that S includes all moves of $\mathtt{A}$ from some conjunctive judgement, say, $\mathtt{t}.A_1 \wedge A_2$. Then $\Theta = \langle \Gamma, \mathtt{t}.A_1 \wedge A_2, f\rangle$ for some f, and S includes all moves $\mathtt{t}.A_i$ of $\mathtt{A}$ for any $i \in \{1, 2\}$. The next positions are, for any i: $\langle \Gamma, \mathtt{t}.A_1 \wedge A_2, \mathtt{t}.A_i, f \cup \{k \mapsto k-1\}\rangle$, where k is the index of $\mathtt{t}.A_i$. If we read the tree S upside down, this corresponds to the introduction rule for $\wedge$: from $\Gamma, \mathtt{t}.A_1 \wedge A_2, \mathtt{t}.A_i$ for all $i \in \{1, 2\}$, deduce $\Gamma, \mathtt{t}.A_1 \wedge A_2$.

We sketched three logical rules, introduction for disjunctive judgements, contraction, introduction for conjunctive judgements, which may define all winning strategies for $\mathtt{E}$ in $\mathtt{Seq}(\mathtt{Tarski}(s.A))$. In the case of restriction $\mathtt{EM}_1(s.A)$ of $\mathtt{Seq}(\mathtt{Tarski}(s.A))$ we have to add some "side condition" to the logical interpretation of S, in order to have correct moves in $\mathtt{EM}_1(s.A)$. Besides, with the $\mathtt{EM}_1$-restriction to backtracking, there are formulas not available for backtracking in the play, and we have to find some game formulation for this. As we did in [9], we interpret the game rule which forbids backtracking by merging the logical rules with a weakening rule. For instance, from $\Gamma, \mathtt{t}.A_1 \vee A_2, \Delta, \mathtt{t}.A_i$ for some i, we deduce $\Gamma, \mathtt{t}.A_1 \vee A_2, \Delta, \Delta'$, for any Δ'. The formulas in Δ' are the formulas in $\Theta = \langle \Gamma, \mathtt{t}.A_1 \vee A_2, \Delta, \Delta', f \rangle$ not available for backtracking to E in the position $\langle \Gamma, \mathtt{t}.A_1 \vee A_2, \Delta, \mathtt{t}.A_i, f' \rangle$.

We formally define a sequent calculus ILW_1 for the language of judgements, whose proof-trees with conclusion $s.A$ are tree-isomorphic to the winning strategies of $\mathtt{EM}_1(s.A)$.

Definition 8 ($\mathbf{ILW}_1$). *Assume $\{J_i | i \in I\}$ is the set of immediate sub-judgements of the disjunctive judgement D and of the conjunctive judgement C. Let J be any judgement.*

$$\frac{\Gamma, D, \Delta, J_i}{\Gamma, D, \Delta, \Delta'}\,(disj) \qquad (D \text{ negative or } \Delta \text{ negative})$$

$$\frac{\Gamma, C, J_i \quad (\forall i)}{\Gamma, C}\,(conj)$$

$$\frac{\Gamma, J, \Delta, J}{\Gamma, J, \Delta, \Delta'}\,(cont) \qquad (J \text{ negative or } \Delta \text{ negative})$$

A simple derived rule is: $\overline{\Gamma, J, \Delta'}\,(axiom)$ (for any J true atomic)

Indeed, an atomic true judgement J is some conjunctive judgement C having an empty family $\{J_i | i \in I\}$ of immediate sub-judgements. Therefore we may infer Γ, J, J from $(\forall i)$, then Γ, J, Δ' by $(cont)$ with $\Delta = \emptyset$.

The cut rule for ILW_1 may be formulated as follows:

$$\frac{\Gamma, \mathtt{t}.A \quad \Delta, \mathtt{f}.A, \Sigma}{\Gamma, \Delta, \Sigma}\,(cut)$$

For any formula different from cut, we call the *active* formula of a rule the formula in the conclusion of the rule which is inferred by the rule. A sequent is an ordered list and there is no explicit Exchange rule. However, we hide Exchange rule through the fact that the active formula, if it is disjunctive, may be in any position in the sequent. Identity rule is trivially derivable. Cut rule is admissible as well, but this result, not at all trivial, will be proved at the end of this paper. For every judgement $s.A$, every proof of $s.A$ in ILW_1 is tree-isomorphic to some winning strategy for E on $\mathtt{EM}_1(s.A)$, and conversely, every winning strategy for E on $\mathtt{EM}_1(s.A)$ is isomorphic to some proof of $s.A$. This relation is similar to the Curry-Howard isomorphism between simply typed λ-terms and proofs of intuitionistic logic of implication, or to the proof/strategy isomorphism between infinitary classical proofs of arithmetic and winning strategy for Coquand's games defined by Herbelin [20].

Definition 9 (The surjection φ). *Assume $s.A$ is any judgement and $\Gamma = \langle s_0.A_0, \ldots, s_{l-1}.A_{l-1} \rangle$ is any non-empty sequent. Let Π be any proof of $s.A$. Let $\alpha = \langle \Gamma_0, \ldots, \Gamma_n \rangle$ be any branch of Π of length $n+1$. For all $i \in [1, n]$, we set:*

1. *$g_\alpha(i) = k$ if the active judgement of Γ_i has index $k \geq 0$ in Γ_i.*

2. $f_\alpha(i) =$ *the last* $j \in [0, i-1]$ *such that* $\texttt{length}(\Gamma_j) = g_\alpha(i) + 1$.

3. $\texttt{last}(\langle s_0.A_0, \ldots, s_{l-1}.A_{l-1}\rangle) = s_{l-1}.A_{l-1}$ *and* $\texttt{Last}(\alpha) = \langle \texttt{last}(\Gamma_0), \ldots, \texttt{last}(\Gamma_n)\rangle$

4. $\varphi(\alpha) = \langle \texttt{Last}(\alpha), f_\alpha\rangle$ *and* $\varphi(\Pi) = \{\varphi(\alpha) | \alpha \in \Pi\}$.

Both a proof tree and a strategy can be seen as trees. We call a map tree-isomorphic when it preserves the root and the order. The next theorem shows a tree-isomorphic map between trees of proofs and trees of strategies.

Theorem 1 (Proof/strategy isomorphism). *Let s.A be any judgement. The map* φ *defined above is a surjection between (possibly non-recursive) proof-trees* Π *of s.A in ILW*$_1$ *and winning strategies for* E *in the game* $\texttt{EM}_1(s.A)$*, such that* Π *and* $\varphi(\Pi)$ *are tree-isomorphic, and if* Π *is recursive then* $\varphi(\Pi)$ *is.*

Proof. The proof is by induction over the length of a branch of Π, as in [9]. □

6 The Infinitary Arithmetic IPA$^-$ for the Subclassical Logic $\texttt{EM}_1$

In the rest of the paper we prove that $\texttt{EM}_1(s.A)$ gives a sound and complete semantics for IPA$^-$. Since Theorem 1 shows that $\texttt{EM}_1(s.A)$ gives a sound and complete semantics for ILW$_1$, it is enough to show the equivalence between ILW$_1$ and IPA$^-$. This equivalence also gives us the cut elimination theorem for ILW$_1$, because the cut elimination theorem for IPA$^-$ has been already proved in our previous paper [23].

The inference rules of IPA$^-$, taken from [23], are those of two-sided infinitary sequent calculus for classical arithmetic, with two important restrictions. First, there is no explicit Exchange rule in IPA$^-$. Secondly, if a rule of IPA$^-$ infers some formula on the right-hand side of the sequent, then the rule (with the only exception of the Weakening rule) infers the *last* formula of the sequent. The part of the Excluded Middle we may prove in IPA$^-$ is exactly $\texttt{EM}_1$. By this we mean: IPA$^-$ derives $\vdash A$ if and only if HA + recursive ω-rule derives $\texttt{EM}_1 \vdash A$. The sequents of IPA$^-$ have the form $\Gamma_0, -, \Gamma_1, -, \ldots, -, \Gamma_n \vdash A_1, \ldots, A_n$, where $-$ is a separation symbol. The intended meaning of $\Gamma_0, -, \Gamma_1, -, \ldots, -, \Gamma_n \vdash A_1, \ldots, A_n$ is "either $\Gamma_0, \ldots, \Gamma_i \vdash A_i$ for some i, or $\Gamma_0 \vdash \emptyset$". The intended meaning of $-$ is the logical constant "true".

$$\frac{}{\Gamma, A \vdash \Delta}\ (\textit{Axiom L}) \quad (A \text{ a false atomic formula})$$

$$\frac{}{\Gamma \vdash \Delta, A}\ (\textit{Axiom R}) \quad (A \text{ a true atomic formula})$$

$$\frac{\Gamma, - \vdash \Delta, A \wedge B, A \quad \Gamma, - \vdash \Delta, A \wedge B, B}{\Gamma \vdash \Delta, A \wedge B} (\wedge R)$$

$$\frac{\Gamma_1, A \wedge B, \Gamma_2, A \vdash \Delta}{\Gamma_1, A \wedge B, \Gamma_2 \vdash \Delta} (\wedge L1) \quad \frac{\Gamma_1, A \wedge B, \Gamma_2, B \vdash \Delta}{\Gamma_1, A \wedge B, \Gamma_2 \vdash \Delta} (\wedge L2)$$

$$\frac{\Gamma, - \vdash \Delta, A \vee B, A}{\Gamma \vdash \Delta, A \vee B} (\vee R1) \quad \frac{\Gamma, - \vdash \Delta, A \vee B, B}{\Gamma \vdash \Delta, A \vee B} (\vee R2)$$

$$\frac{\Gamma_1, A \vee B, \Gamma_2, A \vdash \Delta \quad \Gamma_1, A \vee B, \Gamma_2, B \vdash \Delta}{\Gamma_1, A \vee B, \Gamma_2 \vdash \Delta} (\vee L)$$

$$\frac{\Gamma, A \vdash \Delta, A \rightarrow B}{\Gamma \vdash \Delta, A \rightarrow B} (\rightarrow R1) \quad \frac{\Gamma, - \vdash \Delta, A \rightarrow B, B}{\Gamma \vdash \Delta, A \rightarrow B} (\rightarrow R2)$$

$$\frac{\Gamma_1, A \rightarrow B, \Gamma_2, - \vdash \Delta, A \quad \Gamma_1, A \rightarrow B, \Gamma_2, B \vdash \Delta}{\Gamma_1, A \rightarrow B, \Gamma_2 \vdash \Delta} (\rightarrow L)$$

$$\frac{\Gamma, - \vdash \Delta, \forall x A, A[m/x] \quad \text{(for all } m)}{\Gamma \vdash \Delta, \forall x A} (\forall R) \quad \frac{\Gamma_1, \forall x A, \Gamma_2, A[m/x] \vdash \Delta}{\Gamma_1, \forall x A, \Gamma_2 \vdash \Delta} (\forall L)$$

$$\frac{\Gamma, - \vdash \Delta, \exists x A, A[m/x]}{\Gamma \vdash \Delta, \exists x A} (\exists R) \quad \frac{\Gamma_1, \exists x A, \Gamma_2, A[m/x] \vdash \Delta \quad \text{(for all } m)}{\Gamma_1, \exists x A, \Gamma_2 \vdash \Delta} (\exists L)$$

$$\frac{\Gamma \vdash \Delta}{\Gamma, - \vdash \Delta, A} (weak\ R) \quad \frac{\Gamma \vdash \Delta}{\Gamma, A \vdash \Delta} (weak\ L)$$

Let $\sharp_- \Gamma$ denote the number of occurrences of the symbol $-$ in Γ. We have shown the cut elimination theorem for IPA$^-$ in [23] for the following cut rule, with the side condition: $\sharp_- \Sigma_1 = |\Sigma_2|$:

$$\frac{\Gamma_1, - \vdash \Gamma_2, A \quad \Delta_1, A, \Sigma_1 \vdash \Delta_2, \Sigma_2}{\Gamma_1, \Delta_1, \Sigma_1 \vdash \Gamma_2, \Delta_2, \Sigma_2} (cut)$$

Theorem 2 ([23]). *The cut elimination theorem holds for IPA$^-$.*

Since the system IPA$^-$ is a two sided version of ILW$_1$, we have the following translation between them and the equivalence proposition.

Definition 10. *The sequent Γ in ILW$_1$ is translated into the sequent $(\Gamma)^2$ in IPA$^-$ as follows:*

(1) the translation $()^2$ of the empty sequent is the empty sequent $\vdash$.
(2) If $(\Sigma)^2$ is $\Gamma \vdash \Delta$, then $(\Sigma, t.A)^2$ is $\Gamma, - \vdash \Delta, A$ and $(\Sigma, f.A)^2$ is $\Gamma, A \vdash \Delta$.

Proposition 1. *The statement that Γ is provable in ILW$_1$ is equivalent to the statement that $(\Gamma)^2$ is provable in IPA$^-$.*

This proposition is proved by induction on the definition of provability. By combining the equivalence between the existence of a recursive winning strategy of $\mathtt{EM}(s.A)$ and the provability of $s.A$ in ILW_1 (Theorem 1), and the equivalence between the provability $s.A$ in ILW_1 and $(s.A)^2$ in IPA^- (Proposition 1), we have our main theorem.

Theorem 3 (Main Theorem). *The game* $\mathtt{EM}(\mathtt{t}.A)$ *is a sound and complete semantics of* IPA^-. *That is, (1) the statement that* $-\vdash A$ *is provable in* IPA^- *is equivalent to the statement that* $\mathtt{EM}(\mathtt{t}.A)$ *has a recursive winning strategy, (2) the statement that* $A \vdash$ *is provable in* IPA^- *is equivalent to the statement that* $\mathtt{EM}(\mathtt{f}.A)$ *has a recursive winning strategy.*

Since the cut rule in ILW_1 is the same as the translation of that in IPA^-, we have the following theorem. This is proved by (1) a proof in ILW_1 with the cut rule is given, (2) Proposition 1 translates it into a proof in IPA^- with the cut rule, (3) the cut elimination theorem (Theorem 2) transforms it into a proof in IPA^- without the cut rule, (4) Proposition 1 translates it into a proof in ILW_1 without the cut rule.

Theorem 4. *The cut rule is admissible in* ILW_1 *and in game semantics* $\mathtt{EM}(s.A)$.

7 Conclusion and Future Work

We introduced an operator $\mathtt{Seq}(G)$, adding to a game G the possibility for the first player to backtrack to a previous position, in a more general way. We applied this operator to the Tarski game associated to a judgement $s.A$, and we showed that the resulting game $\mathtt{Seq}(\mathtt{Tarski}(s.A))$ may be restricted in order to represent various subclassical logic, like intuitionism and $\mathtt{EM}_1$-logic. We believe that the same method may be applied to produce models of functional languages extended with exception (this feature corresponds through the proof/strategy isomorphism to $\mathtt{EM}_1$-logic) or continuations (this feature corresponds through the proof/strategy isomorphism to full classical logic).

In a forthcoming paper we hope to interpret cut-elimination through dialogue in game semantics, as done by Coquand, Hyland and Ong. In a play with dialogue, both players may backtrack and there is a visibility relation for each player. Visibility in a position $\langle \alpha, f \rangle$ of a play with dialogue, with $\alpha = \langle p_0, \ldots, p_n \rangle$ and $f : [1, n] \to [0, n-1]$, is a binary relation on $[0, n]$ defined by induction over $i \in [0, n]$. No index is visible from 0. The player making the move of index $i+1$ has visible i, and all indexes he may see from i. The other player may see $f^k(i)$, for the first $k > 0$ such that $p_{f^k(i)} <_1 p_i$, and all indexes he may see from $f^k(i)$. A player may only answer to a visible position, and if both player want to backtrack then the next player is randomly selected. The results which we hope to prove are the following: every strategy on $\mathtt{Seq}(s.A)$ has a canonical extension to a strategy on plays with dialogue, and if we let play two well-founded strategies we obtain a well-founded tree of possible plays. To say otherwise: the resulting play, with a random selection of the next player when both strategies want to backtrack, is always finite if both strategies are terminating.

References

1. Akama, Y., Berardi, S., Hayashi, S., Kohlenbach, U.: An Arithmetical Hierarchy of the Law of Excluded Middle and Related Principles. In: LICS 2004, pp. 192–201 (2004)
2. Aschieri, F., Berardi, S.: A Realization Semantics for EM1-Arithmetic. In: Curien, P.-L. (ed.) TLCA 2009. LNCS, vol. 5608, pp. 20–34. Springer, Heidelberg (2009)
3. Aschieri, F.: Learning, Realizability and Games in Classical Arithmetic. Ph. D. thesis, Torino (2011)
4. Aschieri, F.: Learning Based Realizability for HA + EM1 and 1-Backtracking Games: Soundness and Completeness. APAL Special Issue for CL&C 2010 (2010)
5. Berardi, S., De'Liguoro, U.: Interactive Realizers: A New Approach to Program Extraction. ACM Trans. Comput. Log. 13(2), 11 (2012)
6. Berardi, S., Coquand, T., Hayashi, S.: Games with 1-backtracking. Ann. Pure Appl. Logic 161(10), 1254–1269 (2010)
7. Berardi, S., Tatsuta, M.: Positive Arithmetic Without Exchange Is a Subclassical Logic. In: Shao, Z. (ed.) APLAS 2007. LNCS, vol. 4807, pp. 271–285. Springer, Heidelberg (2007)
8. Berardi, S., De'Liguoro, U.: Toward the interpretation of non-constructive reasoning as non-monotonic learning. IC 207(1), 63–81 (2009)
9. Berardi, S., Yamagata, Y.: A Sequent Calculus for Limit ComputableMathematics. APAL 153(1-3), 111–126 (2008)
10. Berardi, S.: Semantics for Intuitionistic Arithmetic Based on Tarski Games with Retractable Moves. In: Della Rocca, S.R. (ed.) TLCA 2007. LNCS, vol. 4583, pp. 23–38. Springer, Heidelberg (2007)
11. Berardi, S.: Interactive Realizability. Summer School Chambéry, June 14-17 (2011), `http://www.di.unito.it/~stefano/Berardi-RealizationChambery-13Giugno2011.ppt`
12. Curien, P.-L., Herbelin, H.: Abstract machines for dialogue games. CoRR abs/0706.2544 (2007)
13. Coquand, T.: A semantics of evidence for classical arithmetic (preliminary version). In: 2nd Workshop on Logical Frameworks, Edinburgh (1991)
14. Coquand, T.: A semantics of evidence for classical arithmetic. Journal of Symbolic Logic 60, 325–337 (1995)
15. Gold, E.M.: Limiting Recursion. Journal of Symbolic Logic 30, 28–48 (1965)
16. Hyland, M., Ong, L.: On full abstraction for PCF. Information and Computation 163(2), 285–408 (2000)
17. Hayashi, S.: Mathematics based on incremental learning, excluded middle and inductive inference. Theoretical Computer Science 350(1), 125–139 (2006)
18. Hayashi, S., Nakata, M.: Towards Limit Computable Mathematics. In: Callaghan, P., Luo, Z., McKinna, J., Pollack, R. (eds.) TYPES 2000. LNCS, vol. 2277, pp. 125–144. Springer, Heidelberg (2002)

19. Hayashi, S.: Can Proofs be animated by games? Fundamenta Informaticae 77(4), 331–343 (2007)
20. Herbelin, H.: Sequents qu'on calcule: de l'interpretation du calcul des sequents. Ph. D. thesis, University of Paris VII (1995)
21. Hodges, W.: Logic and games. In: The Stanford Encyclopedia of Philosophy, Winter (2004), http://plato.stanford.edu/archives/win2004/entries/logic-games/
22. Felscher, W.: Lorenzen's game semantics. In: Handbook of Philosophical Logic, vol. 5, 2nd edn., pp. 115–145. Kluwer Academic Publishers (2002)
23. Tatsuta, M., Berardi, S.: Non-Commutative Infinitary Peano Arithmetic. In: Proceedings of CSL 2011, pp. 538–552 (2011), Preliminary version: http://www.di.unito.it/~stefano/CSL2011-NonCommutativePeanoArithmetic.pdf

Realizability for Peano Arithmetic with Winning Conditions in HON Games

Valentin Blot

Laboratoire de l'Informatique et du Parallélisme
ENS Lyon - Université de Lyon
UMR 5668 CNRS ENS-Lyon UCBL INRIA
46, allée d'Italie
69364 Lyon cedex 07 - France
valentin.blot@ens-lyon.fr

Abstract. We build a realizability model for Peano arithmetic based on winning conditions for HON games. First we define a notion of winning strategies on arenas equipped with winning conditions. We prove that the interpretation of a classical proof of a formula is a winning strategy on the arena with winning condition corresponding to the formula. Finally we apply this to Peano arithmetic with relativized quantifications and give the example of witness extraction for Π_2^0-formulas.

1 Introduction

Realizability is a technique to extract computational content from formal proofs. It has been widely used to analyze intuitionistic systems (for e.g. higher-order arithmetic or set theory), see [1] for a survey. Following Griffin's computational interpretation of Peirce's law [2], Krivine developed in [3–5] a realizability for second-order classical arithmetic and Zermelo-Fraenkel set theory.

On the other hand, Hyland-Ong game semantics provide precise models of various programming languages such as PCF [6] (a similar model has simultaneously been obtained in [7]), also augmented with control operators [8] and higher-order references [9]. In these games, plays are interactions traces between a program (player P) and an environment (opponent O). A program is interpreted by a strategy for P which represents the interactions it can have with any environment.

In this paper, we devise a notion of realizability for HON general games based on winning conditions on plays. We show that our model is sound for classical Peano arithmetic and allows to perform extraction for Π_2^0-formulas.

HON games with winning conditions on plays have been used in e.g. [10] for intuitionistic propositional logic with fixpoints. Our winning conditions can be seen as a generalization of the ones of [10] in order to handle full first-order classical logic, while [10] only deals with totality. Our witness extraction is based on a version of Friedman's trick inspired from Krivine [4]. Classical logic is handled similarly to the unbracketed game model of PCF of [8].

M. Hasegawa (Ed.): TLCA 2013, LNCS 7941, pp. 77–92, 2013.

We start from the cartesian closed category of single-threaded strategies which contains the unbracketed and non-innocent strategies used to model control operators and references. We use a category of continuations in the coproduct completion of [11], so that the usual flat arena of natural numbers in HON games is indeed in the image of a negative translation. Our realizability is then obtained by equipping arenas with winning conditions on plays.

The paper is organized as follows. Section 2 recalls the game semantics framework and how to interpret $\lambda\mu$-calculus in it. Section 3 defines the notion of winning strategies. Section 4 contains the definition of our realizability relation and its adequacy for classical logic. Section 5 applies our realizability model to Peano arithmetic and shows witness extraction for Π_2^0-formulas.

2 HON Games

Our realizability model is based on the Hyland-Ong-Nickau games [6] with no bracketing or innocence constraint, so as to model control operators and references [8, 9]. We consider single-threaded strategies in order to have a cartesian closed category.

2.1 Arenas and Strategies

Definition 1 (Arena). *An* arena *is a countable forest of* moves. *Each move is given a polarity O (for Opponent) or P (for Player or Proponent):*

- *A root is of polarity O.*
- *A move which is not a root has the inverse polarity of that of his parent.*

A root of an arena is also called an initial move. We will often identify an arena with its set of moves.

Definition 2 (Justified sequence). *Given an arena $\mathcal{A}$, we define a* justified sequence *on $\mathcal{A}$ to be a word s (finite or infinite) of $\mathcal{A}$ together with a partial* justifying function $f : |s| \rightharpoonup |s|$ *such that:*

- *If $f(i)$ is undefined, then s_i is an initial move.*
- *If $f(i)$ is defined, then $f(i) < i$ and s_i is a child of $s_{f(i)}$.*

We denote the empty justified sequence by ϵ. Remark here that by definition of the polarity, if $f(i)$ is undefined (s_i is initial), then s_i is of polarity O, and if $f(i)$ is defined, then s_i and $s_{f(i)}$ are of opposite polarities. Also, $f(0)$ is never defined, and so s_0 is always an initial O-move. A justified sequence is represented for example as:

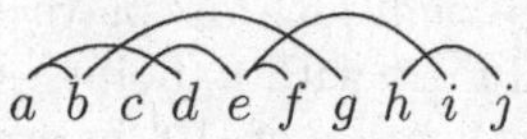

A subsequence of a justified sequence s is a subword of s together with a justifying function defined accordingly. In particular if a move a points to a move b in the original sequence and if a is in the subsequence but b is not, then the pointer from a is left undefined. For example the following sequence is a subsequence of the one above:

$$a\ b\ e\ f\ g\ i$$

If $\mathcal{A}$ is an arena, X is a subset of $\mathcal{A}$ and s is a justified sequence on $\mathcal{A}$, then $s_{|X}$ is the subsequence of s consisting of the moves of s which are in X.
In a sequence s, a move s_j is hereditarily justified by a move s_i if s_i is initial and for some n, $f^n(j) = i$.

Definition 3 (Thread). *If s is a justified sequence on $\mathcal{A}$ and if s_i is initial, then the* thread associated *to s_i is the subsequence of s consisting of the moves hereditarily justified by s_i. The* set of threads *of s,* Threads(s)*, is the set of threads associated to the initial moves of s.*

For example we have:

$$\text{Threads}\left(a\ b\ c\ d\ e\ f\ g\ h\ i\ j\right) = \left\{a\ b\ d\ g; c\ e\ f\ i; h\ j\right\}$$

Warning. Note that a thread is a justified sequence which may not be alternating, so our definition of thread differs from the usual one.

By extension a justified sequence s will be called a thread if it contains exactly one thread (i.e. Threads(s) $= \{s\}$). Remark that Threads(ϵ) $= \emptyset$ and so ϵ is not a thread.
A P-sequence (resp. O-sequence) is a sequence ending with a P-move (resp. a O-move). Write $t \sqsubseteq s$ if t is a prefix of s, i.e. t is a prefix of s as a word and their justifying functions coincide (this is a particular case of subsequence). Write $t \sqsubseteq_P s$ (resp. $t \sqsubseteq_O s$) if t is a P-prefix (resp. O-prefix) of s, i.e. $t \sqsubseteq s$ and t is a P-sequence (resp. O-sequence).

Definition 4 (Play). *A* play *s on $\mathcal{A}$ is an alternating justified sequence of $\mathcal{A}$, i.e., for any i, s_{2i} is a O-move and s_{2i+1} is a P-move. We denote the set of plays of $\mathcal{A}$ by $\mathcal{P}_\mathcal{A}$.*

A play on an arena is the trace of an interaction between a program and a context, each one performing an action alternatively. A P-play (resp. O-play) is a play which is a P-sequence (resp. O-sequence).

Definition 5 (Strategy). *A* strategy *σ on $\mathcal{A}$ is a P-prefix-closed set of finite P-plays on $\mathcal{A}$ such that:*

- *σ is deterministic: if sm and sm' are in σ, then $m = m'$.*
- *σ is single-threaded: for any P-play s, $s \in \sigma \Leftrightarrow$ Threads(s) $\subseteq \sigma$.*

Our notion of single-threadedness matches the usual one of thread-independence (see e.g. [9]). Remark also that a strategy always contains the empty play ϵ since $Threads(\epsilon) = \emptyset$.

2.2 Cartesian Closed Structure

The constructions we use will sometimes contain multiple copies of the same arena (for example $\mathcal{A} \to \mathcal{A}$), so we distinguish the instances with superscripts (for example $\mathcal{A}^{(1)} \to \mathcal{A}^{(2)}$).

Let $\mathcal{U}$ be the empty arena and $\mathcal{V}$ be the arena with only one (opponent) move. If $\mathcal{A}$ and $\mathcal{B}$ are arenas consisting of the trees $\mathcal{A}_1 \ldots \mathcal{A}_p$ and $\mathcal{B}_1 \ldots \mathcal{B}_q$, then the arenas $\mathcal{A} \to \mathcal{B}$ and $\mathcal{A} \times \mathcal{B}$ can be represented as follows:

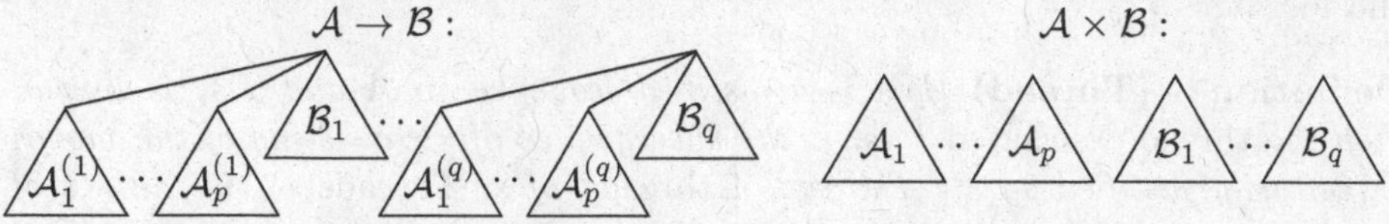

The constructions described here define a cartesian closed category whose objects are arenas and morphisms are strategies. Details of the construction can be found in [12]. In the following this category will be denoted as $\mathcal{C}$.

These definitions of arenas will be used to associate arenas to the following simple types:

Definition 6 (Simple types). *The* simple types *are defined by the following grammar, where ι ranges over a set of base types:*

$$T, U \ := \ \iota \mid \mathsf{void} \mid \mathsf{unit} \mid T \times U \mid T \to U$$

We suppose given an object $[\![\iota]\!]$ of $\mathcal{C}$ for each base type ι, and we associate to each simple type T an object $[\![T]\!]$ of $\mathcal{C}$ as follows:

$$[\![\mathsf{void}]\!] = \mathcal{V} \qquad [\![\mathsf{unit}]\!] = \mathcal{U} \qquad [\![T \times U]\!] = [\![T]\!] \times [\![U]\!] \qquad [\![T \to U]\!] = [\![U]\!]^{[\![T]\!]}$$

Since $\mathcal{C}$ is cartesian closed, we use the syntax of λ-calculus to define strategies from other strategies. In order to distinguish this notation from the $\lambda\mu$-terms of Sect. 2.3 we use a bold lambda $\boldsymbol{\lambda}$.

2.3 Interpretation of the Call-by-Name $\lambda\mu$-Calculus

We map classical proofs to strategies using the interpretation of call-by-name $\lambda\mu$-calculus in categories of continuations described in [13]. In order to make explicit the double negation translation of the base types, we base the model on the category of continuations $R^{\mathrm{Fam}(\mathcal{C})}$, where the response category $\mathrm{Fam}(\mathcal{C})$ is a variant of the coproduct completion described in [11] applied to the category $\mathcal{C}$ defined in Sect. 2.2:

Definition 7 ($\mathrm{Fam}(\mathcal{C})$)**.** *The objects of* $\mathrm{Fam}(\mathcal{C})$ *are families of objects of $\mathcal{C}$ indexed by at most countable sets, and a morphism from $\{A_i \mid i \in I\}$ to $\{B_j \mid j \in J\}$ is a function $f : I \to J$ together with a family of morphisms of $\mathcal{C}$ from A_i to $B_{f(i)}$, for $i \in I$.*

Remark here that we differ from [11] because $\mathcal{C}$ doesn't have weak coproducts nor all small products, and the families are countable. Thus $\mathrm{Fam}(\mathcal{C})$ is not bicartesian closed, but since $\mathcal{C}$ is cartesian closed and has countable products, Fam($\mathcal{C}$) is still a distributive category with finite products and coproducts, and has exponentials of all singleton families. The empty product and terminal object is the singleton family $\{1\}$, the empty sum and initial object is the empty family $\{\}$, and:

$$\{A_i \mid i \in I\} \times \{B_j \mid j \in J\} = \{A_i \times B_j \mid (i,j) \in I \times J\}$$
$$\{A_i \mid i \in I\} + \{B_j \mid j \in J\} = \{C_k \mid k \in I \uplus J\} \text{ where } C_k = \begin{cases} A_k \text{ if } k \in I \\ B_k \text{ if } k \in J \end{cases}$$
$$\{B_0\}^{\{A_i \mid i \in I\}} = \{\Pi_{i\in I} B_0^{A_i}\}$$

We fix once and for all:

$$R = \{\mathcal{V}\} = \{[\![\mathsf{void}]\!]\}$$

which is an object of Fam($\mathcal{C}$) as a singleton family. R has all exponentials as stated above. Note that the canonical morphism $\delta_A : A \to R^{(R^A)}$ is a mono.

The category of continuations $R^{\mathrm{Fam}(\mathcal{C})}$ is the full subcategory of Fam($\mathcal{C}$) consisting of the objects of the form R^A. The objects of $R^{\mathrm{Fam}(\mathcal{C})}$ are singleton families, and $R^{\mathrm{Fam}(\mathcal{C})}$ is isomorphic to $\mathcal{C}$. We will consider that objects and morphisms of $R^{\mathrm{Fam}(\mathcal{C})}$ are arenas and strategies and we will use the vocabulary defined at the end of Sect. 2.2 on $R^{\mathrm{Fam}(\mathcal{C})}$ also.

Interpreting the Call-by-Name $\lambda\mu$-Calculus. The types of $\lambda\mu$-calculus are the simple types of Definition 6. Let k^T range over a set of typed constants and x^T (resp. α^T) range over a countable set of variables (resp. names) for each type T. The grammar of $\lambda\mu$-terms is the following:

$$M, N \ := \ k^T \mid x^T \mid * \mid \langle M, N\rangle \mid \pi_1 M \mid \pi_2 M \mid \lambda x^T.M \mid MN \mid \mu\alpha^T.M \mid [\alpha]M$$

The typing rules can be found in [13], where our unit is their $\top$, our $\times$ is their $\wedge$ and our void is their $\bot$. For instance, the Law of Peirce is the type of the following term (we omit the type annotation of the variables).

$$\lambda x.\mu\alpha.[\alpha]s(\lambda y.\mu\beta.[\alpha]y) : ((T \to U) \to T) \to T \qquad (1)$$

This $\lambda\mu$-term will be denoted cc.

We follow [13] to interpret call-by-name $\lambda\mu$-calculus in $R^{\mathrm{Fam}(\mathcal{C})}$. In particular if M is a $\lambda\mu$-term of type T with free variables in $\{x_1^{T_1}, \ldots, x_n^{T_n}\}$, then its interpretation is a morphism $[\![M]\!]$ from $[\![T_1]\!] \times \ldots \times [\![T_n]\!]$ to $[\![T]\!]$. This morphism coincides with the interpretation of the call-by-name CPS translation of M (defined in [13]) in the cartesian closed category $R^{\mathrm{Fam}(\mathcal{C})}$. See [13] for details. As stated in [13], if the call-by-name CPS translations of two terms are $\beta\eta$-equivalent, then their interpretations are the same.

In the following we will drop the double brackets for the interpretation of simple types.

3 Winning Conditions on Arenas

We will now define our notion of realizability. We equip arenas with winning conditions on threads. Realizers are then winning strategies, intuitively strategies which threads are all winning.

It is well-known that preservation of totality by composition of strategies is problematic in game semantics. Luckily we do not need to preserve totality, but only winningness. We thus do not impose any totality condition on strategies, but when it turns to the definition of winning threads, we have to take into account all maximal threads, including both infinite and odd-length threads. This leads to the notion of winning strategy proposed in Definition 12.

In order to define the notion of winning condition on an arena we introduce the notion of P-subthread and O-subthread:

Definition 8 (P-subthread, O-subthread). *If t is a thread and u is a subsequence of t which is a thread, then u is a:*

- *P-subthread of t if when m^O points to n^P in t and $n^P \in u$, then $m^O \in u$,*
- *O-subthread of t if when m^P points to n^O in t and $n^O \in u$, then $m^P \in u$.*

Now we can define the notion of winning condition on an arena:

Definition 9 (Winning condition). *A* winning condition *on $\mathcal{A}$ is a set $\mathcal{W}$ of threads on $\mathcal{A}$ such that:*

- *If t is a thread on $\mathcal{A}$ and if some P-subthread of t is in $\mathcal{W}$, then $t \in \mathcal{W}$.*
- *If $t \in \mathcal{W}$ then all the O-subthreads of t are in $\mathcal{W}$.*

A justified sequence s on the arena $\mathcal{A}$ equipped with the winning condition $\mathcal{W}$ is said to be winning *if* $\mathrm{Threads}(s) \subseteq \mathcal{W}$.

Our notion of winning sequence can be seen as a generalization of the one defined in [10]. In order to obtain a realizability model of first-order logic, the notion of winning sequence is non-trivial and there can be odd-length sequences which are winning and even-length sequences which are losing.
Remark that if t is a thread on $\mathcal{A} \to \mathcal{B}$, then $t_{|\mathcal{B}}$ is a thread on $\mathcal{B}$, so $t_{|\mathcal{B}}$ is winning iff $t_{|\mathcal{B}} \in \mathcal{W}_{\mathcal{B}}$, and if t is a thread on $\mathcal{A} \times \mathcal{B}$, then t is either a thread on $\mathcal{A}$, either a thread on $\mathcal{B}$.

Definition 10 (Arrow and product of winning conditions). *If $W_{\mathcal{A}}$ and $W_{\mathcal{B}}$ are sets of threads on the arenas $\mathcal{A}$ and $\mathcal{B}$, then we define:*

$$W_{\mathcal{A}\to\mathcal{B}} = \{t \text{ thread on } \mathcal{A} \to \mathcal{B} \mid \mathrm{Threads}(t_{|\mathcal{A}}) \subseteq W_{\mathcal{A}} \;\Rightarrow\; t_{|\mathcal{B}} \in W_{\mathcal{B}}\}$$

$$W_{\mathcal{A}\times\mathcal{B}} = \left\{ t \text{ thread on } \mathcal{A} \times \mathcal{B} \;\middle|\; \begin{array}{l} t \text{ thread on } \mathcal{A} \Rightarrow t \in W_{\mathcal{A}} \\ t \text{ thread on } \mathcal{B} \Rightarrow t \in W_{\mathcal{B}} \end{array} \right\}$$

Lemma 1. *If $W_{\mathcal{A}}$ and $W_{\mathcal{B}}$ are winning conditions on $\mathcal{A}$ and $\mathcal{B}$, then $W_{\mathcal{A}\to\mathcal{B}}$ is a winning condition on $\mathcal{A} \to \mathcal{B}$ and $W_{\mathcal{A}\times\mathcal{B}}$ is a winning condition on $\mathcal{A} \times \mathcal{B}$.*

Winning Strategies. In order to define what is a winning strategy, we use a notion of augmented plays of a strategy inspired from [14]:

Definition 11 (Augmented play). *If σ is a strategy on $\mathcal{A}$, then s is an* augmented play *of σ if one of the following holds:*

- *$s \in \sigma$, or*
- *$s \in \mathcal{P}_\mathcal{A}$ is such that $\forall t \sqsubseteq_P s, t \in \sigma$ and $\forall t \in \sigma, s \not\sqsubseteq t$.*

In particular, in the second case of the above definition, s is either a O-sequence, either an infinite sequence (in which case $s \sqsubseteq t \Leftrightarrow s = t$ and so the second condition, equivalent to $s \notin \sigma$, is always true). Remark that unlike [14], we consider not only odd-length extensions (with an O-move), but also infinite ones.

Definition 12 (Winning strategy). *If σ is a strategy on the arena $\mathcal{A}$ equipped with the winning condition $\mathcal{W}$, then σ is said to be* winning *if all its augmented plays are winning.*

The following lemma will be useful to prove that a strategy σ is winning on $(\mathcal{A}, \mathcal{W})$.

Lemma 2. *If σ is a strategy on $\mathcal{A}$ and if s is an augmented play of σ, then every $t \in Threads(s)$ is an augmented play of σ.*

Using this lemma it is sufficient to prove that every augmented play of σ which is a thread (let us call it an augmented thread of σ) is in $\mathcal{W}_\mathcal{A}$ in order to prove that σ is winning on $(\mathcal{A}, \mathcal{W}_\mathcal{A})$.

We now prove that the winning conditions on the arrow and product are compatible with application and pairing of strategies.

Lemma 3. *If σ is a winning strategy on $(\mathcal{A} \to \mathcal{B}, \mathcal{W}_{\mathcal{A}\to\mathcal{B}})$ and τ is a winning strategy on $(\mathcal{A}, \mathcal{W}_\mathcal{A})$, then $\sigma(\tau)$ is a winning strategy on $(\mathcal{B}, \mathcal{W}_\mathcal{B})$.*

Proof. Let t be an augmented thread of $\sigma(\tau)$. By definition of composition of strategies, there is some augmented play u of σ such that $u_{|\mathcal{A}}$ is an augmented play of τ and $u_{|\mathcal{B}} = t$. Since t is a thread, u is also a thread, so since σ is winning on $\mathcal{A} \to \mathcal{B}$, $u \in \mathcal{W}_{\mathcal{A}\to\mathcal{B}}$. $u_{|\mathcal{A}}$ is an augmented play of τ which is winning on $\mathcal{A}$, so $u_{|\mathcal{B}}$ is winning, and so $t = u_{|\mathcal{B}}$ is a winning thread: $t \in \mathcal{W}_\mathcal{B}$. Therefore $\sigma(\tau)$ is winning. □

Lemma 4. *If σ is a winning strategy on $(\mathcal{A}, \mathcal{W}_\mathcal{A})$ and τ is a winning strategy on $(\mathcal{B}, \mathcal{W}_\mathcal{B})$, then $\langle \sigma, \tau \rangle$ is a winning strategy on $(\mathcal{A} \times \mathcal{B}, \mathcal{W}_{\mathcal{A}\times\mathcal{B}})$.*

Proof. Let t be an augmented thread of $\sigma(\tau)$. By definition of product of strategies, $t_{|\mathcal{A}}$ is an augmented play of σ and $t_{|\mathcal{B}}$ is an augmented play of τ, so since σ and τ are winning, $t_{|\mathcal{A}}$ and $t_{|\mathcal{B}}$ are winning, and so $t \in \mathcal{W}_{\mathcal{A}\times\mathcal{B}}$. Therefore $\langle \sigma, \tau \rangle$ is winning. □

The following technical lemma on the interpretation of cc will be useful.

Lemma 5. *If t is an augmented thread of* $[\![\mathrm{cc}]\!]$ *on the arena* $((T \to U) \to T) \to T$ *(written* $((T^{(1)} \to U) \to T^{(2)}) \to T^{(3)}$*), then the threads of* $t_{|T^{(1)}}$ *and* $t_{|T^{(2)}}$ *are P-subthreads of* $t_{|T^{(3)}}$.

It follows easily from this lemma and Lemma 1 that for any winning conditions W_T and W_U, $[\![\mathrm{cc}]\!]$ is winning on the arena

$$\Big(((T \to U) \to T) \to T, W_{((T\to U)\to T)\to T)}\Big)$$

Remark on the Arrow on Winning Conditions. Let $\mathcal{A}$, $\mathcal{B}$ be arenas equipped with winning conditions $W_{\mathcal{A}}$, $W_{\mathcal{B}}$. We define here a strategy σ on $\mathcal{A} \to \mathcal{B}$ such that for any winning strategy τ on $\mathcal{A}$, $\sigma(\tau)$ is winning on $\mathcal{B}$, but σ is not winning on $\mathcal{A} \to \mathcal{B}$. Hence the arrow on winning conditions differs from the usual Kleene realizability arrow (see [1]).
We choose $\mathcal{A}$ and $\mathcal{B}$ to be the same arena $\mathcal{Q}$ consisting of one root with three children $\sharp$, $\flat$ and $\natural$, equipped with the winning condition

$$\mathcal{W}_{\mathcal{Q}} = \{q^O a_1^P a_2^P \ldots \mid \exists i, a_i \in \{\sharp, \natural\}\}$$

where the threads may be finite or infinite. We define a strategy σ on $\mathcal{Q} \to \mathcal{Q}$ such that for any τ winning on $(\mathcal{Q}, \mathcal{W}_{\mathcal{Q}})$, $\sigma(\tau)$ is winning on $(\mathcal{Q}, \mathcal{W}_{\mathcal{Q}})$, but σ is not winning on $(\mathcal{Q} \to \mathcal{Q}, \mathcal{W}_{\mathcal{Q}\to\mathcal{Q}})$. σ is the innocent strategy defined by the views:

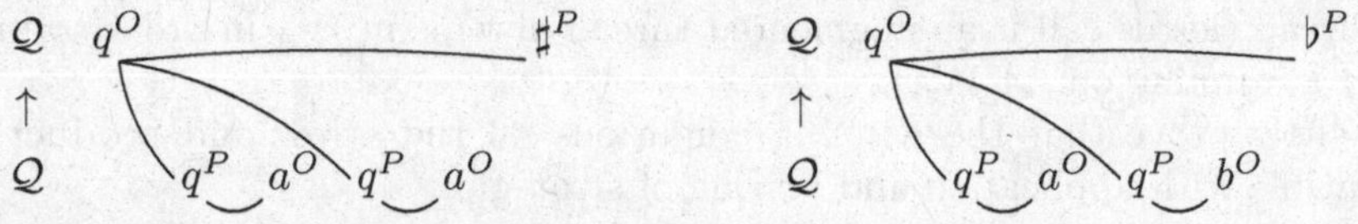

where a and b are distinct moves. The interaction with any single threaded strategy will produce the left view, and so the projection $q^O \sharp^P$ will be winning, but the right view (which will never happen in an interaction with a single-threaded strategy) with $a = \sharp$ and $b = \natural$ is losing, so σ is losing.

4 First-Order Logic

We define a realizability model for first-order classical logic with possibilities of witness extraction. For that the proposition $\bot$ will be mapped to an arena ι in general different from $\mathcal{V}$. Its associated winning condition will be a parameter of the model, in the spirit of [4].

Let x range over a countable set of variables, f range over a set of function symbols with fixed finite arity and P range over a set of predicate symbols with fixed finite arity. First-order terms and formulas are defined by the following grammar:

$$a, b \ := \ x \mid f(a_1, \ldots, a_n)$$

$$A, B \ := \ P(a_1, \ldots, a_n) \mid \top \mid A \wedge B \mid A \Rightarrow B \mid \forall x A \mid \bot$$

In the following we use syntactic sugar for the negation of formulas: $\neg A \equiv A \Rightarrow \bot$ and for the existential: $\exists x A \equiv \neg\forall x \neg A$. We fix a countable first-order structure interpreting the terms of our logic, that is a countable set E together with an interpretation $f^E : E^n \to E$ for each function symbol. The interpretation is extended to every closed term: if a is a closed term of the logic, then a^E denotes its interpretation in the first-order structure, so a^E is an element of E.

4.1 Realizability

We let $\perp\!\!\!\perp$ be an arbitrary subset of E. We consider simple types with a type constant P^* for each predicate P and a type constant ι to interpret $\bot$. We can map any first-order formula A to such a simple type A^* as follows:

$$(P(a_1, \ldots, a_n))^* = P^* \qquad \top^* = \mathsf{unit} \qquad (A \wedge B)^* = A^* \times B^*$$

$$(A \Rightarrow B)^* = A^* \to B^* \qquad (\forall x A)^* = A^* \qquad \bot^* = \iota$$

Remark that the type $\bot^*$ is not the type void because the associated arena would be too small to hold informational content.

Recall that we omit the double bracket notation for the arenas, so a type T also denotes the associated arena. We suppose that for each atomic predicate P, the type P^* comes with its associated arena. We fix the arena associated to ι to be $R^{(R^{\mathbf{E}})}$, where $\mathbf{E} = \{\mathcal{U}_e \mid e \in E\}$ is the countable family of empty arenas (and $R = \{\mathcal{V}\}$). Hence ι is the usual flat arena for the set E.

Let us suppose we associate to each predicate $P(a_1, \ldots, a_n)$ with $a_1, \ldots, a_n$ closed first-order terms a winning condition $\mathcal{W}_{P(a_1,\ldots,a_n)}$ on the arena P^*. We can then define for each closed first-order formula A a winning condition $\mathcal{W}_A$ on the arena A^*. The winning conditions $\mathcal{W}_{A\wedge B}$ and $\mathcal{W}_{A\Rightarrow B}$ are as in Definition 10, and we let:

$$\mathcal{W}_\top = \emptyset \qquad \mathcal{W}_{\forall x A} = \bigcap_{a \text{ closed}} \mathcal{W}_{A[a/x]} \qquad \mathcal{W}_\bot = \{q^O m_1^P m_2^P \ldots \mid \exists i, m_i \in \perp\!\!\!\perp\}$$

Note that these are indeed winning conditions. For $\mathcal{W}_\top$, the empty set is a winning condition on $\mathcal{U}$ which is the empty arena with no thread. For $\mathcal{W}_{\forall x A}$, it is easy to see that an intersection of winning conditions is a winning condition. For $\mathcal{W}_\bot$, the thread $q^O m_1^P m_2^P \ldots$ (that may be finite or infinite) has only itself as O-subthread and $q^O m_{i_1}^P m_{i_2}^P \ldots$ for $1 \leq i_1 < i_2 \leq \ldots$ as P-subthreads so $\mathcal{W}_\bot$ is a winning condition on ι.

We can now define our notion of realizability:

Definition 13 (Realizability relation). *If A is a closed first-order formula and if σ is a strategy on A^*, then σ realizes A (denoted $\sigma \Vdash A$) if σ is a winning strategy on $(A^*, \mathcal{W}_A)$.*

The following lemma shows that the identity formulas are realized by the corresponding identity strategies.

Lemma 6. *If A is a closed formula, then the identity strategy id_{A^*} on A^* is a realizer for the formula $A \Rightarrow A$.*

Proof. Let $\mathcal{A}^{(1)} \to \mathcal{A}^{(2)}$ denote the arena $A^* \to A^*$. If t is an augmented thread of id_{A^*}, then $t_{|\mathcal{A}^{(1)}} = t_{\mathcal{A}^{(2)}}$, so if $t_{|\mathcal{A}^{(1)}}$ is winning, then $t_{|\mathcal{A}^{(2)}} = t_{\mathcal{A}^{(1)}}$ is also winning, and so $t \in \mathcal{W}_{A^*}$. □

The following result is a consequence of the remark following Lemma 5.

Lemma 7. *If A and B are closed formulas, then* $\mathsf{cc} \Vdash ((A \Rightarrow B) \Rightarrow A) \Rightarrow A$.

4.2 Adequacy for Minimal Classical Logic

We now show that realizability is compatible with deduction in minimal classical logic. Full classical logic is discussed in Sect. 4.3.

Deduction System. Let $\boldsymbol{Ax}$ be a set of closed formulas. We use the following deduction system based on natural deduction with a rule for the law of Peirce, where Γ is a sequence of formulas $A_1, \ldots, A_n$.

$$\frac{}{\Gamma \vdash A}\, A \in \Phi \qquad \frac{}{\Gamma \vdash ((A \Rightarrow B) \Rightarrow A) \Rightarrow A} \qquad \frac{}{\Gamma \vdash A}\, A \subset \boldsymbol{Ax}$$

$$\frac{}{\Gamma \vdash \top} \qquad \frac{\Gamma \vdash A \qquad \Gamma \vdash B}{\Gamma \vdash A \wedge B} \qquad \frac{\Gamma \vdash A \wedge B}{\Gamma \vdash A} \qquad \frac{\Gamma \vdash A \wedge B}{\Gamma \vdash B}$$

$$\frac{\Gamma, A \vdash B}{\Gamma \vdash A \Rightarrow B} \qquad \frac{\Gamma \vdash A \Rightarrow B \qquad \Gamma \vdash A}{\Gamma \vdash B}$$

$$\frac{\Gamma \vdash A}{\Gamma \vdash \forall x A}\, x \notin FV(\Gamma) \qquad \frac{\Gamma \vdash \forall x A}{\Gamma \vdash A[a/x]}$$

Remark that $\bot$ has no associated rule, since the ex-falso rule has a particular status, given the interpretation of $\bot$. This will be discussed in Sect. 4.3.

Translation of Proofs to Strategies. We use $\lambda\mu$-calculus and its interpretation in $R^{\mathrm{Fam}(\mathcal{C})}$ to map a first-order proof to a typed $\lambda\mu$-term which is then interpreted in $R^{\mathrm{Fam}(\mathcal{C})}$ as a strategy.

Assume given a constant k^A of type A^* for each $A \in \boldsymbol{Ax}$. We map a derivation ν of $A_1, \ldots, A_n \vdash A$ to a typed $\lambda\mu$-term ν^* of type A with free variables among $x^{A_1^*}, \ldots, x^{A_n^*}$ as follows:

$$\frac{}{A_1, \ldots, A_n \vdash A_i} \rightsquigarrow x^{A_i^*} \qquad \frac{}{\Gamma \vdash ((A \Rightarrow B) \Rightarrow A) \Rightarrow A} \rightsquigarrow \mathsf{cc}\ (\text{see } (1))$$

$$\frac{\dfrac{\nu}{\Gamma \vdash A} \qquad \dfrac{\nu'}{\Gamma \vdash B}}{\Gamma \vdash A \wedge B} \rightsquigarrow \langle \nu^*, \nu'^* \rangle \qquad \frac{}{\Gamma \vdash \top} \rightsquigarrow * \qquad \frac{}{\Gamma \vdash A}\, A \in \boldsymbol{Ax} \rightsquigarrow k^A$$

$$\frac{\dfrac{\nu}{\Gamma \vdash A \wedge B}}{\Gamma \vdash A} \rightsquigarrow \pi_1 \nu^* \qquad \frac{\dfrac{\nu}{\Gamma \vdash A}}{\Gamma \vdash \forall x A}\, x \notin FV(\Gamma) \rightsquigarrow \nu^* \qquad \frac{\dfrac{\nu}{\Gamma, A \vdash B}}{\Gamma \vdash A \Rightarrow B} \rightsquigarrow \lambda x^{A^*}.\nu^*$$

$$\dfrac{\dfrac{\nu}{\Gamma \vdash A \wedge B}}{\Gamma \vdash B} \rightsquigarrow \pi_2 \nu^* \qquad \dfrac{\dfrac{\nu}{\Gamma \vdash \forall x A}}{\Gamma \vdash A[a/x]} \rightsquigarrow \nu^* \qquad \dfrac{\dfrac{\nu}{\Gamma \vdash A \Rightarrow B} \quad \dfrac{\nu'}{\Gamma \vdash A}}{\Gamma \vdash B} \rightsquigarrow \nu^*(\nu'^*)$$

Adequacy. We now prove that the strategies interpreting the proofs are realizers of the proved formula. If A is a formula and θ an assignment of terms to variables, then $\theta(A)$ denotes A where all the free variables are replaced with their image by θ.

Lemma 8. *Let* $\perp\!\!\!\perp \subseteq E$. *Suppose that we have a realizer for each formula of* $\boldsymbol{Ax}$. *If* ν *is a derivation of the sequent* $\Gamma \vdash A$ *and if* θ *is an assignment of closed first-order terms to variables, then* $[\![\nu^*]\!]$ *is a winning strategy on* $\Gamma^* \to A^*$ *equipped with* $\mathcal{W}_{\theta(\Gamma \Rightarrow A)}$.

Proof (sketch). The case of the variable follows from Lemma 6. That of cc comes from Lemma 5. Product introduction is dealt with using Lemma 4, and arrow elimination using Lemma 3. The other cases are straightforward. □

4.3 Full Classical Logic

In order to get full classical logic we need to add an ex-falso rule. However since the arena $\bot^*$ is not empty (see Sect. 4.1), we restrict ex-falso to a certain class of formulas. We have to ensure that $(\iota, \mathcal{W}_\bot)$ is included in $(A^*, \mathcal{W}_A)$. This means that ι is a subtree of A^*, so a play on ι is in particular a play on A^*, and that $\mathcal{W}_\bot \subseteq \mathcal{W}_A$. We will call these formulas explodable since they satisfy the principle of explosion. We add to our deduction system the following rule:

$$\frac{\Gamma \vdash \bot}{\Gamma \vdash A} \; A \textbf{ explodable}$$

In particular any formula ending with $\bot$ is explodable, where a formula ending with $\bot$ is a formula generated by the grammar:

$$C, D \;:=\; C \wedge D \mid A \Rightarrow C \mid \forall x C \mid \bot$$

where A is any first-order formula (defined in Sect. 4). The corresponding adequacy lemma is immediate from Lemma 8.

4.4 First-Order Logic with Equality

We now show how to handle equality. We suppose that our first-order language contains an inequality predicate $\neq$ of arity 2 interpreted by the simple type ι (see Sect. 4.1). The associated winning condition is:

$$\mathcal{W}_{a \neq b} = \begin{cases} \mathcal{W}_\bot & \text{if } a^E = b^E \\ \text{the set of all threads on } \iota & \text{otherwise} \end{cases}$$

(recall that E is the first-order structure chosen at the beginning of Sect. 4). It is easy then to verify that any formula ending with the predicate $a \neq b$ is explodable. In the following we use the notation $(a = b) \equiv \neg(a \neq b)$. The axioms for equality are:

$$\textbf{(refl)} \quad \forall x(x = x) \qquad\qquad \textbf{(Leib)} \quad \forall x \forall y(\neg A[x] \Rightarrow A[y] \Rightarrow x \neq y)$$

Recall that $\forall x(x = x)$ is only syntactic sugar for $\forall x(x \neq x \Rightarrow \bot)$, and that $\forall x \forall y(\neg A[x] \Rightarrow A[y] \Rightarrow x \neq y)$ is also syntactic sugar for $\forall x \forall y((A[x] \Rightarrow \bot) \Rightarrow A[y] \Rightarrow x \neq y)$.

Lemma 9. *Let* $\perp\!\!\!\perp \subseteq E$.

1. *The identity strategy on* ι, *is a realizer of* **(refl)**.
2. *The identity strategy on* $A^* \to \iota$, *is a realizer of* **(Leib)**.

Proof (sketch). For the first point, we always have $a^E = a^E$, so $\mathcal{W}_{a \neq a} = \mathcal{W}_{\bot}$. Concerning the second point,if a and b are closed first-order terms, if $a^E \neq b^E$ then any thread is winning on $a \neq b$, otherwise if we win on $A[b]$ then we win on $A[a]$, so if we win on $\neg A[a]$ then we win on $\bot$ and therefore on $a \neq b$. □

5 Peano Arithmetic

We now proceed to the realizability interpretation of full Peano arithmetic.

5.1 Definitions

Our first-order language is built from the function symbols 0 of arity 0, S of arity 1 and $+$ and $\times$ of arity 2. The predicate symbols are $\neq$ of arity 2 and nat of arity 1. This choice of function symbols is only for simplicity, and we could choose to have all the symbols of primitive recursive functions.

We also fix the structure interpreting the terms of the logic to be the set of natural numbers $\mathbb{N}$. The symbols 0, S, $+$ and $\times$ are interpreted the standard way. The typed $\lambda\mu$-calculus in which we interpret the proofs has ι as unique base type. All the predicate symbols and $\bot$ are interpreted as ι, and the associated arena in $R^{\mathrm{Fam}(\mathcal{C})}$ is $[\![\iota]\!] = R^{(R^{\mathbf{N}})}$ where $\mathbf{N} = \{\mathcal{U}_n \mid n \in \mathbb{N}\}$ (see Sect. 4.1). Hence the type of natural numbers is interpreted as the negative translation of $\mathbf{N}$. Note that this is the usual flat arena of natural numbers:

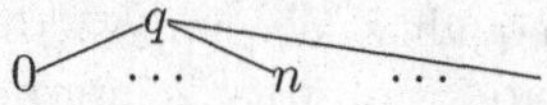

This differs from Laird's interpretation of PCF with control [15], where the base type of natural numbers is interpreted by the arena $(\iota \to \iota) \to \iota$.

The winning conditions for $\bot$ and $a \neq b$ are as in Sects. 4.1 and 4.4, and the winning condition for $\mathsf{nat}(a)$ is:

$$\mathcal{W}_{\mathsf{nat}(a)} = \{q^O n_1^P n_2^P \ldots \mid \exists i, n_i = a^{\mathbb{N}}\}$$

which is a winning condition, using the same arguments as for $\mathcal{W}_\bot$. From this we can check that every formula which contains no $\mathsf{nat}(a)$ predicate at rightmost position is explodable. We use the following syntactic sugar:

$$\forall^{\mathsf{n}} x A \equiv \forall x\, (\mathsf{nat}(x) \Rightarrow A) \qquad \exists^{\mathsf{n}} x A \equiv \neg\forall^{\mathsf{n}} x \neg A \equiv \neg\forall x\, (\mathsf{nat}(x) \Rightarrow \neg A)$$

The relativization A^{n} of a formula is defined as the identity on all constructions except for the quantification: $(\forall x A)^{\mathsf{n}} \equiv \forall^{\mathsf{n}} x A^{\mathsf{n}}$. Note that if a formula does not contain any $\mathsf{nat}(a)$ predicate, then its relativization has no $\mathsf{nat}(a)$ predicate at rightmost position, so it is explodable.

The axioms are the ones for equality (defined in Sect. 4.4) and the universal closures of:

(*Snz*)	$S(x) \neq 0$	**(*Sinj*)**	$x \neq y \Rightarrow S(x) \neq S(y)$
(+0)	$x + 0 = x$	**(*nat*0)**	$\mathsf{nat}(0)$
(+*S*)	$x + S(y) = S(x + y)$	**(*natS*)**	$\mathsf{nat}(x) \Rightarrow \mathsf{nat}(S(x))$
(×0)	$x \times 0 = 0$	**(*nat*+)**	$\mathsf{nat}(x) \Rightarrow \mathsf{nat}(y) \Rightarrow \mathsf{nat}(x + y)$
(×*S*)	$x \times S(y) = x \times y + x$	**(*nat*×)**	$\mathsf{nat}(x) \Rightarrow \mathsf{nat}(y) \Rightarrow \mathsf{nat}(x \times y)$

$$\textbf{(\textit{ind})} \quad A[0] \Rightarrow \forall^{\mathsf{n}} x (A[x] \Rightarrow A[S(x)]) \Rightarrow \forall^{\mathsf{n}} x A[x]$$

We will now define the realizers for these axioms. We first define the strategies computing basic operations and recursion on natural numbers.

In $\mathrm{Fam}(\mathcal{C})$ a morphism from $\top^* = \{\mathcal{U}\}$ to $\mathbf{N} = \{\mathcal{U}_n \mid n \in \mathbb{N}\}$ is given by a function from the singleton set to $\mathbb{N}$ together with a strategy from $\mathcal{U}$ to $\mathcal{U}$. Since there is only one such strategy, such a morphism is given by a natural number. We will call this morphism τ_n. Similarly a morphism from $\mathbf{N}^k$ to $\mathbf{N}$ is given by a function from $\mathbb{N}^k$ to $\mathbb{N}$. This leads to morphisms τ_S, τ_+ and $\tau_\times$ respectively on $\mathbf{N} \to \mathbf{N}$, $\mathbf{N} \to \mathbf{N} \to \mathbf{N}$ and $\mathbf{N} \to \mathbf{N} \to \mathbf{N}$. From these we define the following morphisms of $R^{\mathrm{Fam}(\mathcal{C})}$:

$$\begin{aligned}
\sigma_n &= \boldsymbol{\lambda} k.k\tau_n \ :\ \iota \\
\sigma_S &= \boldsymbol{\lambda} n.\boldsymbol{\lambda} k.n(\boldsymbol{\lambda} n'.k(\tau_S n')) \ :\ \iota \to \iota \\
\sigma_+ &= \boldsymbol{\lambda} m \boldsymbol{\lambda} n.\boldsymbol{\lambda} k.m(\boldsymbol{\lambda} m'.n(\boldsymbol{\lambda} n'.k(\tau_+ m'n'))) \ :\ \iota \to \iota \to \iota \\
\sigma_\times &= \boldsymbol{\lambda} m \boldsymbol{\lambda} n.\boldsymbol{\lambda} k.m(\boldsymbol{\lambda} m'.n(\boldsymbol{\lambda} n'.k(\tau_\times m'n'))) \ :\ \iota \to \iota \to \iota
\end{aligned}$$

The above morphisms correspond to the expected strategies:

Lemma 10. *The strategies σ_n, σ_S, σ_+ and $\sigma_\times$ are the innocent strategies defined by the views:*

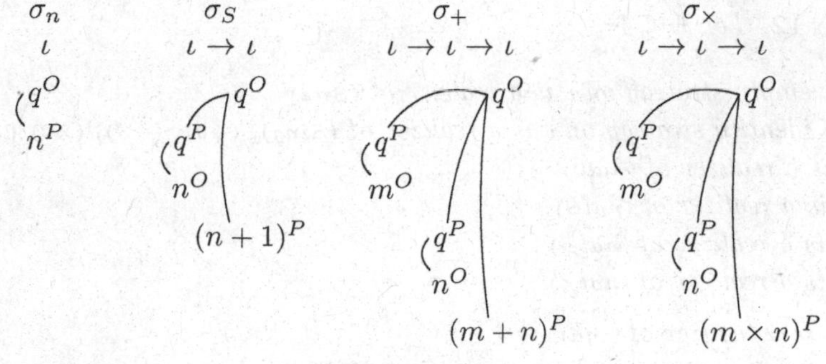

We now move to the definition of ρ^T, the recursor on type T, which is the usual recursor of Gödel's system T. For that we define for each $n \in \mathbb{N}$ and simple type T a strategy ρ_n^T by:

$$\begin{aligned}
&\rho_0^T = [\![\lambda x.\lambda y.x]\!] : T \to (\iota \to T \to T) \to T \\
&\xi^T = [\![\lambda n.\lambda r.\lambda x.\lambda y.y(n)(rxy)]\!] \\
&\qquad\qquad\qquad : \iota \to (T \to (\iota \to T \to T) \to T) \to T \to (\iota \to T \to T) \to T \\
&\rho_{n+1}^T = \xi^T(\sigma_n)(\rho_n^T) : T \to (\iota \to T \to T) \to T
\end{aligned}$$

and we finally define the strategy ρ^T as the innocent strategy which views are:

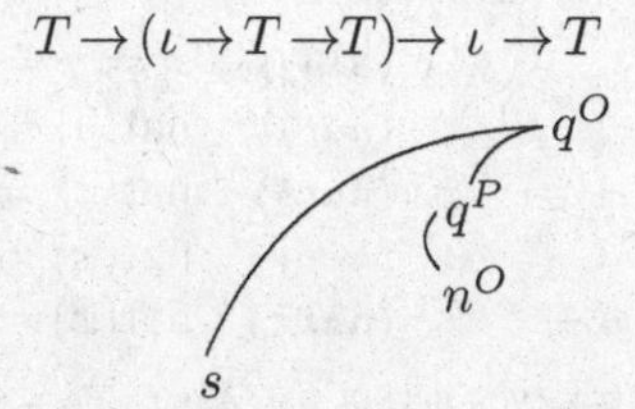

where $q^\frown s$ is a view of ρ_n^T on the subarena $T \to (\iota \to T \to T) \to T$.

We use the following lemma in order to prove the validity of (*ind*):

Lemma 11. *1.* ρ_0^T *is a realizer of* $A[0] \Rightarrow \forall^{\mathsf{n}} x(A[x] \Rightarrow A[S(x)]) \Rightarrow A[0]$

2. ξ^T *is a realizer of:*

$$\forall^{\mathsf{n}} y\Big(\Big(A[0] \Rightarrow \forall^{\mathsf{n}} x(A[x] \Rightarrow A[S(x)]) \Rightarrow A[y]\Big) \\ \Rightarrow A[0] \Rightarrow \forall^{\mathsf{n}} x(A[x] \Rightarrow A[S(x)]) \Rightarrow A[S(y)]\Big)$$

Proof. This is an immediate consequence of Lemma 8, since the strategies ρ_0^T and ξ^T are the interpretations of proofs of the formulas.

5.2 Validity of Axioms

We prove that all the axioms are realized:

Lemma 12. *Let* $\perp\!\!\!\perp \subseteq \mathbb{N}$.

1. *The empty strategy on* ι *is a realizer of* (*Snz*)
2. *The identity strategy on* ι *is a realizer of* (*Sinj*), (+0), (+*S*), (×0) *and* (×*S*)
3. σ_0 *is a realizer of* (*nat*0)
4. σ_S *is a realizer of* (*natS*)
5. σ_+ *is a realizer of* (*nat*+)
6. $\sigma_\times$ *is a realizer of* (*nat*×)
7. ρ^{A^*} *is a realizer of* (*ind*)

Proof (sketch). The cases 1 and 2 are straightforward. We prove cases 3, 4, 5 and 6 using Lemma 10. For 7 we prove by induction on n that:

$$\rho_n^{A^*} \Vdash A[0] \Rightarrow \forall^{\mathsf{n}} x(A[x] \Rightarrow A[S(x)]) \Rightarrow A[\overline{n}]$$

using Lemma 11. We finally prove that ρ^{A^*} is a realizer of the (*ind*) axiom for formula A. Let t be an augmented thread of ρ^{A^*} on the arena

$$A^{(1)} \to \left(\iota^{(1)} \to A^{(2)} \to A^{(3)}\right) \to \iota^{(2)} \to A^{(4)}$$

Let suppose that $t_{|A^{(1)}}$ is winning on $A[0]$ and $t_{|\iota^{(1)} \to A^{(2)} \to A^{(3)}}$ is winning on $\forall^{\mathsf{n}} x(A[x] \Rightarrow A[S(x)])$. We want to prove that $t_{|\iota^{(2)} \to A^{(4)}}$ is winning on $\forall x^{\mathsf{n}} A[x]$, so let a be a closed first-order term, let $n = a^{\mathbb{N}}$ and let suppose that $t_{|\iota^{(2)}}$ is winning on $\mathsf{nat}(a)$. Then there must be some n^O in $t_{|\iota^{(2)}}$. Let u be the subsequence of t consisting of the initial q^O, the following q^P, this n^O and all the moves m of t such that the view obtained immediately after m contains n^O. Then u is a play of $\rho_n^{A^*}$. Since a P-move does not change the current view, the threads of $u_{|A^{(1)}}$ are O-subthreads of $t_{|A^{(1)}}$ (the projection induces an inversion of polarities), so they are winning on $A[0]$, and the threads of $u_{|\iota^{(1)} \to A^{(2)} \to A^{(3)}}$ are O-subthreads of $t_{|\iota^{(1)} \to A^{(2)} \to A^{(3)}}$, so they are winning on $\forall^{\mathsf{n}} x(A[x] \Rightarrow A[S(x)])$. Then by the property on $\rho_n^{A^*}$, $u_{|A^{(4)}}$ is winning on $A[a]$. But $u_{|A^{(4)}}$ is a P-subthread of $t_{|A^{(4)}}$ (no inversion here), so $t_{|A^{(4)}}$ is winning on $A[a]$. □

Theorem 1. *If A is provable in Peano arithmetic then there is a computable strategy σ such that $\sigma \Vdash A^{\mathsf{n}}$.*

5.3 Extraction

We now show that from any provable Π_2^0-formula we can extract a computable witnessing function.

Suppose that we have a proof of $\vdash \forall^{\mathsf{n}} x \exists^{\mathsf{n}} y(a = b)$. We obtain by double-negation elimination a proof of $\vdash \forall^{\mathsf{n}} x(\neg\forall^{\mathsf{n}} y(a \neq b))$, and we map it to a strategy σ such that:

$$\sigma \quad \Vdash \quad \forall^{\mathsf{n}} x(\neg\forall^{\mathsf{n}} y(a \neq b)) \equiv \forall x(\mathsf{nat}(x) \Rightarrow (\forall y(\mathsf{nat}(y) \Rightarrow a \neq b) \Rightarrow \bot)$$

Then if $n \in \mathbb{N}$, $\sigma_n \Vdash \mathsf{nat}(\overline{n})$, so $\sigma(\sigma_n) \Vdash \forall y(\mathsf{nat}(y) \Rightarrow a[\overline{n}/x] \neq b[\overline{n}/x]) \Rightarrow \bot$. Let now fix $\perp\!\!\!\perp = \{m \in \mathbb{N} \mid (a[\overline{n}/x, \overline{m}/y])^{\mathbb{N}} = (b[\overline{n}/x, \overline{m}/y])^{\mathbb{N}}\}$. By a simple disjunction of cases we get

$$id_\iota \Vdash \forall y(\mathsf{nat}(y) \Rightarrow a[\overline{n}/x] \neq b[\overline{n}/x])$$

and therefore $\sigma(\sigma_n)(id_\iota) \Vdash \bot$. Then we can prove that $\sigma(\sigma_n)(id_\iota)$ is some σ_m such that $m \in \perp\!\!\!\perp$. Indeed, if $\sigma(\sigma_n)(id_\iota)$ is the empty strategy then its only augmented play is q^O, which is losing on $\bot$.

6 Conclusion and Future Work

We have built a realizability model for Peano arithmetic using winning conditions on arenas, and have used it in the context of witness extraction for Π^0_2-formulas. Future work will be the comparison of the present model with the game interpretation of classical arithmetic of [16], and with the winning conditions on sequential games of [17] and [18]. Our main goal is to compare two different versions of realizers for the axiom of dependent choices: the modified bar recursion of [19] and the clock of [3].

References

1. Troelstra, A.: Chapter VI Realizability. Studies in Logic and the Foundations of Mathematics 137, 407–473 (1998)
2. Griffin, T.: A Formulae-as-Types Notion of Control. In: POPL, pp. 47–58. ACM Press (1990)
3. Krivine, J.L.: Dependent choice, 'quote' and the clock. Theor. Comput. Sci. 308 (1-3), 259–276 (2003)
4. Krivine, J.L.: Realizability in classical logic. Panoramas et Synthèses 27, 197–229 (2009)
5. Krivine, J.L.: Typed lambda-calculus in classical Zermelo-Frænkel set theory. Arch. Math. Log. 40(3), 189–205 (2001)
6. Hyland, J.M.E., Ong, C.H.L.: On Full Abstraction for PCF: I, II, and III. Inf. Comput. 163(2), 285–408 (2000)
7. Nickau, H.: Hereditarily Sequential Functionals. In: Matiyasevich, Y.V., Nerode, A. (eds.) LFCS 1994. LNCS, vol. 813, pp. 253–264. Springer, Heidelberg (1994)
8. Laird, J.: Full Abstraction for Functional Languages with Control. In: LICS, pp. 58–67. IEEE (1997)
9. Abramsky, S., Honda, K., McCusker, G.: A Fully Abstract Game Semantics for General References. In: LICS, pp. 334–344. IEEE (1998)
10. Clairambault, P.: Least and Greatest Fixpoints in Game Semantics. In: de Alfaro, L. (ed.) FOSSACS 2009. LNCS, vol. 5504, pp. 16–31. Springer, Heidelberg (2009)
11. Abramsky, S., McCusker, G.: Call-by-Value Games. In: Nielsen, M. (ed.) CSL 1997. LNCS, vol. 1414, pp. 1–17. Springer, Heidelberg (1998)
12. Harmer, R.: Games and full abstraction for non-deterministic languages. PhD thesis. Imperial College, London (University of London) (1999)
13. Selinger, P.: Control categories and duality: on the categorical semantics of the lambda-mu calculus. Mathematical Structures in Computer Science 11(2), 207–260 (2001)
14. Melliès, P.A.: Sequential algorithms and strongly stable functions. Theor. Comput. Sci. 343(1-2), 237–281 (2005)
15. Laird, J.: A semantic analysis of control. PhD thesis, University of Edinburgh (1999)
16. Coquand, T.: A Semantics of Evidence for Classical Arithmetic. J. Symb. Log. 60(1), 325–337 (1995)
17. Hyland, J.M.E.: Game semantics. In: Pitts, A.M., Dybjer, P. (eds.) Semantics and Logics of Computation, vol. 14. Cambridge University Press (1997)
18. Melliès, P.A., Tabareau, N.: Resource modalities in game semantics. In: LICS, pp. 389–398. IEEE (2007)
19. Berardi, S., Bezem, M., Coquand, T.: On the Computational Content of the Axiom of Choice. J. Symb. Log. 63(2), 600–622 (1998)

The Resource Lambda Calculus Is Short-Sighted in Its Relational Model

Flavien Breuvart

PPS, UMR 7126, Univ Paris Diderot, Sorbonne Paris Cité, F-75205 Paris, France
breuvart@pps.univ-paris-diderot.fr

Abstract. Relational semantics is one of the simplest and categorically most natural semantics of Linear Logic. The co-Kleisli category MRel associated with its multiset exponential comonad contains a fully abstract model of the untyped λ-calculus. That particular object of MRel is also a model of the resource λ-calculus, deriving from Ehrhard and Regnier's differential extension of Linear Logic and related to Boudol's λ-calculus with multiplicities. Bucciarelli et al. conjectured that model to be fully-abstract also for the resource λ-calculus. We give a counter-example to the conjecture. As a by-product we achieve a context lemma for the resource λ-calculus.

Keywords: Full abstraction, resource λ-calculus, linear logic, nondeterminism.

1 Introduction

Rel. The category Rel of set and relations is known to model Linear Logic, and its construction is canonical from categorical point of view. Indeed, Rel can be seen as the free infinite biproduct completion of the Boolean ring seen as a category with one object and two morphisms (true and false), the conjunction being the identity [13]. The exponential modality ! of linear logic is given by the finite multisets comonad that precisely is the free commutative comonad in Rel [13]. Moreover, despite the biproduct, proofs are morally preserved, *i.e.* the interpretation of cut free proofs is injective up to isomorphism[1] [7].

This multiset comonoid $!A$ of a set A is the set of finite multisets of elements in A. Intuitively a finite multiset in $a \in !A$ is a resource that behaves as $\invamp_{\alpha\in a}\alpha$, *i.e.* like a resource that must be used by a program exactly once per element in a (with multiplicities). This behavior enabling an interesting resource management, it was natural to develop a syntactical counterpart.

Resource λ-Calculus. A restricted version was previously introduced by Boudol in 1993 [1]. Boudol's resource λ-calculus extends the call-by-value λ-calculus with a special resource sensitive application (able to manage finite resources) that involves multisets of affine arguments each one used at most once. Independently from our considerations on Rel, this was seen as a natural

[1] Up to technical details, but the unrestricted injectivity is strongly conjectured.

M. Hasegawa (Ed.): TLCA 2013, LNCS 7941, pp. 93–108, 2013.

way to export resource sensitiveness into the functional setting. However, restricted by a fixed evaluation strategy, it was not fully explored. Later on, Ehrhard and Regnier, working on the implementation of behaviors discovered in Rel, came to a similar calculus, the differential λ-calculus [11], which enjoys many syntactical and semantical properties (confluence, Taylor expansion). In Ehrhard and Regnier's differential λ-calculus the resource-sensitiveness is obtained by adding to the λ-calculus a derivative operation $\frac{\partial M}{\partial x}(N)$ (will be implemented in our notations as the term $M\langle N/x\rangle$, see section 2). This operator syntactically corresponds to a substitution of exactly one occurrence of x by N in M (introducing non determinism on the choice of the substituted occurrence); confluence of the calculus is recovered, then, by performing all the possible choices at once. This linear substitution takes place when β-reducing specific applications where an argument is marked as linear, in order to be used exactly once. We will adopt the syntax of [16] that re-implements improvements from differential λ-calculus into Boudol's calculus, and we will call it resource λ-calculus or $\partial\lambda$-calculus.

MRel. For Rel as for most categorical models of Linear Logic, the interpretation of the exponential modality induces a comonad from which we can construct the Kleisli category that contains a model of the λ-calculus. In the case of Rel, this category, MRel, corresponds to the category whose objects are sets and whose morphisms from A to B are the relations from $\mathcal{N}\langle A\rangle$ (the set of finite multisets over A) to B. It is then a model of both λ and $\partial\lambda$-calculi. This construction being very natural, the reflexive objects of MRel are the most-studied models of the $\partial\lambda$-calculus.

MRel and $\partial\lambda$-Calculus. The depth of the connection between the reflexive objects of MRel and the $\partial\lambda$-calculus is precisely the purpose of our work. More precisely, we investigate the question of the full abstraction of $\mathcal{M}_\infty$, a reflexive object for the $\partial\lambda$-calculus [5]. We also endowed $\partial\lambda$-calculus with a particular choice of reduction that is the may-outer-reduction; this is not the only choice, but this corresponds to the intuition that conducts from Rel to Ehrhard-Regnier's differential calculus. Until now we knew that $\mathcal{M}_\infty$ was adequate for the $\partial\lambda$-calculus [3], *i.e.* that two terms carrying the same interpretations in $\mathcal{M}_\infty$ behave the same way in all contexts. But we did not know anything about the converse, the completeness, and thus about the full abstraction.

Full Abstraction. The full abstraction of $\mathcal{M}_\infty$ has been thoroughly studied. For lack of direct results, the full abstraction has been proved for restrictions and extensions of the $\partial\lambda$-calculus: for the untyped λ-calculus (which is the deterministic and linear-free fragment of the $\partial\lambda$-calculus); for the orthogonal bang-free restriction where the application only accepts bags of linear arguments; and for the extension with tests of [3], an extension with must non-determinism and with operators inspired by 0-ary *par* and *tensor product* that could be added freely in DiLL-proof nets (DiLL for Differential Linear Logic).

These studies were encouraging since they systematically showed MRel to be fully abstract for these calculi ([14] for untyped λ-calculus , [4] for the bang-free restriction and [3] for resource λ-calculus with tests). Therefore Bucciarelli *et.al.* [3] conjectured a full abstraction for the $\partial\lambda$-calculus.

The Counter-Example. The purpose of this article is to set out a highly unexpected counter-example to this conjecture. We will see how an untyped fixpoint and a may non-deterministic sum can combine to produce a term $\boldsymbol{A}$ (Equation 7) behaving like an infinite sum $\Sigma_{i\geq 1}\boldsymbol{B}_i$ where every $\boldsymbol{B}_i$ begins with $(i+1)$ λ-abstractions, put its $(i+1)^{th}$ argument in head position but otherwise behave as the identity in applicative contexts with exactly i arguments; that how $\boldsymbol{A}$ can be thought to have an arbitrary number of λ-abstractions. Such a term can thus look for an argument further than the length of any bounded applicative context. There lies the immediate interest of achieving a context lemma (which have not been done for this calculus, yet) in order to prove that the observational equivalence is so short-sighted. This will refute the inequational full abstraction since the relational semantics can sublimate this short-sightedness. More concretely we will see that $\boldsymbol{A}$ is observationally above the identity but not denotationally. It is not difficult, then, to refute the equational full abstraction.

We proceed in this order. Section 2 present the $\partial\lambda$-calculus and its properties. Section 3 describes MRel and its reflexive object $\mathcal{M}_\infty$, and see how it is related to $\partial\lambda$-calculus. Section 4 gives our results with the context lemma followed by the counter-example (Theorem 8). We will also discuss the generality of this counter-example in the conclusion and explain how it is representative of an unhealthy interaction between untyped fixpoints and *may-non-determinism* that can be reproduced in other calculi like the *may-non-deterministic* extension of λ-calculus.

Notation: We denote $\mathcal{N}\langle A\rangle$ for the set of finite multisetes of elements in the set A.

2 Syntax

2.1 $\partial\lambda$-Calculus

In this section we give some background on the $\partial\lambda$-calculus, a lambda calculus with resources. The grammar of its syntax is the following:

(terms)	Λ :	$L, M, N ::=$	$x \mid \lambda x.M \mid M\,P$	
(bags)	Λ^b :	$P, Q ::=$	$1 \mid [M] \mid [M^!] \mid P\cdot Q$	
(sums)	$\mathbb{\Lambda}, \mathbb{\Lambda}^b$:		$\mathbb{L}, \mathbb{M} \in \mathcal{N}\langle\Lambda\rangle$	$\mathbb{P}, \mathbb{Q} \in \mathcal{N}\langle\Lambda^b\rangle$

Fig. 1. Grammar of the $\partial\lambda$-calculus

The $\partial\lambda$-calculus extends the standard λ-calculus in two directions. First, it is a non deterministic λ-calculus. The argument of an application is a superposition of inputs, called *bag of resources* and denoted by a multiset in multiplicative notation (namely $P\cdot Q$ is the disjoint union of P and Q). Symmetrically, the result of a reduction step is a superposition of outputs denoted by a multiset in additive notation (namely $\mathbb{L}+\mathbb{M}$ is the disjoint union of $\mathbb{L}$ and $\mathbb{M}$). We also have empty multisets, expressing an absence of available inputs (denoted by 1) or of results (denoted by 0).

Second, the $\partial\lambda$-calculus distinguishes between *linear* and *reusable* resources. The formers will never suffer any duplication or erasing regardless the reduction strategy. A reusable resource will be denoted by a *banged* term $M^!$ in a bag, e.g. $[N^!, L, L]$ is a bag of two linear occurrences of the resource L and a reusable occurrence of the resource N. We use the notation $N^{(!)}$ whenever we do not set out whether M occurs linearly or not in a bag. A bag with no *banged resources* will be said *linear* and one with only *banged resources* will be said *exponential.*

Finally, keeping all possible results of a reduction step (with multiplicities) into a finite multiset $\Sigma_i M_i$ of outcomes allows to have a confluent rewriting system in such a non-deterministic setting [16].

Small Latin letters $x, y, z, ...$ will range over an infinite set of λ-calculus variables. Capital Latin letters L, M, N (resp. P, Q, R) are meta-variables for terms (resp bags). Initial capital Latin letters E, F will denote indifferently terms and bags and will be called *expressions.* Finally, the meta-variables $\mathbb{L}, \mathbb{M}, \mathbb{N}$ (resp $\mathbb{P}, \mathbb{Q}, \mathbb{R}$) vary over sums (*i.e.* multisets in additive notation) of terms (resp. bags). Bags and sums are multisets, so we are assuming associativity and commutativity of the disjoint union and neutrality of the empty multiset.

Notice that the sum operator is always at the top level of the syntax trees. This is a design choice taken from [16] allowing for a lighter syntax. However, it is sometimes convenient to write sums inside an expression as a short notation for the expression obtained by distributing the sums following the conventions:

$$\begin{array}{ll} \lambda x.(\Sigma_i M_i) := \Sigma_i(\lambda x.M_i) & (\Sigma_i M_i)\ (\Sigma_j P_j) := \Sigma_{i,j}(M_i\ P_j) \\ [(\Sigma_i M_i)^!]\cdot P := [M_1^!, \ldots, M_n^!]\cdot P & [\Sigma_i M_i]\cdot P := \Sigma_i [M_i]\cdot P \end{array}$$

Notice that every construct is (multi)-linear but the bang $(\cdot)^!$, where we apply the linear logic equivalence $[(M+N)^!] = [M^!]\cdot[N^!]$ which is reminiscent of the standard exponential rule $e^{a+b} = e^a \cdot e^b$. Notice moreover that the 0-ary version of those rules also hold.

Since we have two kinds of resources, we need two different substitutions: the usual one, denoted $\{.\}$, and the linear one, denoted $\langle . \rangle$. Supposing that $x \neq y$, and $x \neq z$, and $z \notin \mathrm{FV}(N)$ (FV denoting free variables):

$$\begin{array}{c} x\langle N/x\rangle := N \qquad y\langle N/x\rangle := 0 \qquad (\lambda z.M)\langle N/x\rangle := \lambda z.(M\langle N/x\rangle) \\ (M\ P)\langle N/x\rangle := (M\langle N/x\rangle\ P) + (M\ P\langle N/x\rangle) \\ [M^!]\langle N/x\rangle := [M\langle N/x\rangle, M^!] \qquad [M]\langle N/x\rangle := [M\langle N/x\rangle] \\ (P\cdot Q)\langle N/x\rangle := (P\langle N/x\rangle)\cdot Q + P\cdot(Q\langle N/x\rangle) \qquad 1\langle N/x\rangle := 0 \end{array}$$

Notice that in the above definition we are heavily using the natural convention of the distributing sums. For example, the bag $P = [x^!, y]\langle N/x\rangle$ reduces $P = [x\langle N/x\rangle, x^!, y] + [x^!, y\langle N/x\rangle] = [N, x^!, y] + [x^!, 0] = [N, x^!, y] + 0 = [N, x^!, y]$.

Substitutions enjoy the following commutation properties:

Lemma 1 ([16]). *For an expression E and terms M, N, if $x \notin \mathrm{FV}(N)$ and if $y \notin \mathrm{FV}(M)$ (potentially $x{=}y$) then:*

$$E\langle M/x\rangle\langle N/y\rangle = E\langle N/y\rangle\langle M/x\rangle \qquad E\{(M{+}x)/x\}\langle N/y\rangle = E\langle N/y\rangle\{(M{+}x)/x\}$$
$$E\{(M{+}x)/x\}\{(N{+}y)/y\} = E\{(N{+}y)/y\}\{(M{+}x)/x\}\}$$

Hence the notion of substitution of variables by bags, denoted $\langle\langle s\rangle\rangle$ (where s is a list of substitutions P/x), may be defined as follows (if $x \notin \mathrm{FV}(N) \cup \mathrm{FV}(P)$):

$$M\langle\langle 1/x\rangle\rangle := M\{0/x\} \qquad M\langle\langle [N^!]\cdot P/x\rangle\rangle := M\{(x{+}N)/x\}\langle\langle P/x\rangle\rangle$$
$$M\langle\langle [N]\cdot P/x\rangle\rangle := M\langle N/x\rangle\langle\langle P/x\rangle\rangle \qquad M\langle\langle s_1; s_2\rangle\rangle := M\langle\langle s_1\rangle\rangle\langle\langle s_2\rangle\rangle$$

2.2 Beta and *Outer Reduction*

Reduction is defined essentially as the contextual closure of the β-rule.

$$\frac{}{(\lambda x.M)\ P \to M\langle\langle P/x\rangle\rangle}\,\beta \qquad \frac{M \to \mathbb{M}}{M\ P \to \mathbb{M}\ P}\,\mathrm{left} \qquad \frac{M \to \mathbb{M}}{\lambda x.M \to \lambda x.\mathbb{M}}\,\mathrm{abs}$$
$$\frac{N \to \mathbb{N}}{M\ [N]\cdot P \to M\ [\mathbb{N}]\cdot P}\,\mathrm{lin} \qquad \frac{N \to \mathbb{N}}{M\ [N^!]\cdot P \to M\ [\mathbb{N}^!]\cdot P}\,!$$
$$\frac{M \to \mathbb{M}' \qquad \mathbb{N} \to \mathbb{N}'}{M{+}\mathbb{N} \to \mathbb{M}'{+}\mathbb{N}'}\,s_1 \qquad \frac{M \to \mathbb{M}'}{M{+}\mathbb{N} \to \mathbb{M}'{+}\mathbb{N}}\,s_2$$

Fig. 2. Reduction rules

Rules s_1 and s_2 allow to reduce one or more terms of a sum in a single step (this is used in Theorem 1).
In the following example and all along this article we denote:

$$\boldsymbol{\omega} := \lambda x.x[x^!] \qquad \boldsymbol{I} := \lambda x.x \qquad \boldsymbol{\Delta} := \lambda gu.u\ [(g\ [g^!]\ [u^!])^!] \qquad \boldsymbol{\Theta} := \boldsymbol{\Delta}[\boldsymbol{\Delta}^!]$$

Example 1.

$$\boldsymbol{I}\ [u^!, v^!] \to u{+}v \qquad (\lambda x.y\ [(x\ [y])^!])\ [u, v^!] \to y\ [u\ [y], (v\ [y])^!] \tag{1}$$
$$(\lambda x.x\ [x, x^!])\ [u, v^!] \to (u\ [v, v^!]){+}(v\ [u, v^!]){+}(v\ [v, u, v^!]) \tag{2}$$
$$u\ [\boldsymbol{I}\ 1] \to 0 \qquad u\ [(\boldsymbol{I}\ 1)^!] \to u\ 1 \tag{3}$$
$$\boldsymbol{\omega}\ [\boldsymbol{\omega}^!] \to \boldsymbol{\omega}\ [\boldsymbol{\omega}^!] \qquad \boldsymbol{\omega}\ [\boldsymbol{\omega}] \to 0 \tag{4}$$
$$\boldsymbol{\Theta}\ [v^!] \to^2 (v\ [(\boldsymbol{\Theta}\ [v^!])^!]) \tag{5}$$
$$\boldsymbol{\Theta}\ [u, v^!] \to^2 (u\ [(\boldsymbol{\Theta}\ [v^!])^!]) + (v\ [(\boldsymbol{\Theta}[u, v^!]), (\boldsymbol{\Theta}[v^!])^!]) \tag{6}$$

As customary, a notion of convergence will be used for relating the operational and denotational semantics of the $\partial\lambda$-calculus.

In this paper, we consider the *may-outer convergence* of [16]. The attribute *may* refers to an angelic notion of non-determinism, hence $M+N$ will converge whenever at least one of the two converges. Indeed, the demonic (must) convergence is also of great interest, however it is harder to deal with (see [17]), in fact the demonic non-determinism does not interact well with the Taylor expansion, which is a crucial tool in our analysis (Section 2.3). Moreover, the attribute *outer* refers to the fact that we reduce only redexes not under the scope of a bang. This turns out to be the analogous of the head-reduction in the λ-calculus.

Definition 1 (*onf* and *monf*). *A term is in* outer-normal form, onf *for short, iff it has no redexes but under a* !, *that is a term of the form:*

$$\lambda x_1, \ldots, x_m.y\ [N_{1,1}^{(!)}, \ldots, N_{1,k_1}^{(!)}] \cdots [N_{n,1}^{(!)}, \ldots, N_{n,k_n}^{(!)}]$$

Where every $N_{i,j}^{(!)}$ *are either* banged *or in* outer-normal form.
A sum of terms is in may-outer-normal form, monf *for short, iff at least one of its addends is in* outer-normal form *(in particular* 0 *is not a* monf*).*

This notion generalizes the one of head-normal form of the untyped lambda calculus. Asking for linear terms of a bag to be in *monf* is a way of expressing that $x\ [\boldsymbol{\omega}\ [\omega^!]]$ diverges while $x\ [(\boldsymbol{\omega}\ [\omega^!])^!]$ is an *onf*. *Monf*s correspond to *may-solvability* [17] in the same way as head-normal-forms correspond to solvability in untyped λ-calculus. From previous examples only contracta of (3.1), (4.1) and (4.2) are not *monf*, and only (3.2)'s redex is.
The restricted reduction leading to the (principal) *monf* of a term is the following:

Definition 2. *The* outer reduction, *denoted* $\rightarrow_o$ *is defined by the rules of Figure 2 but the rule* !, *which is omitted. We denote by* $\rightarrow^*$ *and* $\rightarrow_o^*$ *the reflexive and transitive closures of* $\rightarrow$ *and* $\rightarrow_o$*, respectively.*

In the Example 1, all reductions but the (3.2) are *outer reductions.*

Lemma 2 ([16]). *If* $M \rightarrow^* \mathbb{M}$ *and* $\mathbb{M}$ *is in* monf, *then there exists a* monf $\mathbb{N}$ *such that* $M \rightarrow_o^* \mathbb{N} \rightarrow^* \mathbb{M}$. *Thus the convergence to a* monf *and the* outer convergence *to a* monf *coincide.*

We will write $M \Downarrow_n$ if there exists a *monf* $\mathbb{M}$ and an *outer reduction* sequence from M to $\mathbb{M}$ of length at most n. We will write $M \Downarrow$ if there exists n such that $M \Downarrow_n$ and say that M *outer converges*. Finally we will write $M \Uparrow$ for the *outer-divergence* on M.
The two rules s_1 and s_2 of Figure 2 allow the followings:

Theorem 1. *If* $M \rightarrow \mathbb{M}_1$ *and* $M \rightarrow \mathbb{M}_2$ *for* $\mathbb{M}_1, \mathbb{M}_2 \neq 0$, *then there is* $\mathbb{N}$ *such that* $\mathbb{M}_1 \rightarrow \mathbb{N}$ *and* $\mathbb{M}_2 \rightarrow \mathbb{N}$.

Corollary 1. *If* $M \Downarrow_{n+1}$ *and* $M \rightarrow_o \mathbb{N}$ *then exists* $N \in \mathbb{N}$ *such that* $N \Downarrow_n$.

This is due to the trivial divergence of the case $M = 0$. Notice moreover that M is not a sum.

2.3 Taylor Expansion

A natural restriction of the $\partial\lambda$-calculus is the fragment $\partial\lambda^\ell$ which is obtained by removing the bang construction $[M^!]$ in Figure 1. This restriction has a very limited computational power, for instance it enjoys the following theorem.

Theorem 2 ([Folklore]). *The reduction $\to$ in $\partial\lambda^\ell$ is strongly normalizing.*

Proof. We set an order $\sqsubseteq$ on the finite multisets of terms generated by $\mathbb{M} \sqsubseteq \mathbb{N}$ if $\mathbb{M} = \mathbb{M}'+\mathbb{L}$, $\mathbb{N} = \mathbb{N}'+\mathbb{L}$ and there exists $N \in \mathbb{N}'$ such that for all $M \in \mathbb{M}'$, the inequality $|M| \leq |N|$ (where $|M|$ is the structural size of M) holds. Then, $\to$ is strictly decreasing in this well founded order. □

The main interest of $\partial\lambda^\ell$ comes with the Taylor expansion. The Taylor expansion of a λ-term M has been developed in [11,12] and it recalls the usual decomposition of an analytic function:

$$f(x) = \sum_{n=0}^{\infty} \frac{1}{n!} D^n(f)(0)x^n$$

In this paper, we are interested only in the support of the Taylor expansion of a $\partial\lambda$-term M defined in [11,12], *i.e.* in the set M^o of the $\partial\lambda^\ell$-terms appearing in the Taylor expansion of M with non-null coefficient. Such a set can be defined as follows.

Definition 3. *The Taylor expansion E^o of an expression E is a (possibly infinite) set of linear expressions defined by structural induction:*

$$(\lambda x.M)^o := \{\lambda x.M' | M' \in M^o\} \qquad (M\ P)^o := \{M'P' | M' \in M^o,\ P' \in P^o\}$$
$$[M]^o := \{[M'] | M' \in M^o\} \qquad (P{\cdot}Q)^o := \{P'{\cdot}Q' | P' \in P^o,\ Q' \in Q^o\}$$
$$[M^!]^o := \{[M_1, \ldots, M_n] | n \geq 0,\ M_1, \ldots, M_n \in M^o\} \quad 1^o := \{1\} \quad x^o := \{x\}$$

In the following we use set inclusion for comparing a finite multiset $\mathbb{N}$ with a set $\mathbb{M}^o$. This mean that the support (*i.e.* the set of element appearing in $\mathbb{N}$ with a nonzero multiplicity) of $\mathbb{N}$ is a subset of $\mathbb{M}^o$.

Lemma 3. *For any sum $\mathbb{M}$ and for any $\mathbb{N} \subseteq \mathbb{M}^o$, if $\mathbb{N}$ converges to a normal form $\mathbb{N}'$ then there exists $\mathbb{M}'$ such that $\mathbb{M} \to^* \mathbb{M}'$ and $\mathbb{N}' \subseteq \mathbb{M}'^o$.*

Proof. By induction on the length of the longest path of reduction for $\mathbb{N} \to^* \mathbb{N}'$ (indeed such a path exists by Theorem 2). The case $\mathbb{N} = \mathbb{N}'$ is trivial. We thus asume that $\mathbb{N} \to \mathbb{N}'' \to^* \mathbb{N}'$. Notice that $\mathbb{N} \to \mathbb{N}''$ does not use the rule s_1, otherwise there would be a longest reduction sequence from $\mathbb{N}$ to $\mathbb{N}'$ which contradicts the hypothesis that we have already choosen the largest path of reduction. This means that $\mathbb{N} \to \mathbb{N}''$ reduces a single redex. This redex being the image of a redex in $\mathbb{M}$ (not necessarily an outer redex), we can perform the corresponding reduction $\mathbb{M} \to \mathbb{M}''$. And by reducing all corresponding redexes on $\mathbb{N}''$ (the duplictions from the Taylor expension), we have $\mathbb{N}'' \to^* \mathbb{N}''' \subseteq \mathbb{M}''^o$. Then we conclude by induction hypothesis. □

3 Model

3.1 Categorical Construction of the Model

We recall the interpretation of the $\partial\lambda$-calculus into the reflexive object $\mathcal{M}_\infty$ of MRel used in [3]. MRel is the Cartesian closed category resulting from the co-Kleisli construction associated with the multiset exponential comonad of the category Rel of sets and relations, which is a well-known model of Linear Logic (and Differential Linear Logic). We refer to [9] for a detailed exposition, here we briefly present MRel and the object $\mathcal{M}_\infty$.

The objects of MRel are the sets. Its morphisms from A to B are the relations from the set of the finite multi-sets of A, namely $\mathcal{N}\langle A\rangle$, to the set B; *i.e.* $\mathbf{MRel}(A,B) := \mathcal{P}(\mathcal{N}\langle A\rangle\times B)$.

The composition of $g \in \mathbf{MRel}(B,C)$ and $f \in \mathbf{MRel}(A,B)$ is given by $f;g = \{(a,\gamma) \in \mathcal{N}\langle A\rangle\times C \mid \exists(a_1,\beta_1),...,(a_n,\beta_n)\in f,\ a{=}\Sigma_i a_i \text{ and } ([\beta_1,\ldots,\beta_n],\gamma)\in g\}$

The identities are $\mathrm{id}_A := \{([\alpha],\alpha)|\alpha \in A\}$. Given a family $(A_i)_{i\in I}$, its Cartesian product is $\&_{i\in I} A_i := \{(i,\alpha)|i{\in}I, \alpha{\in}A_i\}$; with the projections $\pi_i := \{([(i,\alpha)],\alpha)|\alpha \in A_i\}$. The terminal object is the empty set. And the exponential object internalizing $\mathbf{MRel}(A,B)$ is $A{\Rightarrow}B := \mathcal{N}\langle A\rangle \times B$. Then the adjuction $\mathbf{MRel}(A \,\&\, B, C) \simeq \mathbf{MRel}(A, B{\Rightarrow}C)$ holds since $\mathcal{N}\langle \&_{i\leq n} A_i\rangle \simeq \prod_{i\leq n}\mathcal{N}\langle A_i\rangle$.

The reflexive object we choose is the simplest stratified object[2] of [14]. It can be recursively defined by (see [5]):

$$\mathcal{M}_0 := \emptyset \qquad \mathcal{M}_{n+1} := \mathcal{N}\langle \mathcal{M}_n\rangle^{(\omega)} \qquad \mathcal{M}_\infty := \bigcup_n \mathcal{M}_n$$

Where $\mathcal{N}\langle M\rangle^{(\omega)}$ is the list of almost everywhere empty multisets over M. Its elements can be generated by:

(elements)	$\mathcal{M}_\infty$:	$\alpha,\beta,\gamma ::=$	$* \mid a{::}\alpha$
(multisets)	$\mathcal{M}_\infty^b$:	$a,b,c ::=$	$[\alpha_1,\ldots,\alpha_n]$

Where $*$, the unique element of $\mathcal{M}_1$, namely the infinite list of empty multisets, enjoys the equation:

$$* = []{::}*$$

The linear morphisms $\boldsymbol{app} \in \mathbf{MRel}(\mathcal{M}_\infty, \mathcal{M}_\infty{\Rightarrow}\mathcal{M}_\infty)$ and $\boldsymbol{abs} \in \mathbf{MRel}(\mathcal{M}_\infty{\Rightarrow}\mathcal{M}_\infty, \mathcal{M}_\infty)$ are defined by:

$$\boldsymbol{app} := \{([a{::}\alpha],(a,\alpha))|(a,\alpha) \in \mathcal{M}_\infty\} \quad \boldsymbol{abs} := \{([(a,\alpha)],a{::}\alpha)|(a,\alpha) \in \mathcal{M}_\infty\}$$

One can easily check that $\boldsymbol{abs};\boldsymbol{app} = \mathrm{Id}_{\mathcal{M}_\infty\Rightarrow\mathcal{M}_\infty}$ (and even $\boldsymbol{app};\boldsymbol{abs} = \mathrm{Id}_{\mathcal{M}_\infty}$).

We could have interpreted the terms of the $\partial\lambda$-calculus by using the categorical structure of MRel. However, we prefer to give a description of such an

[2] Any other stratified object will also be subject to the counter-example since they share the crucial element $*$.

interpretation, using a non-idempotent intersection type system, following [10]. This type system has been introduced in [8].

The usual grammar of non-idempotent intersection types corresponds exactly to the grammar of $\mathcal{M}_\infty$. The cons operator (::) replaces the arrow and the multisets notation replaces the intersection notation. We will use the second one for uniformity consideration. The multisets of $\mathcal{M}_\infty^b$ will be denoted multiplicatively.

A *typing context* is a finite partial function from variables into multisets in $\mathcal{M}_\infty^b$, we denote $(x_i : a_i)_{i\in I}$ the context associating x_i to a_i for $i \in I$. We have two kinds of typing judgments, depending whether we type terms or bags: the former are typed by elements in $\mathcal{M}_\infty$ and the latter by multisets in $\mathcal{M}_\infty^b$.

$$\frac{\Gamma \vdash M : \alpha}{x:1, \Gamma \vdash M : \alpha} \qquad \frac{\Gamma \vdash P : a}{x:1, \Gamma \vdash P : a} \qquad \frac{}{x:[\alpha] \vdash x : \alpha} \qquad \frac{\Gamma \vdash M : \alpha}{\Gamma \vdash M + \mathbb{M} : \alpha}$$

$$\frac{\Gamma, x : a \vdash M : \alpha}{\Gamma \vdash \lambda x.M : a{::}\alpha} \qquad \frac{(x_i : a_i)_{i\in I} \vdash M : b{::}\alpha \qquad (x_i : a'_i)_{i\in I} \vdash P : b}{(x_i : a_i{\cdot}a'_i)_{i\in I} \vdash M\ P : \alpha}$$

$$\frac{}{\vdash 1 : 1} \qquad \frac{(x_i : a_i)_{i\in I} \vdash P : b \qquad (x_i : a'_i)_{i\in I} \vdash Q : c}{(x_i : a_i{\cdot}a'_i)_{i\in I} \vdash P{\cdot}Q : b{\cdot}c}$$

$$\frac{(x_i : a_i)_{i\in I} \vdash L : \beta}{(x_i : a_i)_{i\in I} \vdash [L] : [\beta]} \qquad \frac{(x_i : a_i^j)_{i\in I} \vdash L : \beta_j \qquad \text{for } j \leq m}{(x_i : \Pi_{j\leq m} a_i^j)_{i\in I} \vdash [L^!] : [\beta_1, \ldots, \beta_m]}$$

The usual presentation of the interpretation can be recovered with:

$$[\![\mathbb{M}]\!]^{x_1,\ldots,x_n} := \{((a_1,\ldots,a_n),\beta) | (x_i : a_i)_i \vdash \mathbb{M} : \beta\} \quad \in \mathbf{MRel}(\&_{i=1}^{n} \mathcal{M}_\infty, \mathcal{M}_\infty)$$

$$[\![\mathbb{P}]\!]^{x_1,\ldots,x_n} := \{((a_1,\ldots,a_n),b) | (x_i : a_i)_i \vdash \mathbb{P} : b\} \quad \in \mathbf{MRel}(\&_{i=1}^{n} \mathcal{M}_\infty, \mathcal{M}_\infty^b)$$

Theorem 3. *If* $\mathbb{M} \to \mathbb{N}$ *then* $[\![\mathbb{M}]\!]^{x_1,\ldots,x_n} = [\![\mathbb{N}]\!]^{x_1,\ldots,x_n}$.

An important characteristic of this model that seems to make it particularly suitable for our original purpose is that it models the Taylor expansion:

Theorem 4 ([15]). *For any term* M, $[\![M]\!]^{\bar{x}} = \bigcup_{N\in M^\circ} [\![N]\!]^{\bar{x}}$.

3.2 Observational Order and Adequacy

A first important result relating syntax and semantic is the sensibility theorem, a corollary of [3], but here reproved focussing on the role of the Taylor expansion.

Theorem 5. $\mathcal{M}_\infty$ *is sensible for* may-outer-convergence *of the $\partial\lambda$-calculus,* i.e.

$$\forall M, \quad M \Downarrow \;\Leftrightarrow\; [\![M]\!] \neq \emptyset.$$

Proof. The left-to-right side is trivial since any *monf* has a non-empty interpretation.

Conversely, assume $(\bar{a},\alpha) \in [\![M]\!]^{\bar{x}}$, by Theorem 4 there exists $N \in M^o$ such that $(\bar{a},\alpha) \in [\![N]\!]^{\bar{x}}$. Any single term of $\partial\lambda^\ell$-calculus converges either to 0 or to a normal form $N_0 + \mathbb{N}$ (by Theorem 2). Since $[\![0]\!] = \emptyset$, N converges into a normal form. By applying Lemma 3, we thus have $M \to^* M_0 + \mathbb{M}$ with $N_0 \in M_0^o$. Since the Taylor expansion conserves every redexes, M_0 is *outer-normal* and M is *may-outer converging.* □

Corollary 2. *A term may-outer converges iff one of the elements of its Taylor expansion may-outer converges:* $M\Downarrow \Leftrightarrow \exists N{\in}M^o; N\Downarrow$. *Equivalently, a term* may-outer diverges *iff any element of its Taylor expansion reduces to 0.*

Proof. For any closed term M, using Theorems 4 and 5:
$M\Downarrow \;\Leftrightarrow_{\text{th5}}\; [\![M]\!]\neq\emptyset \;\Leftrightarrow_{\text{th4}}\; \exists N{\in}M^o; [\![N]\!]\neq\emptyset \;\Leftrightarrow_{\text{th5}}\; \exists N{\in}M^o; N\Downarrow.$ □

In the following we use contexts, *i.e.* terms with holes that will be filled by terms. Contexts can be described by the grammar:

(contexts)	$\Lambda(\!\vert.\vert\!)$:	$C(\!\vert.\vert\!) ::=$	$(\!\vert.\vert\!) \mid M \mid \lambda x.C(\!\vert.\vert\!) \mid C(\!\vert.\vert\!)\, P(\!\vert.\vert\!)$
(bag-contexts)	$\Lambda^b(\!\vert.\vert\!)$:	$P(\!\vert.\vert\!) ::=$	$[C_1(\!\vert.\vert\!)^{(!)}, \ldots, C_n(\!\vert.\vert\!)^{(!)}]$

We define the notions of observational preorder and equivalence using as basic observation the *may-outer-convergence* of terms. This is not the only possibility (*must* or *inner* declensions); we discuss this issue in the conclusion.

Definition 4. *We say that a term M is observationally below another term N (denoted $M \leq_o N$), if for all contexts $C(\!|.|\!)$:*

$$C(\!|M|\!)\Downarrow \;\Rightarrow\; C(\!|N|\!)\Downarrow$$

They are observationally equivalent (denoted $M \equiv_o N$) if $M \leq_o N$ and $N \leq_o M$.

Using sensibility we thus assert our adequation.

Theorem 6. *$\mathcal{M}_\infty$ is inequationally adequate for $\partial\lambda$-calculus,*

$$\forall M, N, \quad [\![M]\!] \subseteq [\![N]\!] \;\Rightarrow\; M \leq_o N.$$

Proof. Assume that $[\![M]\!] \subseteq [\![N]\!]$ and $C(\!|M|\!)\Downarrow$. Then since $[\![.]\!]$ is defined by structural induction we have $[\![C(\!|N|\!)]\!] \supseteq [\![C(\!|M|\!)]\!] \neq \emptyset$ and $C(\!|N|\!)\Downarrow$. □

4 Failure of the Full Abstraction

The main result of this paper is the refutation of the full abstraction conjecture:

Conjecture 1 ([3]). $\mathcal{M}_\infty$ is fully abstract for $\partial\lambda$-calculus. *i.e.* the denotational and the observational equivalences are identical:

$$\forall M, N, \quad [\![M]\!] = [\![N]\!] \;\Leftrightarrow\; M \equiv_o N$$

Its refutation (Theorem 8) proceeds as follows. First, we define a term $\boldsymbol{A}$ (Equation 7) and we prove that $\boldsymbol{I} \leq_o \boldsymbol{A}$ (Lemma 7, which uses a context lemma: Theorem 7), but $[\![\boldsymbol{I}]\!] \not\subseteq [\![\boldsymbol{A}]\!]$ (Lemma 9). This results in the refutation of the stronger conjecture:

Conjecture 2 ([3]). $\mathcal{M}_\infty$ is inequationally fully abstract for $\partial\lambda$-calculus. *i.e.* the denotational and the observational preorders are identical:

$$\forall M, N,\ \ [\![M]\!] \subseteq [\![N]\!] \ \Leftrightarrow\ M \leq_o N$$

Only then will we consider the term $\boldsymbol{A'} := \boldsymbol{I}\ [\boldsymbol{A}^!, \boldsymbol{I}^!]$ and prove that $\boldsymbol{A'}$ and $\boldsymbol{A}$ yield a counter-example to Conjecture 1 (Theorem 8).

4.1 Context Lemma

Definition 5. Linear contexts *are contexts with exactly one hole and with this hole in linear position:*

$$\textit{(linear contexts)}\ \ \Lambda(\!|.|\!)_l :\ \ D(\!|.|\!) ::= \ (\!|.|\!)\ \mid\ \lambda x.D(\!|.|\!)\ \mid\ D(\!|.|\!)\ P\ \mid\ M\ [D(\!|.|\!)]\cdot P$$

The applicative contexts *are particular linear contexts of the form* $K(\!|.|\!) = (\lambda x_1 \ldots x_n.(\!|.|\!))\ P_1 \cdots\ P_k$

Lemma 4. *For any term* M *and any bags* P, Q, *there exists a decomposition* $P = P^{l_1}.P^{l_2}.P^e$ *such that* $P^{l_1}.P^{l_2}$ *is linear,* P^e *exponential, and if the convergence* $(M\ Q)\langle\!\langle P/x\rangle\!\rangle \Downarrow_n$ *holds then* $M\langle\!\langle P^{l_1}.P^e/x\rangle\!\rangle\ Q\langle\!\langle P^{l_2}.P^e/x\rangle\!\rangle \Downarrow_n$

Proof. By definition of the may convergence, since $(M\ \ Q)\langle\!\langle P/x\rangle\!\rangle = \Sigma_{P=P^{l_1}.P^{l_2}.P^e} M\langle\!\langle P^{l_1}.P^e/x\rangle\!\rangle\ Q\langle\!\langle P^{l_2}.P^e/x\rangle\!\rangle$ ☐

Lemma 5 (Linear context lemma). *For any terms* M *and* N, *if there is a linear context* $D(\!|.|\!)$ *such that* $D(\!|M|\!) \Downarrow$ *and* $D(\!|N|\!) \Uparrow$ *then there is an applicative context that does the same.*

Proof. We will prove the following stronger property:
For every terms M, N, every bags $P_1, \ldots, P_{p+q}$, and every variables $x_1,\ldots,x_p \notin \bigcup_{1\leq i\leq p+q} \mathrm{FV}(P_i)$, if $\langle\!\langle s\rangle\!\rangle := \langle\!\langle P_1/x_1;\ldots;P_p/x_p\rangle\!\rangle$ and if a linear context $D(\!|.|\!)$ is such that $(D(\!|M|\!)\langle\!\langle s\rangle\!\rangle\ P_{p+1}\ \cdots\ P_{p+q}) \Downarrow_n$ and $(D(\!|N|\!)\langle\!\langle s\rangle\!\rangle\ P_{p+1}\ \cdots\ P_{p+q}) \Uparrow$ then there exists an applicative context $K(\!|.|\!)$such that $K(\!|M|\!) \Downarrow$ and $K(\!|N|\!) \Uparrow$.
By cases, making induction on the lexicographically ordered pair $(n, D(\!|.|\!))$:

- If $D(\!|.|\!) = (\!|.|\!)$:
  $K(\!|.|\!) = (\lambda x_1, \ldots, x_p(\!|.|\!))\ P_1\ \cdots\ P_{p+q}$
- If $D(\!|.|\!) = \lambda z.D'(\!|.|\!)$:
  - If $q = 0$:
    The hypothesis gives $D'(\!|M|\!)\langle\!\langle s\rangle\!\rangle \Downarrow_n$ and $D'(\!|N|\!)\langle\!\langle s\rangle\!\rangle \Uparrow$, thus we can directly apply our induction hypothesis on $D'(\!|.|\!)$. That gives directly the required $K(\!|.|\!)$.

- Otherwise:
  By assuming that z does not appear in $P_{p+2}, ..., P_{p+q}$:
  The hypothesis and Corollary 1 apply to $D(\!|M|\!)\langle\!\langle s\rangle\!\rangle\ P_{p+1}\cdots P_{p+q}$ gives $(D'(\!|M|\!)\langle\!\langle P_{p+1}/z; s\rangle\!\rangle\ P_{p+2}\cdots P_{p+q}) \Downarrow_{n-1}$. Moreover $(D'(\!|N|\!)\langle\!\langle P_{p+1}/z; s\rangle\!\rangle\ P_2\cdots P_q) \Uparrow$.
  Then the induction hypothesis directly gives the required $K(\!|.|\!)$.

– If $D(\!|.|\!) = L\ [D'(\!|.|\!)]\cdot Q$:
By assuming that $x_i \notin \mathrm{FV}(P_j)$ for $i \leq j \leq p$ and by Lemma 4, there exists, for all $i \leq p$, a decomposition $P_i = P_i^{l_1}\cdot P_i^{l_2}\cdot P_i^e$ such that if $\langle\!\langle s_j\rangle\!\rangle := \langle\!\langle P_1^{l_j}\cdot P_1^e/x_1; ...; P_p^{l_j}\cdot P_p^e/x_p\rangle\!\rangle$ (for all $j \in \{1,2\}$), there is $L' \in L\langle\!\langle s_1\rangle\!\rangle$ with $(L'\ ([D'(\!|M|\!)]\cdot Q)\langle\!\langle s_2\rangle\!\rangle\ P_1\cdots\ P_q) \Downarrow_n$ and $(L'\ ([D'(\!|N|\!)]\cdot Q)\langle\!\langle s_2\rangle\!\rangle\ P_1\cdots\ P_q) \Uparrow$.
Then there are two cases. Either $L' \to_o \mathbb{L}$ and there is $L'' \in \mathbb{L}$ such that $((L''\ [D'(\!|M|\!)\cdot Q)\langle\!\langle s_2\rangle\!\rangle\ P_1\cdots\ P_q) \Downarrow_{n-1}$ (using Corollary 1) that allow us to apply the induction hypothesis that result in the wanted $K(\!|.|\!)$. Or L' is in *outer-normal form*:
- if $L' = \lambda z.L''$:
  Let $D''(\!|.|\!) = L''\langle\!\langle [D'(\!|.|\!)]\cdot Q/z\rangle\!\rangle$.
  We have $(D''(\!|M|\!)\langle\!\langle s_2\rangle\!\rangle\ P_1\cdots\ P_q) \Downarrow_{n-1}$ and $(D''(\!|N|\!)\langle\!\langle s_2\rangle\!\rangle\ P_1\cdots\ P_q) \Uparrow$.
  Then we can apply our induction hypothesis on $D'(\!|.|\!)$ that is still a linear context since $D'(\!|.|\!)$ was not under a "!". This gives directly the required applicative context.
- if $L' = y\ Q_1\cdots Q_r$ with $y \neq x_i$ for all i:
  There exists, for all $i \leq p$, a multiset $P_i^{l_3} \subseteq P_i^{l_2}$ such that $D'(\!|M|\!)\langle\!\langle P_1^{l_3}\cdot P_1^e/x_1; ...; P_p^{l_3}\cdot P_p^e/x_p\rangle\!\rangle \Downarrow_n$ and $D'(\!|N|\!)\langle\!\langle P_1^{l_3}\cdot P_1^e/x_1; ...; P_p^{l_3}\cdot P_p^e/x_p\rangle\!\rangle \Uparrow$. Then we can apply the induction hypothesis on $D'(\!|.|\!)$ and obtain the wanted $K(\!|.|\!)$.

– If $D(\!|.|\!) = D'(\!|.|\!)\ Q$:
By Lemma 4, there exists $P_i^{\ell_1}\cdot P_i^{\ell_1}\cdot P_i^e = P_i$ such that, if we denote $\langle\!\langle s_j\rangle\!\rangle := \langle\!\langle P_1^{\ell_i}\cdot P_1^e/x_1; ...; P_p^{\ell_i}\cdot P_p^e/x_p\rangle\!\rangle$:

$$(D'(\!|M|\!)\langle\!\langle s_1\rangle\!\rangle\ Q\langle\!\langle s_2\rangle\!\rangle\ P_{p+1}\ \cdots\ P_{p+q}) \Downarrow_n \qquad (D'(\!|N|\!)\langle\!\langle s_1\rangle\!\rangle\ Q\langle\!\langle s_2\rangle\!\rangle\ P_{p+1}\ \cdots\ P_{p+q}) \Uparrow$$

The induction hypothesis on $D'(\!|.|\!)$ (with $Q\langle\!\langle s_2\rangle\!\rangle$ seen as one of the P_i's) results in the required applicative context. □

Theorem 7 (Context lemma). *For any terms M and N, if there is a context $C(\!|.|\!)$ such that $C(\!|M|\!) \Downarrow$ and $C(\!|N|\!) \Uparrow$ then there is an applicative context that does the same.*

Proof. Let $C(\!|.|\!)$ be such a context.
Let $\{x_1, ..., x_n\} = \mathrm{FV}(M) \cup \mathrm{FV}(N)$ be the free variables of M and N.
Let $L = \lambda u.C(\!|u\ [x_1^!]\ \cdots\ [x_n^!]|\!)$, $D(\!|.|\!) = \lambda x_1...x_n(\!|.|\!)$ and $C'(\!|.|\!) = L\ [D(\!|.|\!)^!]$.
Notice that $C'(\!|M|\!) \to^* C(\!|M|\!)$ and $C'(\!|N|\!) \to^* C(\!|N|\!)$. Hence, the hypothesis and Lemma 9 gives $C'(\!|M|\!) \Downarrow$ and $C'(\!|N|\!) \Uparrow$. Moreover, we have that $C'(\!|M|\!) = \bigcup_{n\geq 0}(L\ [D(\!|M|\!)^n)^o$; thus, by applying twice Corollary 2 we have an $n \in \mathcal{N}$ such

that $L\ [D(\!|M|\!)^n] \Downarrow$. Also, since $(L\ [D(\!|N|\!)^n])^o \subseteq C'(\!|N|\!)^o$, the same corollary and the hypothesis $C'(\!|N|\!) \Uparrow$ gives $L\ [D(\!|N|\!)^n] \Uparrow$
Since $L\ [D(\!|N|\!)^k, D(\!|M|\!)^{n-k}]$ converges for $k = 0$ and diverges for $k = n$ there exists $k_0 < n$ such that it converges for $k = k_0$ and diverges for $k = K_0+1$. Thus by applying Lemma 5 on the *linear context* $C''(\!|.|\!) = L\ [D(\!|N|\!)^{k_0}, D(\!|M|\!)^{n-k_0-1}, D(\!|.|\!)]$ we can conclude. □

4.2 Counter Example

We first exhibit a term $\boldsymbol{A}$ that is observationally above the identity $\boldsymbol{I}$, but whose interpretation will not contain $[*]{::}*$ in order to break Conjecture 2. We would like to have $\boldsymbol{A}$ somehow respecting:

$$\boldsymbol{A} \simeq \Sigma_{n\geq 1}\boldsymbol{B}_n \quad \text{with for } n \geq 1: \quad \boldsymbol{B}_n = \lambda v_1 \ldots v_n w.w\ [\boldsymbol{I}\ [v_1^!]\ [v_2^!]\ \cdots\ [v_n^!]]$$

This term will converge on any applicative context that converges on the identity (take $\boldsymbol{B}_n$ with n greater than the number of applications), and thus is observationally above the identity. On the other side, its semantic will be independent to the semantics of the identity since none of the $[\![B_i]\!]$ contains $[*]{::}* \in [\![\boldsymbol{I}]\!]$. Such an infinite sum $\Sigma_{n\geq 1}\boldsymbol{B}_n$ does not exists in our syntax so we have to represent it by using a fix point combinator and a bag of linear and non-linear resources. We define:

$$\boldsymbol{A} := \boldsymbol{\Theta}\ [\boldsymbol{G}, \boldsymbol{F}^!] \tag{7}$$

where $\boldsymbol{G}$ and $\boldsymbol{F}$ are defined by:

$$\boldsymbol{G} := \lambda uvw.w\ [\boldsymbol{I}\ [v^!]] \qquad\qquad \boldsymbol{F} := \lambda uv_1v_2.u\ [\boldsymbol{I}\ [v_1^!]\ [v_2^!]]$$

$\boldsymbol{A}$ seems quite complex, but, it can be seen as a non deterministic *while* that recursively apply $\boldsymbol{F}$ until it chooses (non-deterministically) to apply $\boldsymbol{G}$, giving one of the $\boldsymbol{B}_i$:

Lemma 6.

1. $\boldsymbol{G}[x^!] \to_o^* \boldsymbol{B}_1$
2. *For all* i, $\boldsymbol{F}\ [\boldsymbol{B}_i^!] \to_o^* \boldsymbol{B}_{i+1}$
3. $\boldsymbol{A} \equiv_\beta \boldsymbol{B}_1 + \boldsymbol{F}[\boldsymbol{A}]$

In particular, for every $i \geq 1$, *we have* $\boldsymbol{A} \equiv_\beta \boldsymbol{F}^i[\boldsymbol{A}] + \Sigma_{j=1}^{i-1}\boldsymbol{B}_j$, *where* $\boldsymbol{F}^1[\boldsymbol{A}] := \boldsymbol{F}[\boldsymbol{A}]$ *and* $\boldsymbol{F}^{i+1} := \boldsymbol{F}^i[\boldsymbol{F}[\boldsymbol{A}]]$

Proof. Item 1 is trivial. Item 2 is just a one-step unfolding of $\boldsymbol{\Theta}$. Item 3 is obtained via the reduction $\boldsymbol{A} \to^* (\boldsymbol{G}[(\boldsymbol{\Theta}[\boldsymbol{F}^!])^!]) + (\boldsymbol{F}[\boldsymbol{A}, (\boldsymbol{\Theta}[\boldsymbol{F}^!])^!]) \equiv_\beta \boldsymbol{B}_1 + (\boldsymbol{F}[\boldsymbol{A}])$ the last step using the linearity of $\boldsymbol{F}$ on its first variable (thus in a context of the kind $[U, V^!]$ only U matters). □

Lemma 7. *For all contexts* $C(\!|.|\!)$ *of the* $\partial\lambda$*-calculus, if* $C(\!|\boldsymbol{I}|\!)$ *converges then* $C(\!|\boldsymbol{A}|\!)$ *converges,* i.e. $\boldsymbol{I} \leq_o \boldsymbol{A}$

Proof. Let $C(\!|.|\!)$ be a context that converges on $\boldsymbol{I}$.

With the context lemma (Theorem 7), and since neither $\boldsymbol{I}$ nor $\boldsymbol{A}$ has free variables, we can assume that $C(\!|.|\!) = (\!|.|\!)\ P_1 \cdots P_k$ (where $P_1, ..., P_k$ are bags). Thus by Lemma 6, we have $\boldsymbol{A} \to^* \boldsymbol{C}_k + \boldsymbol{B}_k$ with $\boldsymbol{C}_k := \boldsymbol{F}^k[\boldsymbol{A}^!]+\Sigma_{j=1}^{k-1}\boldsymbol{B}_j$ and the following converges:

$$C(\!|\boldsymbol{A}|\!) \to^* C(\!|\boldsymbol{C}_k|\!) + \lambda w.w\ [\boldsymbol{I}\ P_1 \cdots P_k] = C(\!|\boldsymbol{C}_k|\!) + \lambda w.w\ [C(\!|\boldsymbol{I}|\!)] \qquad \square$$

We will now compare $\boldsymbol{A}$ and $\boldsymbol{I}$ at the denotational level.

Lemma 8. *We have*

$$[\![\boldsymbol{A}]\!] = \bigcup_i [\![\boldsymbol{B}_i]\!]$$

Proof. $[\![\boldsymbol{A}]\!] \supseteq \bigcup_i [\![\boldsymbol{B}_i]\!]$ is a corollary of the Lemma 6 (the interpretation is stable by reduction), so we have to prove that $[\![\boldsymbol{A}]\!] \subseteq \bigcup_i [\![\boldsymbol{B}_i]\!]$:
Let $\alpha \in [\![\boldsymbol{A}]\!]$. By Theorem 4, there exists $M \in \boldsymbol{A}^o$ such that $\alpha \in [\![M]\!]$. By Theorem 2: $M \to^* \mathbb{N}$, with every element of $\mathbb{N}$ outer-normal. And trivially there is $N \in \mathbb{N}$ such that $\alpha \in [\![N]\!]$. By application of Lemma 3, there exists L such that $\boldsymbol{A} \to^* L + \mathbb{L}$ and $N \in L^o$ (thus $\alpha \in [\![L]\!]$). Since the Taylor expansion conserves all *outer-redexes*, necessary L is outer-normal. We conclude by Lemma 6 that one of the $\boldsymbol{B}_i$ is reducing to L. $\square$

Lemma 9. $[*]::* \notin [\![\boldsymbol{A}]\!]$, *while* $[*]::* \in [\![\boldsymbol{I}]\!]$

Proof. Because of Lemma 8, we just have to prove that $[*]::*$ is not in any $\boldsymbol{B}_i$, which is trivial since the elements of $[\![\boldsymbol{B}_i]\!]$ must be of the form $a_1:: \cdots ::a_i::[a_1:: \cdots ::a_i::\alpha]::\alpha$, for $i \geq 1$. $\square$

Hence, we have refuted the Conjecture 2 concerning the equality between the observational and denotational orders. We will now refute the Conjecture 1:

Theorem 8. *$\mathcal{M}_\infty$ is not fully abstract for the λ-calculus with resources. In particular* $\boldsymbol{A}' := \boldsymbol{I}\ [\boldsymbol{A}^!, \boldsymbol{I}^!] \equiv_o \boldsymbol{A}$ *but* $[*]::* \in [\![\boldsymbol{A}']\!]$ *and* $[*]::* \notin [\![\boldsymbol{A}]\!]$

Proof. Since $\boldsymbol{A}' \to \boldsymbol{A}+\boldsymbol{I}$, we have $\boldsymbol{A}' \geq_o \boldsymbol{A}$ and $\boldsymbol{A}' \leq_o \boldsymbol{A}+\boldsymbol{A} = \boldsymbol{A}$. But in the same time $[\![\boldsymbol{A}']\!] = [\![\boldsymbol{A}]\!] \cup [\![I]\!] \ni [*]::*$ $\square$

5 Conclusion

Literature (on resource sensitive natural constructions from Linear Logic) are especially focussing on two objects, one in the semantical world, $\mathcal{M}_\infty$, and the other in the syntactical one, $\partial\lambda$-calculus. But they appeared not to respect full abstraction.

This unexpected result leads to questions on its generalization. For example, the idea can be applied to refute the full abstraction of $\mathcal{M}_\infty$ for the may-non-deterministic λ-calculus (an extension with a non deterministic operator endowed with a may-convergence operational semantic). Indeed, we can set $\boldsymbol{A}_0 = \lambda x.\boldsymbol{\Theta}\ (\lambda xy.x + \lambda xy.y)$ playing the role of $\boldsymbol{A}$. Such an $\boldsymbol{A}_0$ behaves as the infinite sum $\Sigma_{i=1}^{\infty} \lambda x_1...x_n y.y$, that is a top in its observational order but whose interpretation is not above the identity.

It can even be extended to other models since we can refute the full abstraction of Scott's $\mathcal{D}_\infty$ for the same may-non-deterministic λ-calculus (restriction of $\partial\lambda$-calculus to terms with only banged bags) or the may-must-non-deterministic λ-calculus (λ-calculus with both a may and a must non determinism), using A' in the same way. One can notice that the last case refutes a conjecture of [6].

More generally this counter-example describe the ill-behaved interaction between fixpoints and may-non-determinism that can tests any non-adequacy between the sights of the observation and of the model. We can thus conclude by giving the four key-points that leads to this kind of counter-examples:

- **short-sightedness of the contexts:** Calculi that offer control operators behaving as infinite applicative contexts like the resource λ-calculus with tests [3] are free of these considerations. This traduce the importance of the context lemma in our proof.
- **good sight of the model:** It is our better hope to find a fully abstract model for $\partial\lambda$-calculus but no known interesting algebraic models seems to break this property. Models tend indeed to approximate the condition "for any contexts of any size" into "for any infinite contexts".
- **Untyped fixpoints:** It is the first constructor that is necessary to construct a term that have a non bounded range. Thus, calculi with no fixpoints like the bang-free fragment of $\partial\lambda$-calculus will not suffer such troubles. But those calculi have limited expressive power.
- **may-non-determinism:** The second constructor, that is the most important part and the most interesting one since it can change our view of these calculi. To get rid of this problem without loosing the non determinism one can imagine a finer observation that discriminate the non idempotence of the sum, like the one provided by a probabilistic calculus.

Finally one may be disappointed by the "magic" resolution of Lemma 9. It was unclear, seeing $\boldsymbol{A}$, that this result would arise, and it needed quite a number of nontrivial lemmas. In this point lies a relation with tests mechanisms of [3], in this system $\tau(\!(.)\!)\ \bar{\tau}(\epsilon))$ *outer*-converges on $\boldsymbol{I}$ but not on $\boldsymbol{A}$, the calculus being inequationally fully abstract this gives Lemma 9 for free. That remark was the base of the previous (unpublished but cited) version of this article [2]. From our point of view the relation with tests is even deeper and essential. Indeed the counter-example was discovered naturally from a trial to prove full abstraction from reducing the one from the calculus with tests into the calculus without. This will be subject to an incoming paper.

Acknowledgments. My deepest acknowledgments are for my PhD advisors A.Bucciarelli and M.Pagani whose help was essential for the redaction of this first paper. I also have to thanks a referee that found and corrected a broken lemma.

References

1. Boudol, G.: The lambda-calculus with multiplicities. INRIA Research Report 2025 (1993)
2. Breuvart, F.: On the discriminating power of tests. arXiv preprint arXiv:1205.4691 (2012) (unpublished)
3. Bucciarelli, A., Carraro, A., Ehrhard, T., Manzonetto, G.: Full Abstraction for Resource Calculus with Tests. In: Bezem, M. (ed.) Computer Science Logic (CSL 2011) - 25th International Workshop/20th Annual Conference of the EACSL, Leibniz International Proceedings in Informatics (LIPIcs), vol. 12, pp. 97–111. Schloss Dagstuhl Leibniz-Zentrum fuer Informatik, Dagstuhl (2011)
4. Bucciarelli, A., Carraro, A., Ehrhard, T., Manzonetto, G.: Full Abstraction for the Resource Lambda Calculus with Tests, through Taylor Expansion. Logical Methods in Computer Science 8(4), 1–44 (2012)
5. Bucciarelli, A., Ehrhard, T., Manzonetto, G.: Not Enough Points Is Enough. In: Duparc, J., Henzinger, T.A. (eds.) CSL 2007. LNCS, vol. 4646, pp. 298–312. Springer, Heidelberg (2007)
6. Bucciarelli, A., Ehrhard, T., Manzonetto, G.: A relational model of a parallel and non-deterministic λ-calculus. In: Artemov, S., Nerode, A. (eds.) LFCS 2009. LNCS, vol. 5407, pp. 107–121. Springer, Heidelberg (2008)
7. De Carvalho, D., Tortora de Falco, L.: The relational model is injective for Multiplicative Exponential Linear Logic (without weakenings). Annals of Pure and Applied Logic (2012)
8. De Carvalho, D.: Execution time of lambda-terms via denotational semantics and intersection types. arXiv preprint arXiv:0905.4251 (2009)
9. Ehrhard, T.: A model-oriented introduction to differential linear logic (2011) (accepted)
10. Ehrhard, T.: An application of the extensional collapse of the relational model of linear logic. In: Accepted at CSL 2012 (2012)
11. Ehrhard, T., Regnier, L.: The differential lambda-calculus. Theoretical Computer Science (2004)
12. Ehrhard, T., Regnier, L.: Böhm Trees, Krivine's Machine and the Taylor Expansion of Lambda-Terms. In: Beckmann, A., Berger, U., Löwe, B., Tucker, J.V. (eds.) CiE 2006. LNCS, vol. 3988, pp. 186–197. Springer, Heidelberg (2006)
13. Laird, J., Manzonetto, G., McCusker, G., Pagani, M.: Weighted relational models of typed lambda-calculi (submitted, 2013)
14. Manzonetto, G.: A general class of models of $\mathcal{H}^*$. In: Královič, R., Niwiński, D. (eds.) MFCS 2009. LNCS, vol. 5734, pp. 574–586. Springer, Heidelberg (2009)
15. Manzonetto, G.: What is a Categorical Model of the Differential and the Resource λ-Calculi? Mathematical Structures in Computer Science 22(3), 451–520 (2012)
16. Pagani, M., Tranquilli, P.: Parallel reduction in resource lambda-calculus. In: Hu, Z. (ed.) APLAS 2009. LNCS, vol. 5904, pp. 226–242. Springer, Heidelberg (2009)
17. Pagani, M., Ronchi Della Rocca, S.: Linearity, Non-determinism and Solvability. Fundamenta Informaticae 103(1-4), 173–202 (2010)

Bounding Skeletons, Locally Scoped Terms and Exact Bounds for Linear Head Reduction

Pierre Clairambault

Computer Laboratory, University of Cambridge
pierre.clairambault@cl.cam.ac.uk

Abstract. Bounding skeletons were recently introduced as a tool to study the length of interactions in Hyland/Ong game semantics. In this paper, we investigate the precise connection between them and execution of typed λ-terms. Our analysis sheds light on a new condition on λ-terms, called *local scope*. We show that the reduction of locally scoped terms matches closely that of bounding skeletons. Exploiting this connection, we give upper bound to the length of linear head reduction for simply-typed locally scoped terms. General terms lose this connection to bounding skeletons. To compensate for that, we show that λ-lifting allows us to transform any λ-term into a locally scoped one. We deduce from that an upper bound to the length of linear head reduction for arbitrary simply-typed λ-terms. In both cases, we prove the asymptotical optimality of the upper bounds by providing matching lower bounds.

1 Introduction

In the last two decades there has been a growing interest in the study of *quantitative* or *intensional* aspects of higher-order programs; in particular, the study of their *complexity* has generated a lot of effort. In the context of the λ-calculus, the first result that comes to mind is the work by Schwichtenberg [14], later improved by Beckmann [2], establishing upper bound to the length of β-reduction sequences for simply-typed λ-calculus. In the somewhat related line of work of *implicit complexity*, type systems have been developed to characterize extensionally certain classes of functions, such as polynomial [10] or elementary [8] time. Such systems rely on a soundness theorem establishing that well-typed terms normalize in a certain restricted time, which is itself established using syntactic methods that are specific to the system being studied. This calls for the development of syntax-independent tools to study precisely the execution time of higher-order programs. The present paper is a step towards that goal.

In [4], in the context of Hyland-Ong game semantics, we showed that provided some size information on strategies we could give a bound to the length of their interactions. This was done by annotating each step of an interaction sequence with a finite tree of natural numbers – hereby called a *bounding skeleton*[1] – and showing that progress in the interaction amounts to a simple reduction on the

[1] They were called *agents* in [4].

M. Hasegawa (Ed.): TLCA 2013, LNCS 7941, pp. 109–124, 2013.

bounding skeleton. We gave bounds to the length of this reduction, hence bounding with it the length of the interaction sequence. The strength of this approach is that the games model is syntax-independent: in the variant considered in [4], it accommodates the simply-typed λ-calculus possibly with computational effects such as non-determinism, control, or ground type references. Its key weakness however, is that the direct connection between game-theoretic interaction and execution has only been made explicit [6] for pure simply-typed λ-terms of the form $M\ N_1\ \dots\ N_p$, where M and the N_is are η-long Böhm trees – we will call such terms *game situations*. Although terms can be transformed into game situations (as briefly described in [4]), the transformation is very inefficient and yields bounds that are sub-optimal and not very informative. For bounding skeletons to be a useful tool in complexity analysis, it is crucial to relate them directly to the execution of programs, without the detour by game semantics. Such a connection is non-trivial, as the dynamics of reduction in all generality is much more complicated than for game situations.

In this paper, we develop such a connection. This is done by introducing a new structural condition on terms, *local scope*, that ensures that information only flows locally through redexes, and not remotely through variables shared by distant subterms. We show that the reduction of η-long, locally scoped terms can be directly simulated within bounding skeletons. Using this property and (a small optimization of) our results in [4] on bounding skeletons, we deduce exact bounds to the execution time of locally scoped terms. We also show that the operation of λ-*lifting* [12] transforms arbitrary terms into locally scoped ones, and exploit this transformation to give exact bounds for the execution time of arbitrary simply-typed λ-terms.

Related works. There are multiple approaches to the complexity analysis of higher-order programs, but they seem to separate into two major families. On the one hand, Beckmann [2], extending earlier work by Schwichtenberg [14], gave exact bounds to the maximal length of β-reduction on simply-typed λ-terms. His analysis uses very basic information on the terms (their length, or height, and order), but gives bounds that are in general very rough. On the other hand other groups, including Dal Lago and Laurent [13], De Carvalho [9], or Bernardet and Lengrand [3], use semantic structures (respectively, game semantics, relational semantics, or non-idempotent intersection types) to capture abstractly the precise complexity of particular terms. Their bounds are much more precise on particular terms, but require information on the terms whose extraction is in general as long to obtain as actual execution. The present work belongs to the first family. However, unlike Beckmann and Schwichtenberg the reduction we consider is *linear head reduction*, which is the notion of execution implemented by abstract machines [7] and is therefore much closer to the actual execution of functional programming languages.

Outline. In Section 2, we start by introducing linear head reduction along with bounding skeletons, and recall the main result of [4]. In Section 3, we introduce local scope, show our simulation result and deduce exact bounds on the length of linear head reduction on locally scoped terms. Finally in Section 4, we show

how to use λ-lifting to transform arbitrary terms into locally scoped ones, and deduce exact bounds for linear head reduction on general terms.

2 Preliminaries

In this section, we start by recalling some of the background of this research. The natural starting point is *linear head reduction* [7], which can be seen as a direct implementation on λ-terms of the notion of execution performed by abstract machines. We will then turn to the presentation of *bounding skeletons*: we will recall the results of [4] on the length of their reductions, along with a small improvement.

2.1 Linear Head Reduction

We work here with the simply-typed λ-calculus *à la Church*, *i.e.* the variables are explicitly annotated with types (although we often omit the annotations for the sake of readability). Types are built from a unique atom o and the arrow constructor $\rightarrow$. We suppose that for every type A, there is a constant $*_A : A$ of type A. We will often omit the index and write $*$. As usual, we write $\text{fv}(M)$ for the set of *free variables* of a term M. The typing relation $\Gamma \vdash M : A$ is defined by the usual deduction rules for simply-typed λ-calculus. All the terms considered in this paper are supposed well-typed. Note that our choices – only one atom, each type is inhabited – merely make the presentation simpler and are not strictly required for our results to hold.

This work focuses strongly on *linear substitution*, for which only one variable occurrence is substituted at a time. In this situation, it is convenient to have a distinguished notation for particular *occurrences* of variables. We will use the notations $x_0, x_1, \ldots$ to denote particular occurrences of the same variable x in a term M. When in need of additional variable identifiers, we will use $x^1, x^2, \ldots$. Sometimes, we will still denote occurrences of x by just x when their index is not relevant. Although it is not the focus of this development, we will occasionally also refer to β-reduction: it is the standard rewriting rule on λ-terms, defined by $(\lambda x.M)\ N \rightarrow_\beta M[N/x]$, where $M[N/x]$ is the substitution of *all* occurrences of the variable x by N, applied in any position within M. We write $\equiv_\beta$ for the corresponding equivalence relation. If x_0 is a specific occurrence of x, we will use $M[N/x_0]$ for the substitution of x_0 by N, leaving all other occurrences of x unchanged. We assume Barendregt's convention and consider all terms up to α-equivalence (so, substitution involves renaming of bound variables).

Intuitively, linear head reduction proceeds as follows. We first locate the head variable occurrence, *i.e.* the leftmost variable occurrence in the term M. Then we locate the abstraction, if any, that binds this variable. Then we locate (again if it exists) the subterm N of M in argument position for that abstraction, and we substitute the head occurrence by N. We touch neither the other occurrences of x nor the redex. It is worth noting that locating the argument subterm can be delicate, as it is not necessarily part of a β-redex. For instance in $(\lambda y^A.(\lambda x^B.x_0 M))N_1 N_2$, we want to replace x_0 by N_2, even though N_2 is not directly applied to $\lambda x^B.x_0 M$. Therefore, the notion of redex will be generalized.

Note that a term is necessarily of the form $* \ M_1 \ \dots \ M_n, x_0 \ M_1 \ \dots \ M_n, \lambda x.M$ or $(\lambda x.M) \ M_1 \ \dots \ M_n$. That will be used quite extensively to define and reason on linear head reduction. The **length** of a term M is the number of characters in M, *i.e.* $\mathrm{l}(*) = 1, \mathrm{l}(x_0) = 1, \mathrm{l}(\lambda x.M) = \mathrm{l}(M) + 1, \mathrm{l}(M_1 \ M_2) = \mathrm{l}(M_1) + \mathrm{l}(M_2)$. Its **height** is $\mathrm{h}(*) = 0, \mathrm{h}(x_0) = 1, \mathrm{h}(\lambda x.M) = \mathrm{h}(M), \mathrm{h}(M_1 \ M_2) = \max(\mathrm{h}(M_1), \mathrm{h}(M_2) + 1)$.

Definition 1. *Given a term M, we define its set of* ***prime redexes****. They are written as pairs $(\lambda x, N)$ where N is a subterm of M, and λx is used to denote the (if it exists, necessarily unique by Barendregt's convention) subterm of M of the form $\lambda x.N'$. We define the prime redexes of M by induction on its length, distinguishing several cases depending on the form of M.*

- *If $M = * \ M_1 \ \dots \ M_n$, then M has no prime redex.*
- *If $M = x_0 \ M_1 \ \dots \ M_n$, then M has no prime redex.*
- *If $M = \lambda x.M'$, then M has the prime redexes of M'.*
- *If $M = (\lambda x.M') \ M_1 \ \dots \ M_n$, then the prime redexes of M are $(\lambda x, M_1)$ plus those of $M' \ M_2 \ \dots \ M_n$.*

The **head occurrence** of a term M is the leftmost occurrence of a variable or constant in M. If $(\lambda x, N)$ is a prime redex of M where the head occurrence of M is an occurrence x_0 of the variable x, then the **linear head reduct** of M is $M' = M[N/x_0]$. We write $M \rightarrow_{\mathrm{lhr}} M'$.

Example 1. As an example, we give the linear head reduction sequence of the term $(\lambda f.\lambda x.f \ (f \ x)) \ (\lambda y.y)$.

$$\begin{aligned}(\lambda f.\lambda x.f \ (f \ x)) \ (\lambda y.y) &\rightarrow_{\mathrm{lhr}} (\lambda f.\lambda x.(\lambda z.z) \ (f \ x)) \ (\lambda y.y) \\ &\rightarrow_{\mathrm{lhr}} (\lambda f.\lambda x.(\lambda z.f \ x) \ (f \ x)) \ (\lambda y.y) \\ &\rightarrow_{\mathrm{lhr}} (\lambda f.\lambda x.(\lambda z.(\lambda u.u) \ x) \ (f \ x)) \ (\lambda y.y) \\ &\rightarrow_{\mathrm{lhr}} (\lambda f.\lambda x.(\lambda z.(\lambda u.x) \ x) \ (f \ x)) \ (\lambda y.y)\end{aligned}$$

At this point the reduction stops since the head occurrence is an occurrence of x, and the corresponding abstraction subterm is not part of a prime redex.

We will abbreviate linear head reduction by lhr. It is straightforward to see that lhr is compatible with β-reduction, in the sense that if $M \rightarrow_{\mathrm{lhr}} M'$ we have $M \equiv_\beta M'$. Just as for β-reduction, lhr always terminates on well-typed terms, let us denote by $\mathcal{N}(M)$ the length of the reduction sequence of M. Since redexes for lhr are not necessarily β-redexes, it will be necessary to consider the following generalization of redexes:

Definition 2 (Generalized redex). *The* ***generalized redexes*** *of a term M are the prime redexes of all subterms of M. In particular, all prime redexes are generalized redexes.*

Example 2. Consider the following λ-term:

$$M = (\lambda x.x)\ ((\lambda y.(\lambda z.u))\ v\ w)$$

The only prime redex of M is $(\lambda x, (\lambda y.(\lambda z.u))\ v\ w)$, and it is therefore also a generalized redex. The two other generalized redexes are $(\lambda y, v)$, which is also a β-redex, and $(\lambda z, w)$, which is not.

2.2 Bounding Skeletons

This section focuses on a pivotal notion of this paper, that of a bounding skeleton. Intuitively, it is what is left of a term when all precise dynamic information is forgotten, and only the structural size information necessary to study termination is retained. Formally, a **bounding skeleton** is a finite tree whose nodes and edges are labeled by natural numbers. We write:

$$n[\{d_1\}a_1, \ldots, \{d_p\}a_p] =$$

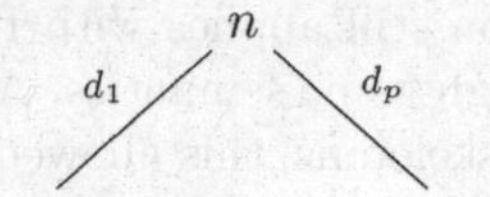

This notion was introduced in [4] where it was extracted from *game semantics*, and more precisely from the notion of *pointing sequence* central to Hyland-Ong games [11], but also appearing crucially in the earlier work of Coquand [5]. Bounding skeletons arise as measures of positions in pointing sequences, progress in the sequence corresponding to reduction of the skeleton. By the operational content of game semantics [6], bounding skeletons can also be seen as measures of terms obtained by lhr from a term of a particular form called a *game situation*. A **game situation** is a term of the form $M\ N_1\ \ldots\ N_n$, where $M : A_1 \to \ldots \to A_n \to o$ and $N_i : A_i$ are closed η-long Böhm trees – the terminology is motivated by the strong geometric correspondence between *η-long Böhm trees* and the *innocent strategies* of [11]. We know by the result of Danos, Herbelin and Regnier [6] that the lhr sequence of $M\ N_1\ \ldots\ N_n$ is in step-by-step correspondence with the game-theoretic interaction between the corresponding strategies $[\![M]\!]$ and $[\![N_i]\!]$. To illustrate how bounding skeletons arise from lhr of game situations, let us suppose for simplicity that $M : (A \to o) \to o$ and that M has the form $\lambda x.x\ M'$ with M' η-long Böhm tree of type A, possibly including x as a free variable. Then we have the lhr step:

$$(\lambda x.x\ M')\ N \to_{\mathrm{lhr}} (\lambda x.N\ M')\ N$$

Therefore, a situation with a closed Böhm tree M applied to a closed Böhm tree N is reduced to a closed Böhm tree N applied to an *open* Böhm tree M', along with an environment associating x to the closed Böhm tree N. In other words:

$$M \star N \to N \star M'^{\{x \mapsto N\}}$$

where the notation used will remain informal, but should be clear nonetheless. This can be represented by the following operation on trees of terms:

$$\begin{array}{c} M \\ | \\ N \end{array} \;\rightarrow\; \begin{array}{c} N \\ | \\ M' \\ | \\ N \end{array}$$

Replacing the terms by some measure of size (which will be made precise later in the paper) and annotating the edges with a measure of their types, these trees give rise to bounding skeletons, and this reduction appears as an instance of the following non-deterministic rule, illustrated in Figure 1.

$$n[\{d_1\}a_1, \ldots, \{d_p\}a_p] \rightarrow_{\mathrm{bs}} a_i \cdot_{d_i - 1} (n-1)[\{d_1\}a_1, \ldots, \{d_p\}a_p]$$

where $n, d_i \geq 1$ and $a \cdot_d b$ denotes the skeleton $n[\{d_1\}a_1, \ldots, \{d_p\}a_q, \{d\}b]$, for two skeletons $a = n[\{d_1\}a_1, \ldots, \{d_p\}a_q]$ and b.

This observation still applies with further reductions of $M\ N$, as we established in [4] through game semantics. Along with bounds to the length of reduction of bounding skeletons, this allowed us to bound the maximum length of lhr sequences starting from game situations. We also gave a bound for regular terms, relying on a very rough translation of arbitrary terms into game situations – as a result, this bound was far from optimal and not very informative. We aim in this paper to study the direct connection of bounding skeletons and syntax outside of game situations and independently on game semantics.

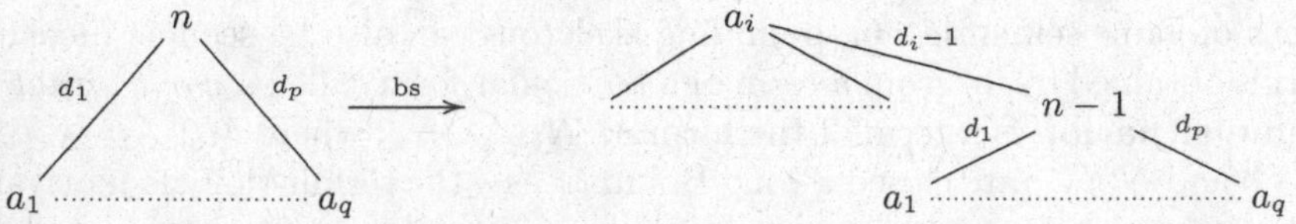

Fig. 1. Rewriting rule on skeletons

Remark 1. Note that reduction is set to only happen in root position, *i.e.* at the root of the tree. Generalizing it to apply deeper leads to pathological behavior. For instance deep reduction does not terminate on the variant without edge labels, whereas the standard (root) reduction does. It is not known whether deep reduction terminates in the presence of edge labels, or to which extent the relationship with syntactic reduction is preserved – the correspondence with game semantics is lost.

The main result of [4] is a bound on the length of reduction for bounding skeletons. The bound we state here is in fact a minor improvement of the result in [4], however the tools and methods to get it are the same. Therefore to save space, we omit the details of the optimization. As for terms, we write $\mathcal{N}(a)$ for the

norm of a bounding skeleton a, *i.e.* the length of its longest reduction sequence. We also write $\max(a)$ for the highest node label in a, $\mathrm{ord}(a)$ for the **order** of a, *i.e.* the highest edge label in a and $\mathrm{depth}(a)$ for the **depth** of a, *i.e.* the maximal depth of a node in a, the root being at depth 1. Here, log denotes the logarithm to base 2 and the tower of exponentials 2^p_n is defined by $2^p_0 = p$ and $2^p_{n+1} = 2^{2^p_n}$.

Theorem 1 (Upper bound). *If* $\mathrm{ord}(a), \mathrm{depth}(a), \max(a) \geq 1$, *then*

$$\mathcal{N}(a) \leq 2^{\mathrm{depth}(a)\log(\max(a)+1)}_{\mathrm{ord}(a)-1}$$

Constructions. In defining the interpretation of terms as bounding skeletons, we will make use of the following constructions. If $(a_i)_{1\leq i\leq n}$ is a finite family of bounding skeletons, then writing $a_i = n_i[\{d_{i,1}\}b_{i,1}, \dots, \{d_{i,p_i}\}b_{i,p_i}]$, we define:

$$\bigsqcup_{i=1}^{n} a_i = (\max_{1\leq i\leq n} n_i) \cdot [\{d_{i,j}\}b_{i,j} \mid 1 \leq i \leq n \ \&\ 1 \leq j \leq p_i]$$
$$\sum_{i=1}^{n} a_i = (\sum_{i=1}^{n} n_i) \cdot [\{d_{i,j}\}b_{i,j} \mid 1 \leq i \leq n \ \&\ 1 \leq j \leq p_i]$$

so, they either take the maximum or the sum of the roots, and simply append all the subtrees of the a^is. In the binary case, we write as usual $+$ for the sum. Finally, each natural number n can be seen as an atomic bounding skeleton $n[]$ without subtrees, still denoted by n. That should never cause any confusion.

Embedding. The norm of a bounding skeleton is unchanged by permutation of subtrees, or merging of identical subtrees, and is only increased by an increase of labels. If $a = n[\{d_1\}a_1, \dots, \{d_p\}a_p]$ and $a' = n'[\{d'_1\}a'_1, \dots, \{d'_{p'}\}a'_{p'}]$, we say that a **embeds** in a', written $a \hookrightarrow a'$, if $n \leq n'$ and for any $i \in \{1, \dots, p\}$ there exists $j \in \{1, \dots, p'\}$ such that $d_i \leq d'_j$ and $a_i \hookrightarrow a'_j$. Then we have:

Lemma 1 (Embedding lemma). *If* $a \hookrightarrow b$, *then* $\mathcal{N}(a) \leq \mathcal{N}(b)$.

We illustrate the reduction in Figure 2, where at each step we emphasize the subtree selected non-deterministically for the next reduction step. For conciseness, we also do not represent the subtrees under a node labeled 0, as they can play no further part in the reduction.

3 Locally Scoped Terms and Bounding Skeletons

This section explains the direct connection between linear head reduction and the reduction of bounding skeletons. We will first introduce *locally scoped terms*, for which this connection holds, then prove their simulation within bounding skeletons, and finally deduce bounds for the length of their reduction.

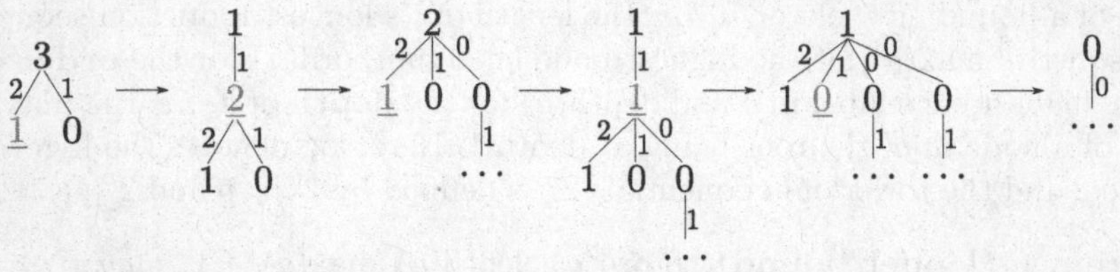

Fig. 2. Example reduction sequence on bounding skeletons

3.1 Locally Scoped Terms

Define inductively a **closure** as an open term M along with an **environment** σ, mapping free variables of M to closures of the same type. We say that a closure M^σ is **hereditarily normal** when M is β-normal and η-long, and when for any $x \in \text{fv}(M)$, the closure $\sigma(x)$ is hereditarily normal. Hereditarily normal closures are very close to bounding skeletons: from a hereditarily normal closure M^σ one can obtain a bounding skeleton having the height of M as root, and the bounding skeletons corresponding to $\sigma(x)$ for $x \in \text{fv}(x)$ as subtrees.

Although we will not make this formal in this paper, our simulation of lhr in bounding skeletons exploits that some terms can be represented as hereditarily normal closures. For instance, take the term:

$$K_1 = (\lambda x^{o \to o}.(\lambda y^{o \to o}.y)\ x)\ (\lambda z^o.z)$$

The term K_1 is faithfully represented by the hereditarily normal closure:

$$y^{y \mapsto x^{x \mapsto \lambda z.z}}$$

From that, we see that K_1 corresponds (ignoring edge labels) to the bounding skeleton $1[1[1]]$. Note in passing that K_1 reduces to $(\lambda x.(\lambda y.x)\ x)\ (\lambda z.z)$, which by the same idea as above corresponds to the hereditarily normal form $x^{x \mapsto \lambda z.z}$, and to the bounding skeleton $1[1]$ – which embeds in the bs-reduct $1[1, 0[1[1]]]$ of $1[1[1]]$, so the lhr reduction of K_1 is accounted for in bounding skeletons.

Unfortunately, this connection does not always work. For instance, take:

$$K_2 = (\lambda x.x\ *)\ (\lambda z.(\lambda y.y)\ z)$$

When trying to represent K_2 as a hereditarily normal closure, we run into the issue that since z is not a closed subterm, there is not way to replace the redex $(\lambda y.y)\ z$ by an environment. Of course, in K_1, we also had a generalized redex $(\lambda y, x)$ where x is not closed. But in K_1, x was *active*, in the sense that we had a redex $(\lambda x, \lambda z.z)$, so we knew how to define the environment on x. On the other hand, z is *passive* in K_2: there is no generalized redex $(\lambda z, N)$. In summary, the issue with K_2 is that there is a generalized redex $(\lambda y, z)$ where z (obviously) contains a passive free variable z, and because of that K_2 cannot be directly represented as a hereditarily normal closure.

Definition 3. *A variable x in M is* **active** *iff it is a free variable or if there is a generalized redex $(\lambda x, N)$ in M. It is* **passive** *otherwise. A term M is* **locally scoped** *(abbreviated l.s.) if for any generalized redex $(\lambda x, N)$ in M all the free variables in N are active in M. Likewise, M is* **strongly locally scoped** *(abbreviated s.l.s.) if for any generalized redex $(\lambda x, N)$ in M, N is closed.*

So, the term K_1 above is locally scoped, but K_2 is not since there is a generalized redex $(\lambda y, z)$ with z passive. Neither of those are strongly locally scoped. Any β-normal term is strongly locally scoped, and so is any term obtained by applications of β-normal forms (such as λ-terms corresponding to terms of combinatory logic). Local scope will be sufficient to ensure that the interpretation to bounding skeletons is a simulation, but the correspondence between terms and bounding skeletons will be tighter for strongly locally scoped terms: for those, the tree structure of the bounding skeleton will match the tree structure of imbricated generalized redexes. Strongly locally scoped terms are not stable under lhr, so we need to develop the full connection on locally scoped terms instead.

Lemma 2. *If M is a locally scoped term* of ground type *and $M \to_{\text{lhr}} M'$, then M' is locally scoped.*

3.2 Interpretation in Bounding Skeletons

Interpretation. The **level** of a type is defined by $\text{lv}(o) = 0$ and $\text{lv}(A \to B) = \max(\text{lv}(A) + 1, \text{lv}(B))$. Likewise, the **level** $\text{lv}(M)$ of a term M is the level of its type. Finally, the **order** $\text{ord}(M)$ of a term M is the maximal $\text{lv}(N)$, for all subterms N of M. Within a term $\Gamma \vdash M : A$ such that $(x : B) \in \Gamma$, we write $\text{lv}_M(x) = \text{lv}(B)$. The term M will generally be obvious from the context, so we will just write $\text{lv}(x)$.

Definition 4. *Let $\Gamma \vdash M : A$ be a term, with a* **bs-environment** *ρ, being defined as a partial function associating to each variable x of Γ on which it is defined a bounding skeleton $\rho(x)$. Then the bounding skeleton $[\![M]\!]_\rho$ is defined by induction on the length of M, as follows:*

$$
\begin{aligned}
[\![* \; M_1 \; \dots \; M_n]\!]_\rho &= 0 \\
[\![x_0 \; M_1 \; \dots \; M_n]\!]_\rho &= 1 + \textstyle\bigsqcup_{i=1}^{n} [\![M_i]\!]_\rho && \textit{if } \rho(x) \textit{ undefined} \\
[\![x_0 \; M_1 \; \dots \; M_n]\!]_\rho &= (1 + \textstyle\bigsqcup_{i=1}^{n} [\![M_i]\!]_\rho) \cdot_{\text{lv}(x)+1} \rho(x) && \textit{if } \rho(x) \textit{ defined} \\
[\![\lambda x.M]\!]_\rho &= [\![M]\!]_\rho \\
[\![(\lambda x.M) \; M_1 \dots \; M_n]\!]_\rho &= [\![M \; M_2 \; \dots \; M_n]\!]_{\rho \cup \{x \mapsto [\![M_1]\!]_\rho\}}
\end{aligned}
$$

We write $[\![M]\!]$ for $[\![M]\!]_\emptyset$.

Measures on terms and their preservation. To estimate lhr on s.l.s. terms, we need to define measures on terms that reflect the geometry of the corresponding bounding skeletons. So instead of the *height*, we have two alternative quantities.

The **depth** depth(M) of a term M is defined by induction on the length of M:

$$\begin{aligned} \mathrm{depth}(*\ M_1\ \ldots\ M_n) &= 1 \\ \mathrm{depth}(x_0\ M_1\ \ldots\ M_n) &= \max_{1\leq i\leq n} \mathrm{depth}(M_i) \\ \mathrm{depth}(\lambda x.M) &= \mathrm{depth}(M) \\ \mathrm{depth}((\lambda x.M)\ M_1\ \ldots\ M_n) &= \max(\mathrm{depth}(M\ M_2\ \ldots\ M_n), \mathrm{depth}(M_1)+1) \end{aligned}$$

Likewise, the **local height** lh(M) of a term M is defined by:

$$\begin{aligned} \mathrm{lh}(*\ M_1\ \ldots\ M_n) &= 0 \\ \mathrm{lh}(x_0\ M_1\ \ldots\ M_n) &= 1 + \max_{1\leq i\leq n} \mathrm{lh}(M_i) \\ \mathrm{lh}(\lambda x.M) &= \mathrm{lh}(M) \\ \mathrm{lh}((\lambda x.M)\ M_1\ \ldots\ M_n) &= \max(\mathrm{lh}(M\ M_2\ \ldots\ M_n), \mathrm{lh}(M_1)) \end{aligned}$$

Then, we have the following lemma:

Lemma 3. *If M is a strongly locally scoped term, then we have:*

$$\mathrm{depth}([\![M]\!]) \leq \mathrm{depth}(M) \qquad \max([\![M]\!]) \leq \mathrm{lh}(M) \qquad \mathrm{ord}([\![M]\!]) \leq \mathrm{ord}(M)$$

Simulation. In order to have our simulation result of linear head reduction into bounding skeletons, we need the additional requirement that the terms being interpreted are **η-long** – it is natural since our tools originate from game semantics, in which strategies are representations of η-long normal forms. As usual, η-expansion is the rule $M \to_\eta \lambda x^A.M\ x$, that applies when M has type $A \to B$ and $x \notin \mathrm{fv}(M)$. Non β-normal η-long terms are often defined as the terms on which any further η-expansion creates a new β-redex. Since we have generalized the notion of redex, we instead define them as the terms for which any η-expansion creates a new generalized redex. Then, η-long terms are stable under lhr. Moreover, we have:

Proposition 1 (Simulation). *Let $\Gamma \vdash M, M' : o$ be η-long locally scoped terms, and suppose $M \to_{\mathrm{lhr}} M'$. Then, there is a such that $[\![M]\!] \to_{\mathrm{bs}} a \hookleftarrow [\![M']\!]$.*

3.3 Bounds for Strongly Locally Scoped Terms

As a first application, we give exact bounds for the maximal length of lhr on strongly locally scoped terms. Formally, we will estimate the following quantity.

$$\mathrm{Lls}_n(h,d) = \max\{\mathcal{N}(M) \mid \mathrm{ord}(M) \leq n\ \&\ \mathrm{lh}(M) \leq h\ \&\ \mathrm{depth}(M) \leq d\ \&\ M \text{ s.l.s.}\}$$

To express our results, we will use some standard notations for comparing growth rates of functions. For functions $f, g : \mathbb{N} \to \mathbb{N}$, we write $f(n) = \Theta(g(n))$ when there exists reals $c_1, c_2 > 0$ and $N \in \mathbb{N}$ such that for all $n \geq N$, $c_1 g(n) \leq f(n) \leq c_2 g(n)$. This is generalized to functions of multiple variables $f, g : \mathbb{N}^p \to \mathbb{N}$

by setting that $f(n_1,\ldots,n_p) = \Theta(g(n_1,\ldots,n_p))$ iff there are $c_1, c_2 > 0$ and $N_i \in \mathbb{N}$ for all $i \in \{1,\ldots,p\}$ such that for all $n_i \geq N_i$ we have $c_1 g(n_1,\ldots,n_p) \leq f(n_1,\ldots,n_p) \leq c_2 g(n_1,\ldots,n_p)$. If $h : \mathbb{N} \to \mathbb{N}$ is another function, we write $f(n_1,\ldots,n_p) = h(\Theta(g(n_1,\ldots,n_p)))$ iff there is a function $\phi : \mathbb{N}^p \to \mathbb{N}$ such that $f(n_1,\ldots,n_p) = h(\phi(n_1,\ldots,n_p))$ and $\phi(n_1,\ldots,n_p) = \Theta(g(n_1,\ldots,n_p))$.

η-long form. Our simulation result only applies to η-long terms. Therefore, in order to obtain the upper bound we first need the following result:

Proposition 2. *If M is a term, then there is an η-long term M' such that:*

$$\begin{array}{ll} \mathrm{lh}(M') \leq \mathrm{lh}(M) + \mathrm{ord}(M) & \quad \mathrm{depth}(M') = \mathrm{depth}(M) \\ \mathrm{ord}(M') = \mathrm{ord}(M) & \quad \mathcal{N}(M') \geq \mathcal{N}(M) \end{array}$$

Moreover if M was s.l.s., M' is still s.l.s..

The proof is mostly direct, but rather long and technical. We show first that η-expansion can only increase the norm and that it preserves strong local scope and the order of terms. Moreover, if η-expansion is restricted so that it does not create new generalized redexes, then it terminates on an η-long form. Besides, restricted η-expansion preserves depth and a variant $\mathrm{lh}'(M)$ of $\mathrm{lh}(M)$ such that $\mathrm{lh}(M) \leq \mathrm{lh}'(M) \leq \mathrm{lh}(M) + \mathrm{ord}(M)$, taking into account the potential size of the variables that are not yet expanded. Details are omitted.

Upper bound. If $\Gamma \vdash M : A_1 \to \ldots \to A_n \to o$ is a s.l.s. term, we first make it of ground type by forming $\Gamma \vdash M \ *_{A_1} \ \ldots \ *_{A_n} : o$ – its norm can only increase, the other quantities stay unchanged and the term is still s.l.s.. By Proposition 2, there is M' η-long, of ground type, and s.l.s. such that $\mathrm{lh}(M') \leq \mathrm{lh}(M) + \mathrm{ord}(M)$, $\mathrm{depth}(M') = \mathrm{depth}(M)$, $\mathrm{ord}(M') = \mathrm{ord}(M)$ and $\mathcal{N}(M') \geq \mathcal{N}(M)$. Along with Lemma 3, Proposition 1 and Theorem 1 this gives the following proposition.

Proposition 3. *Suppose M is a strongly locally scoped term of order at least one. Then, $\mathcal{N}(M) \leq 2_{\mathrm{ord}(M)-1}^{\mathrm{depth}(M)\log(\mathrm{lh}(M)+\mathrm{ord}(M)+1)}$.*

Lower bound. We now set to prove the optimality of this upper bound by exhibiting a family of terms whose reduction length asymptotically reaches it. The family we describe is a variant of one used by Beckmann in [2], constructed by iterated exponentiation of Church numerals. We define higher types for Church integers by setting $A_{-2} = o$ and $A_{n+1} = A_n \to A_n$. Then, writing $\underline{n}_p$ for the Church integer for n of type A_p, we define, for $n, k, p \geq 0$ and $M : A_p$:

$$[n]_p^0(M) = M \qquad\qquad [n]_p^{k+1}(M) = \underline{n}_{p+1}\ [n]_p^k(M)$$

One can immediately check that $[n]_p^k(M) : A_p$ and that for all $q \in \mathbb{N}$, $[n]_p^k(\underline{q}_p) \to_\beta^* \underline{q^{n^k}}_p$. Exploiting this construction we set, for $n, k, p \geq 0$:

$$S_{n,k,p} = [n]_p^k(\underline{2}_p)\ \underline{2}_{p-1}\ \cdots\ \underline{2}_0$$

For which it is immediate to check that for all $n, k, p \geq 0$ we have $S_{n,k,p} \rightarrow_\beta^* \underline{2_p^{2_0^{n^k}}}$. Moreover, by construction of $S_{n,k,p}$, for $n \geq 2$ and $p, k \geq 1$ we have $\mathrm{lh}(S_{n,k,p}) = n+1$, $\mathrm{depth}(S_{n,k,p}) = k+1$ and $\mathrm{ord}(S_{n,k,p}) = p+3$, and $S_{n,k,p}$ is s.l.s.. To deduce a lower bound from this, we need to relate it to lhr using:

Lemma 4. *If $M \rightarrow_\beta^* \underline{n}_0$, then $\mathcal{N}(M\ \mathrm{id}_o) \geq n$, where $\mathrm{id}_o = \lambda x^o.x$.*

Proof. By induction on n, exploiting that lhr preserves β-equivalence.

Theorem 2. *For fixed $n \geq 2$ we have $\mathrm{Lls}_n(h, d) = 2_{n-1}^{\Theta(d \log(h))}$.*

Proof. Let us first consider the case where $n \geq 3$, as $n = 2$ requires a separate construction for the lower bound. Let us fix $h \geq 3$ and $d \geq 2$. By Proposition 3, we already know that $\mathrm{Lls}_n(d, h) \leq 2_{n-1}^{d \log(h+n+1)}$. Moreover, we have $\mathrm{lh}(S_{h-1,d-1,n-3}\ \mathrm{id}_o) = h$ and $\mathrm{depth}(T_{h-1,d-1,n-3}\ \mathrm{id}_o) = d$, and by Lemma 4 we have $\mathcal{N}(T_{h-1,d-1,n-3}\ \mathrm{id}_o) \geq 2_{n-1}^{(d-1)\log(h-1)}$. To summarize:

$$2_{n-1}^{(d-1)\log(h-1)} \leq \mathrm{Lls}_n(d, h) \leq 2_{n-1}^{d \log(h+n+1)}$$

Therefore, with $n \geq 3$ fixed and d, h parameters we have $\mathrm{Lls}_n(h, d) = 2_{n-1}^{\Theta(d \log(h))}$.

For $n = 2$, the upper bound still holds. For $d, p \geq 2$, define:

$$U_{n,d} = \underline{n}_1\ (\underline{n}_1\ \dots (\underline{n}_1\ \mathrm{id}_o) \dots)$$

where there are d copies of $\underline{n}_1$ in total. Then, the term $U_{n,d}$ is s.l.s. and we have $\mathrm{lh}(U_{n,d}) = n+1$, $\mathrm{depth}(U_{n,d}\ \mathrm{id}_o) = d+1$, $\mathrm{ord}(U_{n,d}\ \mathrm{id}_o) = 2$ and $\mathcal{N}(U_{n,d}) \geq n^d = 2^{d \log(n)}$. It follows that $\mathrm{Lls}_2(d, h) = 2^{\Theta(d \log(h))}$.

In particular, reduction length for s.l.s. second-order terms of fixed depth is bounded by a polynomial of degree less than the depth.

4 Exact Bounds for General Terms

4.1 Lambda-Lifting to Strongly Locally Scoped Terms

In order to deduce bounds for general terms, we now describe a transformation taking any λ-term M to a corresponding s.l.s. term M'; this transformation is a variant of the familiar notion of λ-lifting [12], adapted to lift variables through generalized redexes as well as β-redexes.

Take a term $M = \lambda x^A.(\lambda y^A.y)\ x$. Obviously, M is not s.l.s.: indeed there is a prime redex $(\lambda y, x)$ and the subterm x has x free. In order to make the variable x "local", we modify the abstraction subterm $\lambda y.y$ to forward explicitly the variable x. We get the term $M' = \lambda x^A.(\lambda y^{A \rightarrow A}.y\ x)(\lambda x'^A.x')$. The type of y has changed, but not the type of the overall term. Note that the terms M and M' are still β-equivalent, although we are not going to use that explicitly. More importantly, the norm has increased, the order has increased by one, and

$$\frac{y \in \mathrm{fv}(M_1)}{(\lambda x.M)\ M_1\ \ldots\ M_n \to_{\lambda \mathrm{l}} (\lambda x.M[x\ y/x])\ (\lambda y'.M_1[y'/y])\ \ldots\ M_n}$$

$$\frac{M_i \to_{\lambda \mathrm{l}} M_i'}{x_0\ M_1\ \ldots\ M_n \to_{\lambda \mathrm{l}} x_0\ M_1\ \ldots\ M_i'\ \ldots\ M_n} \qquad \frac{M \to_{\lambda \mathrm{l}} M'}{\lambda x.M \to_{\lambda \mathrm{l}} \lambda x.M'}$$

$$\frac{M_1 \to_{\lambda \mathrm{l}} M_1'}{(\lambda x.M)\ M_1\ \ldots\ M_n \to_{\lambda \mathrm{l}} (\lambda x.M)\ M_1'\ \ldots\ M_n}$$

$$\frac{M\ M_2\ \ldots\ M_n \to_{\lambda \mathrm{l}} M'\ M_2'\ \ldots\ M_n'}{(\lambda x.M)\ M_1\ \ldots\ M_n \to_{\lambda \mathrm{l}} (\lambda x.M')\ M_1\ M_2'\ \ldots\ M_n'}$$

Fig. 3. Definition of the λ-lifting expansion $\to_{\lambda \mathrm{l}}$

the other quantities are essentially unchanged. We formalize this construction by the λ-lifting expansion $\to_{\lambda \mathrm{l}}$, defined in Figure 3.

In general $\to_{\lambda \mathrm{l}}$ leaves the type unchanged, although it can change the type of bound variables. Moreover $\to_{\lambda \mathrm{l}}$ terminates, and its normal form is necessarily s.l.s.. Altogether, we have the following result:

Lemma 5. *For any term M, there is a strongly locally scoped M' such that:*

$$\mathrm{lh}(M') \leq \mathrm{lh}(M) + 1 \qquad \mathrm{depth}(M') = \mathrm{depth}(M)$$
$$\mathrm{ord}(M') \leq \mathrm{ord}(M) + 1 \qquad \mathcal{N}(M') \geq \mathcal{N}(M)$$

This is established by a rather lengthy technical proof, studying commutations between $\to_{\lambda \mathrm{l}}$ and $\to_{\mathrm{lhr}}$. Preservation of depth is easy since we do not add generalized redexes, and (relative) preservation of order and local height is established as for $\to_\eta$, by building variants lh′ and ord′ which take into account the potential expansion of variables, that satisfy $\mathrm{lh}(M) \leq \mathrm{lh}'(M) \leq \mathrm{lh}(M) + 1$ and $\mathrm{ord}(M) \leq \mathrm{ord}'(M) \leq \mathrm{ord}(M) + 1$ and are preserved by $\to_{\lambda \mathrm{l}}$.

4.2 Expanding Variables

For non locally scoped terms, the local height and the depth are rather unnatural quantities, and the bounds are not naturally expressed in terms of them. We convert one to the other using another norm-increasing term transformation.

Lemma 6. *For any term M, there exists a term M' such that $M \to_{\eta^*} M'$,*

$$\mathrm{lh}(M') \leq 2 \qquad \mathrm{depth}(M') \leq \mathrm{h}(M)$$
$$\mathrm{ord}(M') = \mathrm{ord}(M) \qquad \mathcal{N}(M') \geq \mathcal{N}(M)$$

The term M' is obtained by replacing each occurrence x_0 in M of a variable $x : A_1 \to \ldots \to A_n \to o$ by its η-expanded form $\lambda {y^1}^{A_1}. \ldots .\lambda {y^n}^{A_n}.x_0\ y^1\ \ldots\ y^n$. Since we have $M \to_{\eta^*} M'$, we already know that $\mathcal{N}(M') \geq \mathcal{N}(M)$, the other inequalities are easily established by induction.

4.3 Exact Bounds for General Terms

In this section, we are interested in estimating the quantity:

$$\mathrm{Lgen}_n(h) = \max\{\mathcal{N}(M) \mid \mathrm{ord}(M) \leq n\ \&\ \mathrm{h}(M) \leq h\}$$

We do that by applying the tools developed earlier to get an upper bound on the length of reduction, and then prove a matching lower bound by providing terms whose length of reduction asymptotically reaches the upper bound.

Upper bound. Starting from a term M, we first expand variables using Lemma 6, then make it s.l.s. using Lemma 5. This gives M' such that:

$$\begin{array}{ll} \mathrm{lh}(M') \leq 3 & \mathrm{depth}(M') = \mathrm{h}(M) \\ \mathrm{ord}(M') \leq \mathrm{ord}(M) + 1 & \mathcal{N}(M') \geq \mathcal{N}(M) \end{array}$$

By applying Proposition 3, we get:

Proposition 4. *Suppose M is a term. Then,* $\mathcal{N}(M) \leq 2_{\mathrm{ord}(M)}^{\mathrm{h}(M)\log(\mathrm{ord}(M)+5)}$.

Lower bound. We provide a lower bound matching asymptotically the upper bound offered by Proposition 4. The construction is essentially the same as the one used in [2] for the lower bound in terms of height.

For $p \geq 1$ and $k \geq 0$, we define $b_0^p = \underline{2}_p$ and $b_{k+1}^p = \lambda x^{A_{p-1}}.b_k^p\ (b_k^p\ x)$. Then, we set:

$$B_k^p = b_k^p\ \underline{2}_{p-1}\ \cdots\ \underline{2}_0$$

Note that this term is *not* s.l.s.. By standard arithmetic of Church numerals, we have that for any $p \geq 1, k \geq 0$, $B_k^p \to_\beta^* \underline{2_{p+2}^k}_0$. By Lemma 4 it follows that $\mathcal{N}(B_k^p\ \mathrm{id}_o) \geq 2_{p+2}^k$. It is direct to check that $\mathrm{ord}(B_k^p) = p+2$ and $\mathrm{h}(B_k^p) = k+3$ (for $k \geq 1$). Therefore, we have:

Theorem 3. *For fixed $n \geq 3$ we have* $\mathrm{Lgen}_n(h) = 2_n^{\Theta(h)}$.

For a term M of height h and order n, Beckmann's results [2] predict that any β-reduction chain of M terminates in less than $2_{n+1}^{\Theta(h)}$ steps. It might seem counter-intuitive that our bound (with linear head reduction) is smaller than Beckmann's (with β-reduction) since we substitute only one occurrence at a time, which is obviously longer. However, Beckmann considers arbitrary β-reduction, not head β-reduction. The possibility of reducing in arbitrary locations of the term unlocks much longer reductions, since higher-order free variables or constants can isolate sections of the term that will never arrive in head position but can still be affected by arbitrary β-reduction. The fact that the length of linear head reduction has the same order of magnitude as head β-reduction is not surprising in the light of Accattoli and Dal Lago's recent result [1] that a similar notion of linear head reduction is quadratically related to head reduction.

5 Conclusion

We have worked out the precise connection between bounding skeletons and syntactic reduction, deducing bounds for linear head reduction in the simply-typed λ-calculus. The analysis uncovers *locally scoped terms*, whose reduction relates closely to game-theoretic interaction. Through this work, we obtain syntax-independent tools to reason on the complexity of programs, hopefully useful in implicit complexity. Although we have only described this connection here for the pure λ-calculus, the connection with games suggest that similar constructions should yield the same results for languages with effects such as control, non-determinism or ground state. In future work we plan plan to generalize these tools to more expressive languages, in particular in the presence of recursion.

Acknowledgment. We gratefully acknowledge the support of the ERC Advanced Grant ECSYM.

References

1. Accattoli, B., Lago, U.D.: On the invariance of the unitary cost model for head reduction. In: Tiwari, A. (ed.) RTA. LIPIcs, vol. 15, pp. 22–37. Leibniz-Zentrum fuer Informatik, Schloss Dagstuhl (2012)
2. Beckmann, A.: Exact bounds for lengths of reductions in typed lambda-calculus. J. Symb. Log. 66(3), 1277–1285 (2001)
3. Bernadet, A., Lengrand, S.: Complexity of strongly normalising λ-terms via non-idempotent intersection types. In: Hofmann, M. (ed.) FOSSACS 2011. LNCS, vol. 6604, pp. 88–107. Springer, Heidelberg (2011)
4. Clairambault, P.: Estimation of the length of interactions in arena game semantics. In: Hofmann, M. (ed.) FOSSACS 2011. LNCS, vol. 6604, pp. 335–349. Springer, Heidelberg (2011)
5. Coquand, T.: A semantics of evidence for classical arithmetic. J. Symb. Log. 60(1), 325–337 (1995)
6. Danos, V., Herbelin, H., Regnier, L.: Game semantics and abstract machines. In: Proceedings of the Eleventh Annual IEEE Symposium on Logic in Computer Science, LICS 1996, pp. 394–405. IEEE (1996)
7. Danos, V., Regnier, L.: How abstract machines implement head linear reduction (2003) (unpublished)
8. Danos, V., Joinet, J.-B.: Linear logic and elementary time. Inf. Comput. 183(1), 123–137 (2003)
9. de Carvalho, D., Pagani, M., de Falco, L.T.: A semantic measure of the execution time in linear logic. TCS 412(20), 1884–1902 (2011)

10. Girard, J.-Y.: Light linear logic. In: Leivant, D. (ed.) LCC 1994. LNCS, vol. 960, pp. 145–176. Springer, Heidelberg (1995)
11. Hyland, J.M.E., Ong, C.-H.L.: On full abstraction for PCF: I, II, and III. Inf. Comput. 163(2), 285–408 (2000)
12. Johnsson, T.: Lambda lifting: Transforming programs to recursive equations. In: Jouannaud, J.-P. (ed.) FPCA 1985. LNCS, vol. 201, pp. 190–203. Springer, Heidelberg (1985)
13. Dal Lago, U., Laurent, O.: Quantitative game semantics for linear logic. In: Kaminski, M., Martini, S. (eds.) CSL 2008. LNCS, vol. 5213, pp. 230–245. Springer, Heidelberg (2008)
14. Schwichtenberg, H.: An upper bound for reduction sequences in the typed λ-calculus. Archive for Mathematical Logic 30(5), 405–408 (1991)

Intersection Type Matching with Subtyping

Boris Düdder, Moritz Martens, and Jakob Rehof

Technical University of Dortmund, Faculty of Computer Science
{boris.duedder,moritz.martens,jakob.rehof}@cs.tu-dortmund.de

Abstract. Type matching problems occur in a number of contexts, including library search, component composition, and inhabitation. We consider the intersection type matching problem under the standard notion of subtyping for intersection types: Given intersection types τ and σ, where σ is a constant type, does there exist a type substitution S such that $S(\tau)$ is a subtype of σ? We show that the matching problem is NP-complete. NP-hardness holds already for the restriction to atomic substitutions. The main contribution is an NP-algorithm which is engineered for efficiency by minimizing nondeterminism and running in PTIME on deterministic input problems. Our algorithm is based on a nondeterministic polynomial time normalization procedure for subtype constraint systems with intersection types. We have applied intersection type matching in optimizations of an inhabitation algorithm.

1 Introduction

By *intersection type matching* we understand the following decision problem:

Given intersection types τ and σ, where σ does not contain any type variables, is there a type substitution S with $S(\tau) \leq_{\mathbf{A}} \sigma$?

Here the relation $\leq_{\mathbf{A}}$ is the standard theory of subtyping for intersection types as defined in [1].

Generally, type matching may refer to a range of decision problems spanning from various forms of type equivalence [2] to problems involving substitution and subtyping [3]. The *subtype matching* problem considered in this paper is a special case of the subtype *satisfiability* problem, to decide for given τ and σ whether there exists a substitution S such that $S(\tau) \leq S(\sigma)$, where $\leq$ is a subtyping relation (a partial order on types). *Equational* matching problems of the form $\exists S.\ S(\tau) \sim \sigma$, where $\sim$ is some equivalence relation, is a special case of *unification* modulo an equational theory (see, e.g., [4]). Typical applications of type matching include component retrieval and composition [3,2], where τ might for example be the parametric type of a component F and σ the expected type in a usage context C, and where the matching condition $S(\tau) \leq \sigma$ ensures that $C[F]$ is well typed.

Our study of intersection type matching with subtyping is motivated from two perspectives. First, from a systematic standpoint, there is a general need to develop the algorithmic theory of intersection type subtyping. Such a theory is

M. Hasegawa (Ed.): TLCA 2013, LNCS 7941, pp. 125–139, 2013.

sorely missing in the literature (see Sect. 2), and an algorithmic study of matching is one necessary step towards such a theory. Second, from the standpoint of applications, we have applied the results reported in this paper to the problem of inhabitation in systems of combinatory logic with intersection types [5,6] in the context of combinatory logic synthesis (see [7] for an introduction to this research programme). With polymorphic combinators [6] a central step in solving the inhabitation problem $\Gamma \vdash ? : \sigma$ (is there a term with type σ in type environment Γ?) is to decide the condition $S(\tau) \leq \sigma$, where σ is an inhabitation goal and τ is the type of a combinator from Γ. A solution to the matching problem leads to a very substantial optimization of the inhabitation algorithm, since it allows us to filter out uninteresting choices of τ early in the process based on the matching condition. We refer the reader to [8] for details of this application.

We show that the problem of intersection type matching is NP-complete, even when substitutions are restricted to be atomic (mapping type variables to either variables or constants). We present an NP-algorithm which is engineered for efficiency by localizing nondeterminism as much as possible. The core of the algorithm consists in a nondeterministic constraint set normalization procedure. Matching substitutions can be efficiently constructed from normalized constraint systems provided they are consistent. The constraint normalization procedure performs a fine-grained analysis of subtyping constraints generated from the input problem. In the absence of nondeterminism in the input problem, the algorithm operates in PTIME (note that this is obviously far superior to approaches where solutions are simply guessed).

Due to space limitations some proofs have been shortened or left out. They can be found in [8].

2 Related Work

Surprisingly, the algorithmic properties of the standard intersection subtyping relation $\leq_{\mathbf{A}}$, clearly of systematic importance in type theory, do not appear to have been very much investigated in the literature. To the best of our knowledge, there are no tight results for any of the problems, unification, satisfiability, or, indeed, matching, for intersection types under the standard subtyping relation of [1]. This situation is not satisfactory, and with the present paper we take a step towards remedying it. Only recently, a PTIME-procedure for deciding the relation $\leq_{\mathbf{A}}$ itself was given in [5] and is used as a subroutine here (decidability of $\leq_{\mathbf{A}}$ follows from the results of [9], but with an exponential time algorithm).

Subtype satisfiability has been studied in various subtyping theories without intersection. It is particularly useful to compare with results in simple types (an overview can be found in [10]). Satisfiability there is PSPACE-complete when arbitrary partial orders of base types are allowed [11,12], but it is in PTIME over lattices of base types [11]. The PTIME result uses the property that consistent constraint systems are satisfiable, because a satisfying substitution can be constructed from such systems using either one of the lattice operations. A similar property is used here (Lem. 3) to solve the intersection type matching problem, but it is based on a much more intricate constraint normalization procedure. As can indeed be concluded from our NP-completeness result, the presence of intersection changes the problem fundamentally. Let us remark that our technique for solving the matching problem does not transfer to the satisfiability problem with intersection (the reason will become clear in our technical development), and satisfiability could very well lie higher in the complexity hierarchy.

There are various results concerning the complexity of matching and unification in different classes of equational theories (cf. [4] for a general survey), which exhibit similar properties as the type operators $\cap$ and $\rightarrow$ (associativity (A), commutativity (C), and idempotence (I), respectively distributivity (D)) and which are NP-complete. For example, AC- and ACI-matching as well as AC- and ACI-unification are NP-complete [13,14,15]. The techniques used to prove membership in NP are completely different from the approach we follow, though. For example, it is easy to show that the size of a substitution solving an AC-matching problem is bounded by the size of the terms involved [13]. Therefore, simply guessing the right substitution yields an NP-algorithm. As discussed in the beginning of Sect. 5 this is not so simple in the case of intersection type matching. Moreover, this approach foregoes any detailed analysis of the sources of nondeterminism and leads to pragmatically suboptimal algorithms. The algorithms deciding AC- and ACI-unification are relatively intricate [15]. They reduce unification to unification in certain commutative semigroups which is known to be NP-complete. This reduction does not appear to be possible in our setting, however.[1] There are semi-decidability results concerning an altogether different notion of unification for intersection types [16,17] where unification is considered along with other operations that can be used to characterize principal typings with intersection types.

3 Preliminaries

Type expressions, ranged over by τ, σ, etc., are defined by $\tau \ ::= \ a \,|\, \tau \rightarrow \tau \,|\, \tau \cap \tau$ where $a, b, p, q \ldots$ range over *atoms* comprising of *type constants*, drawn from a finite set $\mathbb{A}$ including the constant ω, and *type variables*, drawn from a disjoint denumerable set $\mathbb{V}$ ranged over by α, β, etc. We let $\mathbb{T}$ denote the set of all types.

As usual, types are taken modulo commutativity ($\tau \cap \sigma = \sigma \cap \tau$), associativity ($(\tau \cap \sigma) \cap \rho = \tau \cap (\sigma \cap \rho)$), and idempotency ($\tau \cap \tau = \tau$). As a matter of notational convention, function types associate to the right, and $\cap$ binds stronger than $\rightarrow$.

A type $\tau \cap \sigma$ is said to have τ and σ as *components*. For an intersection of several components we sometimes write $\bigcap_{i=1}^{n} \tau_i$ or $\bigcap_{i \in I} \tau_i$ or $\bigcap\{\tau_i \mid i \in I\}$, where the empty intersection is identified with ω.

The standard [1] intersection type *subtyping* relation $\leq_{\mathbf{A}}$ is the least preorder (reflexive and transitive relation) on $\mathbb{T}$ generated by the following set $\mathbf{A}$

---

[1] We also note that the results on unification mentioned above do not directly encompass the complexity of intersection type unification modulo the ACID equational theory induced by $\leq_{\mathbf{A}}$.

of axioms:

$$\sigma \leq_{\mathbf{A}} \omega, \quad \omega \leq_{\mathbf{A}} \omega \to \omega, \quad \sigma \cap \tau \leq_{\mathbf{A}} \sigma, \quad \sigma \cap \tau \leq_{\mathbf{A}} \tau, \quad \sigma \leq_{\mathbf{A}} \sigma \cap \sigma;$$
$$(\sigma \to \tau) \cap (\sigma \to \rho) \leq_{\mathbf{A}} \sigma \to \tau \cap \rho;$$
$$\text{If } \sigma \leq_{\mathbf{A}} \sigma' \text{ and } \tau \leq_{\mathbf{A}} \tau' \text{ then } \sigma \cap \tau \leq_{\mathbf{A}} \sigma' \cap \tau' \text{ and } \sigma' \to \tau \leq_{\mathbf{A}} \sigma \to \tau'.$$

We identify σ and τ when $\sigma \leq_{\mathbf{A}} \tau$ and $\tau \leq_{\mathbf{A}} \sigma$. The distributivity properties $(\sigma \to \tau) \cap (\sigma \to \rho) = \sigma \to \tau \cap \rho$ and $(\sigma \to \tau) \cap (\sigma' \to \tau') \leq_{\mathbf{A}} \sigma \cap \sigma' \to \tau \cap \tau'$ follow from the axioms of subtyping. Note also that $\tau_1 \to \cdots \to \tau_m \to \omega = \omega$. We say that a type τ is *reduced with respect to* ω if it has no subterm of the form $\rho \cap \omega$ or $\tau_1 \to \cdots \to \tau_m \to \omega$ with $m \geq 1$. It is easy to reduce a type with respect to ω, by applying the equations $\rho \cap \omega = \rho$ and $\tau_1 \to \cdots \to \tau_m \to \omega = \omega$ left to right.

A type of the form $\tau_1 \to \cdots \to \tau_m \to a$, where $a \neq \omega$ is an atom, is called a *path of length m*. A type τ is *organized* if it is a (possibly empty) intersection of paths (those are called *paths in* τ). Every type τ is equal to an organized type $\overline{\tau}$, computable in polynomial time, with $\overline{a} = a$, $\overline{\tau \cap \sigma} = \overline{\tau} \cap \overline{\sigma}$, and $\overline{\tau \to \sigma} = \bigcap_{i \in I}(\tau \to \sigma_i)$ where $\overline{\sigma} = \bigcap_{i \in I} \sigma_i$. Note that premises in an organized type do not have to be organized, i.e., organized types are not necessarily *normalized* as defined in [9] (in contrast to organized types, the normalized form of a type may be exponentially large in the size of the type).

A *substitution* is a function $S : \mathbb{V} \to \mathbb{T}$, such that S is the identity everywhere but on a finite subset of $\mathbb{V}$. Whenever we consider a substitution S and a type variable α such that $S(\alpha)$ is not explicitly defined we assume $S(\alpha) = \alpha$. A substitution S is tacitly lifted to a function on types, $S : \mathbb{T} \to \mathbb{T}$, by homomorphic extension.

The following property, probably first stated in [1], is often called *beta-soundness*. Note that the converse is trivially true.

Lemma 1. *Let a and a_j, for $j \in J$, be atoms.*

1. *If $\bigcap_{i \in I}(\sigma_i \to \tau_i) \cap \bigcap_{j \in J} a_j \leq_{\mathbf{A}} a$ then $a = a_j$, for some $j \in J$.*
2. *If $\bigcap_{i \in I}(\sigma_i \to \tau_i) \cap \bigcap_{j \in J} a_j \leq_{\mathbf{A}} \sigma \to \tau$, where $\sigma \to \tau \neq \omega$, then the set $H = \{i \in I \mid \sigma \leq_{\mathbf{A}} \sigma_i\}$ is nonempty and $\bigcap\{\tau_i \mid i \in H\} \leq_{\mathbf{A}} \tau$.*

We will need three specializations of this lemma to organized types:

Lemma 2.

1. *Let $\tau = \bigcap_{i \in I} \tau_{i,1} \to \ldots \to \tau_{i,m_i} \to p_i$ be an organized type and let $\sigma = \sigma_1 \to \ldots \to \sigma_m \to p$ be a path. We have $\tau \leq_{\mathbf{A}} \sigma$ if and only if there is an $i \in I$ with $m_i = m$, $\sigma_j \leq_{\mathbf{A}} \tau_{i,j}$ for all $j \leq m$, and $p_i = p$.*
2. *Let $\tau = \bigcap_{i \in I} \tau_{i,1} \to \ldots \to \tau_{i,m_i} \to p_i$ be an organized type and let $\sigma = \sigma_1 \to \ldots \to \sigma_m \to \rho$ be a type with $\rho \neq \omega$. We have $\tau \leq_{\mathbf{A}} \sigma$ if and only if there is a nonempty subset $I' \subseteq I$ such that for all $i \in I'$ and all $1 \leq j \leq m$ we have $\overline{\sigma_j} \leq_{\mathbf{A}} \overline{\tau_{i,j}}$ and such that $\bigcap_{i \in I'} \tau_{i,m+1} \to \ldots \to \tau_{i,m_i} \to p_i \leq_{\mathbf{A}} \rho$.*
3. *Let $\tau = \tau_1 \to \ldots \to \tau_m \to \rho$ be a type and let $\sigma = \sigma_1 \to \ldots \to \sigma_n \to p$ be a path with $m \leq n$. We have $\tau \leq_{\mathbf{A}} \sigma$ if and only if for all $1 \leq j \leq m$ we have $\sigma_j \leq_{\mathbf{A}} \tau_j$ and $\rho \leq_{\mathbf{A}} \sigma_{m+1} \to \ldots \to \sigma_n \to p$.*

Using Lem. 1, the lemmas are proved by induction over m. We conclude by formally defining intersection type matching:

Definition 1. *Let $\tau, \sigma \in \mathbb{T}$ be types and let $\tau \leq \sigma$ be a formal type constraint.*

Let $C = \{\tau_1 \leq \sigma_1, \ldots, \tau_n \leq \sigma_n\}$ be a set of type constraints such that for every i it is the case that σ_i or τ_i does not contain any type variables. We say that C is matchable *if there is a substitution $S : \mathbb{V} \to \mathbb{T}$ such that for all i we have $S(\tau_i) \leq_{\mathbf{A}} S(\sigma_i)$. We say that S* matches *C.*

CMATCH *denotes the decision problem whether a given set of constraints C is matchable.* cMATCH *denotes the decision problem whether a given constraint $\tau \leq \sigma$ where σ does not contain any type variables is matchable.*

We sometimes denote CMATCH and cMATCH as matching problems. Note that in $S(\tau) \leq_{\mathbf{A}} S(\sigma)$ at least one of the two types does not contain variables, i.e., we have $S(\sigma) = \sigma$ or $S(\tau) = \tau$. If it is known that σ does not contain any variables we write $S(\tau) \leq_{\mathbf{A}} \sigma$ (and analogously for τ). Note that we use $\leq$ to denote a formal constraint whose matchability is supposed to be checked whereas $\tau \leq_{\mathbf{A}} \sigma$ states that τ is a subtype of σ.

4 NP-Hardness of Intersection Type Matching

We show that intersection type matching is NP-hard by defining a reduction $\mathcal{R}$ from 3SAT to CMATCH such that any formula F in 3CNF is satisfiable if and only if $\mathcal{R}(F)$ is matchable. Let $F = c_1 \wedge \ldots \wedge c_m$ where for each i we have $c_i = L_i^1 \vee L_i^2 \vee L_i^3$ and each L_i^j is either a propositional variable x or a negation $\neg x$ of such a variable. For all propositional variables x occurring in F we define two fresh type variables called α_x and $\alpha_{\neg x}$. Furthermore, we assume the two type constants 1 and 0. For a given formula F, let $\mathcal{R}(F)$ denote the set containing the following constraints:

1. For all x in F: $\big((1 \to 1) \to 1\big) \cap \big((0 \to 0) \to 0\big) \leq (\alpha_x \to \alpha_x) \to \alpha_x$
2. For all x in F: $(0 \to 1) \cap (1 \to 1) \leq \alpha_x \to 1$
3. For all x in F: $\big((1 \to 1) \to 1\big) \cap \big((0 \to 0) \to 0\big) \leq (\alpha_{\neg x} \to \alpha_{\neg x}) \to \alpha_{\neg x}$
4. For all x in F: $(0 \to 1) \cap (1 \to 1) \leq \alpha_{\neg x} \to 1$
5. For all x in F: $(1 \to 0) \cap (0 \to 1) \leq \alpha_x \to \alpha_{\neg x}$
6. For all c_i: $\alpha_{L_i^1} \cap \alpha_{L_i^2} \cap \alpha_{L_i^3} \leq 1$

It is clear that $\mathcal{R}(F)$ can be constructed in polynomial time.

Theorem 1. *A formula F in* 3CNF *is satisfiable if and only if $\mathcal{R}(F)$ is matchable.*

Proof. For the "only if"-direction let v be a valuation that satisfies F. We define a substitution S_v as follows:

– $S_v(\alpha_x) = v(x)$
– $S_v(\alpha_{\neg x}) = \neg v(x)$

By way of slight notational abuse the right hand sides of these defining equations represent the truth values $v(x)$ and $\neg v(x)$ as types. We claim that S_v matches $\mathcal{R}(F)$. For the first five constraints this is obvious. Consider a clause c_i in F and the corresponding constraint in the sixth group of constraints: because $v(F) = 1$ there is a literal L_i^j with $v(L_i^j) = 1$. Thus, $S_v(\alpha_{L_i^j}) = 1$ and the constraint corresponding to c_i is matched.

For the "if"-direction, from a substitution S matching $\mathcal{R}(F)$ we construct a satisfying valuation v_S for F. We define $v_S(x) = S(\alpha_x)$, and show that v_S is well-defined and satisfies F. Consider a type variable α_x. Using Lem. 1 it is not difficult to show that S can only match the first constraint if $S(\alpha_x) \in \{0, 1, \omega\}$. The second constraint, however, will not be matched if $S(\alpha_x) = \omega$. It is matched by the instantiations $S(\alpha_x) = 0$ and $S(\alpha_x) = 1$, though. Thus, the first two constraints make sure that $S(\alpha_x) \in \{0, 1\}$. The same argument, using the third and fourth constraint, shows $S(\alpha_{\neg x}) \in \{0, 1\}$. These two observations can be used together with the fact that the fifth constraint is matched for x to show that $S(\alpha_x) = 1$ if and only if $S(\alpha_{\neg x}) = 0$ and vice versa. We conclude that v_S is well-defined. In order to show that it satisfies F we need to show that for every clause c_i there is a literal L_i^j with $v_S(L_i^j) = 1$. Because S matches $\mathcal{R}(F)$ we have $S(\alpha_{L_i^1}) \cap S(\alpha_{L_i^2}) \cap S(\alpha_{L_i^3}) \leq_{\mathbf{A}} 1$. Lemma 1 states that the type on the left-hand side must have a component which is equal to 1. We already know that each of the three variables is instantiated either to 0 or 1. Thus, at least one of them must be instantiated to 1. Therefore, $v_S(c_i) = 1$ and v_S satisfies F.

We immediately get the following corollary:

Corollary 1. CMATCH *is* NP*-hard.*

Exploiting co- and contravariance, a set C of constraints can be transformed into a single constraint c such that C is matchable if and only if c is matchable. This yields a reduction from CMATCH to cMATCH, hence the following corollary:

Corollary 2. cMATCH *is* NP*-hard.*

Remark 1. Notice that the lower bound holds even when restricting the matching problem to atomic substitutions (mapping variables to atoms), since S_v as constructed above only uses such substitutions. Furthermore, note that the source of nondeterminism in the matching problem stems from the fact that in the first case of Lem. 1 one has to choose $j \in J$ whereas in the second case a nonempty subset of I has to be chosen. Both cases are reflected in the constraints defined for the reduction $\mathcal{R}$. A nondeterministic choice as in the first case of the lemma has to be made in the constraints of the sixth kind whereas the second case arises in the other constraints resulting in the choice whether a type variable should be substituted by 0 or 1.

5 NP-Membership of Intersection Type Matching

We show that CMATCH and cMATCH are in NP. Interestingly, the NP upper bound is quite challenging. In a first approach, one could try to polynomially bound the size of substitutions and then guess one nondeterministically. However, proving such a bound turns out to be complicated. Moreover, the nondeterminism exhibited by such an algorithm would completely ignore the sources of nondeterminism that were identified in Rem. 1 and would be pragmatically very suboptimal. Instead, we exploit the special structure of paths and organized types and attempt to minimize nondeterminism. Recall that every type can be organized in polynomial time. Thus, we may assume that all types occurring in a set of constraints are organized. Using the cases of Lem. 2, we successively decompose the set of constraints until we arrive at a set of constraints, that are basic in a certain sense. This process exhibits nondeterministic choices *only* in exactly the same manner as in the lemma with one exception: the case where $\tau \leq \sigma$ is matched by a substitution S with $S(\sigma) = \omega$ is not treated by the lemma and must be dealt with by a separate nondeterministic choice. For a set of basic constraints it is then possible to define a notion of consistency which is similar to the notion of ground consistency defined in [11]. Then, we can use intersections to construct a matching substitution for a consistent set of basic constraints. Since it can be shown that a set of basic constraints resulting from a successful run of the algorithm is matchable if and only if the original set of constraints was matchable, we arrive at an efficient nondeterministic polynomial-time algorithm deciding intersection type matching.

5.1 Algorithm

Our algorithm is shown in the figure Alg. 1 below. It will be seen that Alg. 1 heavily exploits the rather regular structure of organized types. We take a short digression to discuss an alternative strategy, that normalizes [9] types in a set of constraints to be matched and uses directed acyclic graphs (DAGs) to represent these types. Even though normalization may lead to an exponentially large syntax tree this is not per se a problem with regard to the polynomial bound, because a sharing representation of the syntax trees by DAGs only needs polynomial space. We do not, however, follow this approach for two reasons. First, it can be seen that such an approach would need exactly the same decomposition strategy for the sets of constraints that arise in Alg. 1. Furthermore, a DAG-approach requires a normalization of the types in advance. In general, this will cause applications of the distributivity law to types that are eliminated from the set of constraints according to some nondeterministic choices. On the other hand, our algorithm only organizes types in an argument position if necessary, i.e., if a constraint containing such a type is added to the set of constraints. In this case the distributivity law is *only* applied to the target position of paths that are components of the top-level intersection of the type in question. This reflects the structure of Lem. 1 which reduces the test whether a type is a subtype of an arrow-type to a subtype-test for certain types which are arguments, respectively targets, of top-level arrow-types. Thus, there is no reason to apply the distributivity law to types which are on some lower levels of the syntax tree.[2]

Definition 2. *We call* $\tau \leq \sigma$ *a* basic *constraint if either* τ *is a type variable and* σ *does not contain any type variables or* σ *is a type variable and* τ *does not contain any type variables.*

Definition 3. *Let* C *be a set of basic constraints.*

Let α *be a variable occurring in* C*. Let* $\tau_i \leq \alpha$ *for* $1 \leq i \leq n$ *be the constraints in* C *where* α *occurs on the right hand side of* $\leq$ *and let* $\alpha \leq \sigma_j$ *for* $1 \leq j \leq m$ *be the constraints in* C *where* α *occurs on the left hand side of* $\leq$*. We say that* C *is* consistent with respect to α *if for all* i *and* j *we have* $\tau_i \leq_{\mathbf{A}} \sigma_j$*.*

We say that C *is* consistent *if it is consistent with respect to all variables occurring in* C*.*

Lemma 3. *Let* C *be a set of basic constraints. The set* C *can be matched if and only if it is consistent.*

Proof. The implication from left to right is easy and left out. For the direction from right to left, let C be a set of basic constraints and let $\alpha \leq \sigma_j$ for $1 \leq j \leq m$ be the constraints in C where α occurs on the left hand side of $\leq$. We define the substitution S_C by $S_C(\alpha) = \bigcap_{j=1}^{m} \sigma_j$, for every variable α occurring in C. Recall that empty intersections equal ω (thus, S_C is well-defined even if $m = 0$). It can easily be verified that S_C matches C, if C is consistent.

The definition $S_C(\alpha) = \bigcap_{j=1}^{m} \sigma_j$ corresponds to the technique used by Tiuryn [11] for satisfiability in simple types over lattices of type constants, but generalized to arbitrary constant types σ_j (and not just type constants). It is important here that we can treat variables independently since the basic constraints in C do not contain any variables on one side of $\leq$ (hence the types $\bigcap_{j=1}^{m} \sigma_j$ contain no variables). The proof technique would not work for the satisfiability problem with intersection types, where the types on *both* sides of $\leq$ may contain variables.

Alg. 1 below is the nondeterministic procedure mentioned above that decomposes the constraints in the set C until we arrive at a set of basic constraints. According to the lemma above, it suffices to check whether this set of constraints is consistent. A few issues that are not made explicit in the algorithm should be addressed:

---

[2] Even with respect to a decision procedure for $\leq_{\mathbf{A}}$ the compact representation of normalized types by DAGs does not help: In [9] subtyping for normalized types $\tau = \bigcap_{i \in I} \tau_i$ and $\sigma = \bigcap_{j \in J} \sigma_j$ is characterized by $\tau \leq_{\mathbf{A}} \sigma \Leftrightarrow \forall j \exists i\ \tau_i \leq_{\mathbf{A}} \sigma_j$. Even if the normalized types were represented using DAGs the characterization requires for all j the identification of an index i with $\tau_i \leq_{\mathbf{A}} \sigma_j$. Checking this subtype relation, however, again requires a similar choice of certain components in the types. Repeating this argument, it can be seen that in order to check the characterization above one has to choose certain paths in the DAGs. The number of paths in a DAG however is still exponential if the original normalized type is of exponential size. Since a PTIME-procedure deciding $\leq_{\mathbf{A}}$ is known [5], clearly this approach using DAGs does not lead to an improvement.

1. Memoization is used to make sure that no constraint is added to C more than once.
2. Failing choices always return **false**.
3. The reduction with respect to ω in line 6 means, in particular, that, unless they are syntactically identical to ω, neither τ nor σ contain any types of the form $\rho_1 \to \ldots \to \rho_m \to \omega$ as top-level components.
4. Line 14 subsumes the case $\sigma = \omega$ if $I = \emptyset$. Then c is simply removed from C, and no new constraints are added.
5. We assume that the cases in the **switch**-block are mutually exclusive and checked in the given order. Thus, we know, for example, that for the two cases in lines 19 and 30 σ is *not* an intersection and therefore a path. Thus, we may fix the notation in line 17. Note, though, that the index set I for τ may be empty.
6. In line 21 the choice between the two cases is made nondeterministically. The first option covers the case where $I_2 \cup I_3 = \emptyset$, i.e., all paths in τ are shorter than σ.[3] The only possibility to match such a constraint is to make sure that $S(\sigma) = \omega$, which is only possible if $S(a) = \omega$. Thus, a must be a variable. Clearly, a cannot be a constant different from ω, and it cannot be ω either, because then σ would have been reduced to ω in line 6 and the case in line 8 would have been applicable.
7. The following example shows that even if $I_2 \cup I_3 \neq \emptyset$ it must be possible to choose the first option in line 21: the constraint set $\{a' \leq \beta, b \to a \leq \beta \to \alpha\}$ is matchable according to the substitution $\{\alpha \mapsto \omega, \beta \mapsto a'\}$. If the algorithm were not allowed to choose the option in line 22 it would have to choose the option in the following line which would result in the constraint set $\{a' \leq \beta, a \leq \alpha, \beta \leq b\}$. This set is clearly not matchable and the algorithm would incorrectly return **false**.
8. We assume that the nondeterministic choice in line 21 is made deterministically, choosing the first option whenever $I_2 \cup I_3 = \emptyset$. In this case choosing the second option would always result in **false**.

Nondeterminism results *only* from line 21 and lines 23, 27, and 31. We emphasize that the nondeterminstic choice in line 21 differs from the other cases in that it is the only occurrence of nondeterminism which is not structural in the sense that it can be traced to Lem. 2. Indeed, the first option of this choice covers the case that the type on the right hand side of $\leq_{\mathbf{A}}$ is of the form $\sigma_1 \to \ldots \to \sigma_m \to \omega$ which is excluded in the lemma. This situation can arise if a substitution maps a variable in the target of σ to ω. The other three occurrences of nondeterminism can directly be traced to Lem. 2: the choice of $I' \subseteq I$ in line 23 corresponds to Lem. 2.2 whereas the choice of *one* index in the other two lines can be traced back to Lem. 2.1.

The following example illustrates why it is indeed necessary to choose an index *set* in line 23 as opposed to a single index as in the other two cases: the constraint set $C = \{(a \to b) \cap (a \to p) \leq a \to \alpha, \alpha \leq b \cap p\}$ is matchable with the substitution $\{\alpha \mapsto b \cap p\}$. If, in line 23, the algorithm were only allowed to choose a *single* index from I' this would result in the addition of the new constraints $a \leq a$ and *either*

[3] In particular, $\tau = \omega$ is allowed if furthermore $I_1 = \emptyset$, as well.

Algorithm 1. Match(C)

1: *Input:* $C = \{\tau_1 \leq \sigma_1, \ldots, \tau_n \leq \sigma_n\}$ such that for all i at most one of σ_i and τ_i contains variables. Furthermore, all types have to be organized.
2: *Output:* **true** if C can be matched otherwise **false**
3:
4: **while** $\exists$ nonbasic constraint in C **do**
5:     choose a nonbasic constraint $c = (\tau \leq \sigma) \in C$
6:     reduce τ and σ with respect to ω
7:     **switch**
8:         **case:** c does not contain any variables
9:             **if** $\tau \leq_A \sigma$ **then**
10:                 $C := C \backslash \{c\}$
11:             **else**
12:                 **return false**
13:             **end if**
14:         **case:** $\sigma = \bigcap_{i \in I} \sigma_i$
15:             $C := C \backslash \{c\} \cup \{\tau \leq \sigma_i | i \in I\}$
16:
17:         write $\tau = \bigcap_{i \in I} \tau_{i,1} \to \ldots \to \tau_{i,m_i} \to p_i$, $\sigma = \sigma_1 \to \ldots \to \sigma_m \to a$
18:         write $I_1 = \{i \in I | m_i < m\}$, $I_2 = \{i \in I | m_i = m\}$, $I_3 = \{i \in I | m_i > m\}$
19:         **case:** σ contains variables, τ does not contain any variables
20:             **if** $a \in \mathbb{V}$ **then**
21:                 **choose** case 1 or 2:
22:                     1. $C := C \backslash \{c\} \cup \{\omega \leq a\}$
23:                     2. **choose** $\emptyset \neq I' \subseteq I_2 \cup I_3$
24:                         $C := C \backslash \{c\} \cup \{\overline{\sigma_j} \leq \overline{\tau_{i,j}} | i \in I', 1 \leq j \leq m\} \cup$
25:                         $\{\bigcap_{i \in I'} \tau_{i,m+1} \to \ldots \to \tau_{i,m_i} \to p_i \leq a\}$
26:             **else**
27:                 **choose** $i_0 \in I_2$
28:                 $C := C \backslash \{c\} \cup \{\overline{\sigma_j} \leq \overline{\tau_{i_0,j}} | 1 \leq j \leq m\} \cup \{p_{i_0} \leq a\}$
29:             **endif**
30:         **case:** τ contains variables, σ does not contain any variables
31:             **choose** $i_0 \in I_1 \cup I_2$
32:             $C := C \backslash \{c\} \cup \{\overline{\sigma_j} \leq \overline{\tau_{i_0,j}} | 1 \leq j \leq m_{i_0}\} \cup \{p_{i_0} \leq \sigma_{m_{i_0}+1} \to \ldots \to \sigma_m \to a\}$
33:     **end switch**
34: **end while**
35: **if** C is consistent **then**
36:     **return true**
37: **else**
38:     **return false**
39: **end if**

$b \leq \alpha$ *or* $p \leq \alpha$. Thus, the resulting set of constraints would be $\{a \leq a, b \leq \alpha, \alpha \leq b \cap p\}$, for example. This set of constraints is not matchable, and it is easy to see that the algorithm would incorrectly return **false** even though the initial set of constraints was matchable. The reason is that, if σ is a path whose target is a variable, a substitution may cause σ to not be a path any more.

As mentioned above, nondeterminism is localized in the algorithm, leading to efficiency. Thus, if for some input C none of the cases above ever arises or, in fact, the choices are deterministic because $I_2 \cup I_3 = \emptyset$, respectively, the index sets to choose from in the other cases are always singleton sets, then the algorithm becomes a PTIME-procedure.

In general, a most general matching substitution is not unique: the constraint set $C = \{\alpha \cap \beta \leq a\}$ is matched by the two substitutions $S_1 = \{\alpha \mapsto a, \beta \mapsto \beta\}$ and $S_2 = \{\alpha \mapsto \alpha, \beta \mapsto a\}$, for example. It is clear that neither of the two substitutions can be obtained from instantiating the other. Furthermore, there is no substitution more general than S_1 and S_2 that still matches C (this substitution would be the identity which clearly does not match C). Note that S_1 and S_2 can directly be traced to the nondeterminstic choice of either α or β in line 31. Thus, only $\alpha \leq a$ or $\beta \leq a$ replaces the original constraint in C. The least upper bound construction from the proof of Lem. 3 then results in S_1 respectively S_2.

Lemma 4. *Algorithm 1 operates in nondeterministic polynomial time.*

Proof. (Sketch) It is clear that the algorithm terminates because in every iteration of the **while**-loop the height of the types involved in a constraint is reduced. Thus, if the algorithm does not return **false**, C only consists of basic constraints after a finite number of iterations.

The algorithm terminates in nondeterministic polynomial time, because (as can be shown by a detailed case analysis) every type occurring in a nonbasic constraint considered during one execution of the **while**-loop is a subterm of a type occurring in an initial constraint (we call such types *initial* types) or a subterm of an organized argument of an initial type. The number of subterms of an initial type is linear. The number of arguments of an initial type is also linear. Organizing each of these linearly many arguments causes only a polynomial blowup. Therefore, it is clear that the number of subterms of an organized argument of an initial type is polynomial. Let k denote the number of subterms of the initial types plus the number of subterms of organized arguments of the initial types. The total number of nonbasic constraints that the algorithm considers is therefore bounded by k^2 which, in turn, is polynomial in the size of C. Furthermore, it can also be seen that every constraint that is considered is of polynomial size. The consistency check and the check in line 9 require checking the truth of constraints $\tau_i \leq \sigma_j$ not containing any type variables, which means deciding the relation $\leq_{\mathbf{A}}$. As is shown in [5] this can be done in PTIME.

The statement of the previous lemma might come as a surprise since the execution of the **while**-loop requires a repeated organization of the arguments of the occurring types. It can be asked why this repeated organization does not result in a normalization [9] of the types involved. As mentioned before, this could cause an exponential blowup in the size of the type. The reason why this problem does not occur is the fact that this organization is interleaved with decomposition steps. We illustrate this by the following small example. We inductively define two families of types:

$$\tau_0 = a_0 \cap b_0 \qquad\qquad \sigma_0 = \alpha_0$$
$$\tau_l = \tau_{l-1} \to (a_l \cap b_l) \qquad\qquad \sigma_l = \sigma_{l-1} \to \alpha_l$$

The size of τ_n in normalized form is exponential in n. However, if the algorithm processes the constraint $\tau_n \leq \sigma_n$ only a polynomial number of new constraints (of polynomial size) are constructed: First, the types have to be organized. We obtain $(\tau_{n-1} \to a_n) \cap (\tau_{n-1} \to b_n) \leq \sigma_n$. In the first iteration of the **while**-loop the case in line 20 applies and the nondeterministic choice in line 21 may be resolved in such a way that a subset of components of the toplevel intersection of $(\tau_{n-1} \to a_n) \cap (\tau_{n-1} \to b_n)$ has to be chosen. In order to maximize the size of C we choose both components which forces the construction of the constraints $a_n \cap b_n \leq \alpha_n$, $\overline{\sigma_{n-1}} \leq \overline{\tau_{n-1}}$, and $\overline{\sigma_{n-1}} \leq \overline{\tau_{n-1}}$. The last two constraints are the same, however, and therefore the memoization of the algorithm makes sure that this constraint is only treated once. In the next step the case in line 14 applies (note that $\overline{\tau_{n-1}}$ is a top-level intersection). This leads to the construction of the constraints $\overline{\sigma_{n-1}} \leq \tau_{n-2} \to a_{n-1}$ and $\overline{\sigma_{n-1}} \leq \tau_{n-2} \to b_{n-1}$. For both constraints the same rule applies and causes a change of C according to line 32. This leads to the construction of the basic constraints $\alpha_{n-1} \leq a_{n-1}$ and $\alpha_{n-1} \leq b_{n-1}$ as well as to the construction of $\overline{\sigma_{n-2}} \leq \overline{\tau_{n-2}}$ and $\overline{\sigma_{n-2}} \leq \overline{\tau_{n-2}}$. Then, the same argument as above can be repeated.

We conclude that the doubling of the arguments of the τ_l that occurs in the normalization (and which eventually causes the exponential blowup if repeated) does not occur in the algorithm, because the types involved are decomposed such that the arguments and targets are treated separately. This implies that the arguments cannot be distinguished any more such that the new constraints coincide and are only added once.

The proof of Lem. 4 relies on the fact that every new nonbasic constraint added to C only contains types that are essentially subterms of an initial type. A new intersection which possibly does *not* occur as a subterm in any of the initial types has to be constructed in line 25, though. Since, in principle, this new intersection represents a subset of an index set, it is not clear that there cannot be an exponential number of such basic constraints. However, this construction of new intersections only happens as a consequence to the nonbasic constraint that is treated there. As noted above there can be at most a polynomial number of nonbasic constraints and therefore new intersections can also only be introduced a polynomial number of times.

5.2 Correctness

We consider correctness of the algorithm, i.e., we have to show that it can return **true** if and only if the original set of constraints can be matched. For soundness we first state the following lemma:

Lemma 5. *Let C be a set of constraints and let C' be a set of constraints that results from C by application of one of the cases of Alg. 1.*

Every substitution that matches C' also matches C.

The proof consists of a case analysis. To give the reader an idea we give a proof sketch which discusses one of the cases in more detail.

Proof. For all cases C' results from C by removing the constraint c and possibly by further adding some new constraints. Assuming we have a substitution S that matches C', it suffices to show that S satisfies c in order to show that it also satisfies C.

The argument is not difficult for the cases occurring because of lines 10, 15, and 22. The interesting cases occur because of lines 24, 28, and 32. We will discuss the case in line 24 in detail. The other cases are argued similarly. We then have $\sigma = \sigma_1 \to \ldots \to \sigma_m \to a$, with $a = \alpha$ a type variable, and $\tau = \bigcap_{i \in I} \tau_i$ with $\tau_i = \tau_{i,1} \to \ldots \to \tau_{i,m_i} \to p_i$. In this case C' is constructed by adding $\overline{\sigma_j} \leq \overline{\tau_{i,j}}$ for all $i \in I'$ and all $1 \leq j \leq m$ and $\bigcap_{i \in I'} \tau_{i,m+1} \to \ldots \to \tau_{i,m_i} \to p_i \leq \alpha$. Since S matches C' we know $S(\overline{\sigma_j}) \leq_{\mathbf{A}} \overline{\tau_{i,j}}$ and $\bigcap_{i \in I'} \tau_{i,m+1} \to \ldots \to \tau_{i,m_i} \to p_i \leq_{\mathbf{A}} S(\alpha)$. We want to show $\tau \leq_{\mathbf{A}} S(\sigma)$. We write $S(\sigma) = \sigma'_1 \to \ldots \to \sigma'_m \to \rho$ where $S(\sigma_j) = \sigma'_j$ and $S(\alpha) = \rho$. Using $S(\overline{\sigma_j}) \leq_{\mathbf{A}} \overline{\tau_{i,j}}$, it can be shown[4] that $\overline{\sigma'_j} \leq_{\mathbf{A}} \overline{\tau_{i,j}}$. We may apply the "if"-part of Lem. 2.2 to conclude $\tau \leq_{\mathbf{A}} S(\sigma)$.

Corollary 3. *Algorithm 1 is sound.*

Proof. Assume that the algorithm returns **true**. This is only possible in line 36 if the algorithm leaves the **while**-loop with a consistent set C of basic constraints. By the "if"-direction of Lem. 3, C is matchable. Using Lem. 5, an inductive argument shows that *all* sets of constraints considered in the algorithm during execution of the **while**-loop are matchable. This is in particular true for the initial set of constraints.

For completeness we need the following lemma:

Lemma 6. *Let C be matchable and $c \in C$.*

There exists a set of constraints C' such that C' results from C by application of one of the cases of Alg. 1 to c and C' is matchable.

Again we give a sketch of the case analysis necessary for the proof:

Proof. We show that no matter which case applies to c, a choice can be made that results in a matchable set C'. In particular, it must be argued that there is a choice that does not result in **false**.

As above, note that for all cases C' results from C by removing c and by possibly adding some new constraints. Let S be a substitution that matches C. In order to show that a choice is possible such that S also matches C' it suffices to show that S matches the newly introduced constraints. The argument is not

[4] At this point a technical lemma showing that for a type σ we have $S(\overline{\sigma}) = S(\sigma)$ has to be proven.

difficult for the cases in lines 8 and 14 (and besides there is no nondeterministic choice involved, here). For the cases in lines 19 and 30 subcase-analyses are necessary. The subcases different from the one in line 22 (explained below) are similar to the following:

We consider the case in line 30. We have $\tau = \bigcap_{i\in I} \tau_i$ where $\tau_i = \tau_{i,1} \to \ldots \to \tau_{i,m_i} \to p_i$ and $\sigma = \sigma_1 \to \ldots \to \sigma_m \to a$, and we know $S(\tau) \leq_{\mathbf{A}} \sigma$. We organize $S(\tau)$ and we write $\overline{S(\tau)} = \bigcap_{h\in H} \rho_{h,1} \to \ldots \to \rho_{h,n_h} \to b_h$. Using Lem. 2.1, we conclude that there exists $h_0 \in H$ with $b_{h_0} = a$, $n_{h_0} = m$, and $\sigma_l \leq_{\mathbf{A}} \rho_{h_0,l}$ for all $1 \leq l \leq m$. Note that with $\pi_{h_0} = \rho_{h_0,1} \to \cdots \to \rho_{h_0,m} \to a$ this implies $\pi_{h_0} \leq_{\mathbf{A}} \sigma$. For π_{h_0} there must be an index $j_0 \in I$ such that π_{h_0} occurs as a component in $\overline{S(\tau_{j_0})}$. It is clear that $m_{j_0} \leq m$. Otherwise all paths in this type would be of length greater than m (neither a substitution nor an organization may reduce the length of a path). Thus, j_0 as above is contained in $I_1 \cup I_2$ (cf. line 18), and we may choose $i_0 = j_0$ in line 31.

We have to show that S matches $\overline{\sigma_l} \leq \overline{\tau_{j_0,l}}$ for all $1 \leq l \leq m_{j_0}$ and $p_{j_0} \leq \sigma_{m_{j_0}+1} \to \ldots \to \sigma_m \to a$. Because π_{h_0} is a component in $\overline{S(\tau_{j_0})}$, it is clear that $\pi_{h_0} \leq_{\mathbf{A}} \sigma$ implies $\overline{S(\tau_{j_0})} \leq_{\mathbf{A}} \sigma$ and, thus, also $S(\tau_{j_0}) \leq_{\mathbf{A}} \sigma$. Applying the "only if"-part of Lem. 2.3 to $S(\tau_{j_0,1}) \to \ldots \to S(\tau_{j_0,m_{j_0}}) \to S(p_{j_0}) = S(\tau_{j_0}) \leq_{\mathbf{A}} \sigma = \sigma_1 \to \ldots \to \sigma_m \to a$, we obtain $\sigma_l \leq_{\mathbf{A}} S(\tau_{j_0,l})$ for all $1 \leq l \leq m_{j_0}$ and $S(p_{j_0}) \leq_{\mathbf{A}} \sigma_{m_{j_0}+1} \to \ldots \to \sigma_m \to a$. It can be shown[5] that from $\sigma_l \leq_{\mathbf{A}} S(\tau_{j_0,l})$ we get $\overline{\sigma_l} \leq_{\mathbf{A}} S(\overline{\tau_{j_0,l}})$. Altogether this shows that S matches the newly introduced constraints.

Concerning line 22, if $\tau \leq_{\mathbf{A}} S(\sigma)$ because a is a variable with $S(a) = \omega$ the nondeterminstic choice (line 21) is resolved such that the first option is chosen. Then, it is clear that the newly added constraint $\omega \leq a$ is matched by S.

Corollary 4. *Algorithm 1 is complete.*

Proof. We assume that the initial set C of constraints is matchable. We have to show that there is an execution sequence of the algorithm that results in **true**. Using Lem. 6 in an inductive argument it can be shown that for every iteration of the **while**-loop it is possible to make the nondeterministic choice in such a way that the iteration results in a matchable set of constraints. Thus, there is an execution sequence of the **while**-loop that results in a matchable set of basic constraints. Lemma 3 shows that this set is consistent and therefore the algorithm returns **true**.

Corollaries 2, 3, and 4 and Lem. 4 immediately prove the following theorem:

Theorem 2. cMATCH *and* CMATCH *are* NP*-complete.*

6 Conclusion and Future Work

We have proven the intersection type matching problem to be NP-complete and have provided an algorithm which is engineered for efficiency by reducing

[5] Again using the lemma, showing that $S(\overline{\sigma}) = S(\sigma)$.

nondeterminism as much as possible. Future work includes further experiments to study optimizations of the inhabitation algorithm in [6] based on matching, and studying the satisfiability and unification problems with intersection types.

References

1. Barendregt, H., Coppo, M., Dezani-Ciancaglini, M.: A Filter Lambda Model and the Completeness of Type Assignment. Journal of Symbolic Logic 48(4), 931–940 (1983)
2. Jha, S., Palsberg, J., Zhao, T.: Efficient Type Matching. In: Nielsen, M., Engberg, U. (eds.) FOSSACS 2002. LNCS, vol. 2303, pp. 187–204. Springer, Heidelberg (2002)
3. Zaremski, A., Wing, J.: Signature Matching: A Tool for Using Software Libraries. ACM Trans. Softw. Eng. Methodol. 4(2), 146–170 (1995)
4. Baader, F., Snyder, W.: Unification Theory. In: Handbook of Automated Reasoning, ch. 8, pp. 439–526. Elsevier (2001)
5. Rehof, J., Urzyczyn, P.: Finite Combinatory Logic with Intersection Types. In: Ong, L. (ed.) TLCA 2011. LNCS, vol. 6690, pp. 169–183. Springer, Heidelberg (2011)
6. Düdder, B., Martens, M., Rehof, J., Urzyczyn, P.: Bounded Combinatory Logic. In: Proceedings of CSL 2012. LIPIcs, vol. 16, pp. 243–258. Schloss Dagstuhl (2012)
7. Rehof, J.: Towards Combinatory Logic Synthesis. In: 1st International Workshop on Behavioural Types, BEAT 2013, January 22 (2013), `http://beat13.cs.aau.dk/pdf/BEAT13-proceedings.pdf`
8. Düdder, B., Martens, M., Rehof, J.: Intersection Type Matching and Bounded Combinatory Logic (Extended Version). Technical Report 841, Faculty of Computer Science, TU Dortmund (2012), `http://ls14-www.cs.tu-dortmund.de/index.php/Jakob_Rehof_Publications#Technical_Reports`
9. Hindley, J.R.: The Simple Semantics for Coppo-Dezani-Sallé Types. In: Dezani-Ciancaglini, M., Montanari, U. (eds.) Programming 1982. LNCS, vol. 137, pp. 212–226. Springer, Heidelberg (1982)
10. Rehof, J.: The Complexity of Simple Subtyping Systems. PhD thesis, DIKU, Department of Computer Science (1998)
11. Tiuryn, J.: Subtype Inequalities. In: Proceedings of LICS 1992, pp. 308–315. IEEE Computer Society (1992)
12. Frey, A.: Satisfying Subtype Inequalities in Polynomial Space. Theor. Comput. Sci. 277(1-2), 105–117 (2002)
13. Benanav, D., Kapur, D., Narendran, P.: Complexity of Matching Problems. J. Symb. Comput. 3(1/2), 203–216 (1987)
14. Kapur, D., Narendran, P.: NP-Completeness of the Set Unification and Matching Problems. In: Siekmann, J.H. (ed.) CADE 1986. LNCS, vol. 230, pp. 489–495. Springer, Heidelberg (1986)
15. Kapur, D., Narendran, P.: Complexity of Unification Problems with Associative-Commutative Operators. J. Autom. Reasoning 9(2), 261–288 (1992)
16. Ronchi Della Rocca, S.: Principal Type Scheme and Unification for Intersection Type Discipline. Theor. Comput. Sci. 59, 181–209 (1988)
17. Kfoury, A.J., Wells, J.B.: Principality and Type Inference for Intersection Types Using Expansion Variables. Theor. Comput. Sci. 311(1-3), 1–70 (2004)

A Type-Checking Algorithm for Martin-Löf Type Theory with Subtyping Based on Normalisation by Evaluation

Daniel Fridlender* and Miguel Pagano**

Universidad Nacional de Córdoba
{fridlend,pagano}@famaf.unc.edu.ar

Abstract. We present a core Martin-Löf type theory with subtyping; it has a cumulative hierarchy of universes and the contravariant rule for subtyping between dependent product types. We extend to this calculus the normalisation by evaluation technique defined for a variant of MLTT without subtyping. This normalisation function makes the subtyping relation and type-checking decidable. To our knowledge, this is the first time that the normalisation by evaluation technique has been considered in the context of subtypes, which introduce some subtletics in the proof of correctness of NbE; an important result to prove correctness and completeness of type-checking.

1 Introduction

The usefulness of proof assistants based on type theory depends on how much effort is required to transform a traditional proof in one accepted by the formal system. For example, as mentioned in [6], when one has a proof of some property for monoids, one immediately realises that the property is also valid for groups. It is desirable that our proof assistants permit us to use the same reasoning in their formalisms. In this paper we extend previous works [2] on NbE to MLTT extended with subtyping, which is one way to allow such forms of reasoning in type-theory.

The calculus we consider is a core fragment of MLTT with a cumulative hierarchy of universes and the usual rule for subtyping between dependent product types. Like in previous works, the formal system features explicit substitution and de Bruijn indices; there are however some other differences besides subtyping: it lacks (η) rule, which makes some semantical constructions simpler, and there is another form of judgement for subtyping between valid contexts.

The semantic domain and the normalisation function are essentially like the ones in [2]; as in that work, the main results follow from completeness and correctness of NbE. Correctness is proved using logical relations. To deal with the subtleties of the subsumption rule it is useful to model subtyping by an order

* Partially supported by MinCyT-Córdoba, and SeCyT-UNC.
** Partially supported by CONICET, MinCyT-Córdoba, and SeCyT-UNC.

M. Hasegawa (Ed.): TLCA 2013, LNCS 7941, pp. 140–155, 2013.

between semantical types, which is finer than set-theoretical inclusion. This leads to a proof that logical relations are preserved by subtyping, while keeping the definition of the logical relations unchanged.

Finally we define a decision procedure for subtyping among well-formed types in normal form; thanks to the NbE algorithm it gives rise to a bidirectional type-checking algorithm. Correctness and completeness of algorithmic subtyping and type-checking follow from injectivity of Fun and the analogous property for subtyping: we say that Fun reflects subtyping, if the converse of the contravariant rule for subtyping between dependent product types holds. In the light of [4] and [15] injectivity of Fun is an interesting property in itself.

2 The Calculus

The calculus is presented in Figs. 1, 2 and 3. Contexts are finite sequences of dependent types. There is a cumulative hierarchy of universes U_l for $l \in \mathbb{N}$. Types are built up from universes and their objects, and are closed by the dependent product type Fun, and by substitution. Terms include function application, de Bruijn-like notation for abstraction and variable (q), and substitution. In the case of type U_l terms include also smaller universes and dependent products.

Substitutions are built up from the empty, identity and weakening (p) substitutions by the operation of extension, and closed under composition. The substitution $[r]$ stands for (id, r), so $\Gamma \vdash [r] \colon \Gamma.A$ if $\Gamma \vdash r \colon A$.

Pre-Terms and pre-substitutions are defined by the following grammar:

$$\mathit{Terms} \ni t, r, A, B ::= \mathsf{q} \mid \mathsf{U}_l \mid \mathsf{App}\ t\ r \mid \lambda t \mid \mathsf{Fun}\, A\, B \mid t\,\sigma$$
$$\mathit{Subs} \ni \sigma, \delta ::= \mathsf{p} \mid \mathsf{id} \mid \langle\rangle \mid (\delta, t) \mid \delta\,\sigma$$

The subtyping relation is generated by the cumulative hierarchy of universes and the contravariant rule for subtyping between dependent product types (FUN-STY). This relation, in turn, induces the definition of the subcontext relation.

Notice the presence of the β-rule (BETA-ET) and the ξ-rule (XI-ET), and the absence of (η). Equality rules (FUN-SUBS-ETY), (FUN-SUBS-ET) and (ABS-SUBS-ET) involve the substitution $(\delta\,\mathsf{p}, \mathsf{q})$: if $\Gamma \vdash \delta \colon \Delta$ and $\Delta \vdash A$, then $\Gamma.A\,\delta \vdash (\delta\,\mathsf{p}, \mathsf{q}) \colon \Delta.A$.

We will make free use of symmetry and transitivity, since they can be derived from rules (REFL-ET) and (SYM-TRAN-ET), or the analogous rules for equal substitutions or equal types, in Fig. 2. If $\Gamma \vdash \delta \colon \Delta$ then $\Gamma \vdash \mathsf{U}_l\,\delta = \mathsf{U}_l$ is derivable using rules (U-SUBS-ET) and (CONG-EL-ETY); we call (U-SUBS-ETY) this derived rule.

Notation. We denote with $|\_|$ the length of contexts, and with p^i, the i-fold self-composition of p. We define $\mathsf{p}^0 = \mathsf{id}$ but $t\,\mathsf{p}^0 = t$ rather than $t\,\mathsf{p}^0 = t\,\mathsf{id}$. We say that $\Delta \leqslant^i \Gamma$ if $\Delta \vdash \mathsf{p}^i \colon \Gamma$. Neutral terms and normal forms are characterised by the following grammar.

Definition 1 (Neutral terms and normal forms).

$$\mathsf{Ne} \ni k ::= \mathsf{q}\,\mathsf{p}^i \mid \mathsf{App}\ k\ v \qquad \mathsf{Nf} \ni v, V, W ::= \mathsf{U}_l \mid \mathsf{Fun}\, V\, W \mid \lambda v \mid k$$

Contexts.

$$\frac{}{\diamond \vdash}\ (\text{EMPTY-CTX}) \qquad \frac{\Gamma \vdash \quad \Gamma \vdash A}{\Gamma.A \vdash}\ (\text{EXT-CTX})$$

Substitutions.

$$\frac{\Gamma \vdash}{\Gamma \vdash \mathsf{id}\colon \Gamma}\ (\text{ID-SUBS}) \qquad \frac{\Sigma \vdash \delta\colon \Delta \quad \Gamma \vdash \sigma\colon \Sigma}{\Gamma \vdash \delta\sigma\colon \Delta}\ (\text{COMP-SUBS})$$

$$\frac{\Gamma \vdash A}{\Gamma.A \vdash \mathsf{p}\colon \Gamma}\ (\text{FST-SUBS}) \quad \frac{\Gamma \vdash \delta\colon \Delta \quad \Delta \vdash A \quad \Gamma \vdash t\colon A\delta}{\Gamma \vdash (\delta, t)\colon \Delta.A}\ (\text{EXT-SUBS})$$

$$\frac{\Gamma \vdash}{\Gamma \vdash \langle\rangle\colon \diamond}\ (\text{EMP-SUBS}) \qquad \frac{\Gamma \vdash \delta\colon \Delta \quad \Delta \leq \Theta}{\Gamma \vdash \delta\colon \Theta}\ (\text{SUBS})$$

Types.

$$\frac{\Gamma \vdash}{\Gamma \vdash \mathsf{U}_l}\ (\text{U-F}) \qquad \frac{\Gamma \vdash A\colon \mathsf{U}_l}{\Gamma \vdash A}\ (\text{U-EL})$$

$$\frac{\Gamma \vdash A \quad \Gamma.A \vdash B}{\Gamma \vdash \mathsf{Fun}\, A\, B}\ (\text{FUN-F}) \quad \frac{\Delta \vdash A \quad \Gamma \vdash \delta\colon \Delta}{\Gamma \vdash A\delta}\ (\text{SUBS-TYPE})$$

Terms.

$$\frac{\Gamma \vdash}{\Gamma \vdash \mathsf{U}_l\colon \mathsf{U}_{l+1}}\ (\text{U-U-F}) \quad \frac{\Gamma \vdash A\colon \mathsf{U}_l \quad \Gamma.A \vdash B\colon \mathsf{U}_l}{\Gamma \vdash \mathsf{Fun}\, A\, B\colon \mathsf{U}_l}\ (\text{FUN-U-F})$$

$$\frac{\Gamma \vdash A}{\Gamma.A \vdash \mathsf{q}\colon A\,\mathsf{p}}\ (\text{HYP}) \quad \frac{\Gamma \vdash A \quad \Gamma.A \vdash B \quad \Gamma.A \vdash t\colon B}{\Gamma \vdash \lambda t\colon \mathsf{Fun}\, A\, B}\ (\text{FUN-I})$$

$$\frac{\Gamma \vdash A \quad \Gamma.A \vdash B \quad \Gamma \vdash t\colon \mathsf{Fun}\, A\, B \quad \Gamma \vdash r\colon A}{\Gamma \vdash \mathsf{App}\ t\ r\colon B\,[r]}\ (\text{FUN-EL})$$

$$\frac{\Delta \vdash A \quad \Delta \vdash t\colon A \quad \Gamma \vdash \delta\colon \Delta}{\Gamma \vdash t\delta\colon A\delta}\ (\text{SUBS-TERM})$$

$$\frac{\Gamma \vdash t\colon A \quad \Gamma \vdash A \leq B}{\Gamma \vdash t\colon B}\ (\text{SUB})$$

Subtypes.

$$\frac{\Gamma \vdash A = B}{\Gamma \vdash A \leq B}\ (\text{REFL-STY}) \qquad \frac{\Gamma \vdash}{\Gamma \vdash \mathsf{U}_l \leq \mathsf{U}_{l+1}}\ (\text{U-STY})$$

$$\frac{\Gamma \vdash A' \leq A \quad \Gamma.A \vdash B \quad \Gamma.A' \vdash B \leq B'}{\Gamma \vdash \mathsf{Fun}\, A\, B \leq \mathsf{Fun}\, A'\, B'}\ (\text{FUN-STY})$$

$$\frac{\Gamma \vdash A \leq A' \quad \Gamma \vdash A' \leq A''}{\Gamma \vdash A \leq A''}\ (\text{TRAN-STY})$$

Subcontexts.

$$\frac{}{\diamond \leq \diamond}\ (\text{EMPTY-SCTX}) \quad \frac{\Gamma \leq \Delta \quad \Delta \vdash B \quad \Gamma \vdash A \leq B}{\Gamma.A \leq \Delta.B}\ (\text{EXT-SCTX})$$

Fig. 1. Typing Rules

Equal Substitutions.

$$\frac{\Gamma \vdash \delta\colon \Delta}{\Gamma \vdash \delta = \delta\colon \Delta}\ (\text{REFL-ES}) \qquad \frac{\Gamma \vdash \delta = \delta'\colon \Delta \qquad \Gamma \vdash \delta = \delta''\colon \Delta}{\Gamma \vdash \delta' = \delta''\colon \Delta}\ (\text{SYM-TRAN-ES})$$

$$\frac{\Gamma \vdash \delta\colon \Delta}{\Gamma \vdash \delta\,\mathsf{id} = \delta\colon \Delta}\ (\text{NEUT-R-ES}) \qquad \frac{\Gamma \vdash A}{\Gamma.A \vdash \mathsf{id} = (\mathsf{p},\mathsf{q})\colon \Gamma.A}\ (\text{PQ-ES})$$

$$\frac{\Gamma \vdash \delta\colon \Delta}{\Gamma \vdash \mathsf{id}\,\delta = \delta\colon \Delta}\ (\text{NEUT-L-ES}) \qquad \frac{\Theta \vdash \delta\colon \Delta \qquad \Sigma \vdash \theta\colon \Theta \qquad \Gamma \vdash \sigma\colon \Sigma}{\Gamma \vdash (\delta\,\theta)\,\sigma = \delta\,(\theta\,\sigma)\colon \Delta}\ (\text{ASSOC-ES})$$

$$\frac{}{\diamond \vdash \mathsf{id} = \langle\rangle\colon \diamond}\ (\text{ID-EMP-ES}) \qquad \frac{\Gamma \vdash \delta\colon \Delta \qquad \Delta \vdash A \qquad \Gamma \vdash t\colon A\delta}{\Gamma \vdash \mathsf{p}\,(\delta, t) = \delta\colon \Delta}\ (\text{FST-ES})$$

$$\frac{\Gamma \vdash \delta\colon \Delta}{\Gamma \vdash \langle\rangle\,\delta = \langle\rangle\colon \diamond}\ (\text{ABSORB-ES}) \qquad \frac{\Sigma \vdash \delta\colon \Delta \quad \Delta \vdash A \quad \Sigma \vdash t\colon A\delta \quad \Gamma \vdash \sigma\colon \Sigma}{\Gamma \vdash (\delta, t)\,\sigma = (\delta\,\sigma, t\,\sigma)\colon \Delta.A}\ (\text{EXT-ES})$$

$$\frac{\Gamma \vdash \sigma = \sigma'\colon \Sigma \qquad \Sigma \vdash \delta = \delta'\colon \Delta}{\Gamma \vdash \delta\,\sigma = \delta'\,\sigma'\colon \Delta}\ (\text{CONG-COMP-ES})$$

$$\frac{\Gamma \vdash \delta = \delta'\colon \Delta \qquad \Delta \vdash A \qquad \Gamma \vdash t = t'\colon A\delta}{\Gamma \vdash (\delta, t) = (\delta', t')\colon \Delta.A}\ (\text{CONG-EXT-ES})$$

$$\frac{\Gamma \vdash \delta = \delta'\colon \Delta \qquad \Delta \leq \Theta}{\Gamma \vdash \delta = \delta'\colon \Theta}\ (\text{SUBS-ES})$$

Equal Types.

$$\frac{\Gamma \vdash A}{\Gamma \vdash A = A}\ (\text{REFL-ETY}) \qquad \frac{\Gamma \vdash A = A' \qquad \Gamma \vdash A = A''}{\Gamma \vdash A' = A''}\ (\text{SYM-TRAN-ETY})$$

$$\frac{\Gamma \vdash A}{\Gamma \vdash A\,\mathsf{id} = A}\ (\text{NEUT-ETY}) \qquad \frac{\Delta \vdash A \qquad \Sigma \vdash \delta\colon \Delta \qquad \Gamma \vdash \sigma\colon \Sigma}{\Gamma \vdash (A\,\delta)\,\sigma = A\,(\delta\,\sigma)}\ (\text{ASSOC-ETY})$$

$$\frac{\Gamma \vdash A = A' \qquad \Gamma.A' \vdash B' \qquad \Gamma.A \vdash B = B'}{\Gamma \vdash \mathsf{Fun}\,A\,B = \mathsf{Fun}\,A'\,B'}\ (\text{CONG-FUN-ETY})$$

$$\frac{\Gamma \vdash A = B\colon \mathsf{U}_l}{\Gamma \vdash A = B}\ (\text{CONG-EL-ETY}) \qquad \frac{\Delta \vdash A = A' \qquad \Gamma \vdash \delta = \delta'\colon \Delta}{\Gamma \vdash A\,\delta = A'\,\delta'}\ (\text{CONG-SUBS-ETY})$$

$$\frac{\Delta \vdash A \qquad \Delta.A \vdash B \qquad \Gamma \vdash \delta\colon \Delta}{\Gamma \vdash (\mathsf{Fun}\,A\,B)\,\delta = \mathsf{Fun}\,(A\,\delta)\,(B\,(\delta\,\mathsf{p},\mathsf{q}))}\ (\text{FUN-SUBS-ETY})$$

Fig. 2. Equality Rules for Types and Substitutions

The following three lemmata are easily proved by induction on derivations.

Lemma 1 (Validity of typing judgements). *Every component of a valid judgement is well-formed. For instance, if* $\Gamma \vdash t = t'\colon A$, *then* $\Gamma \vdash$, $\Gamma \vdash A$, $\Gamma \vdash t\colon A$ *and* $\Gamma \vdash t'\colon A$; *if* $\Gamma \vdash \delta\colon \Delta$ *or* $\Gamma \leq \Delta$, *then* $\Gamma \vdash$ *and* $\Delta \vdash$; *etc.*

Lemma 2. *If* $\Delta \vdash A \leq B$ *and* $\Gamma \vdash \delta\colon \Delta$ *then* $\Gamma \vdash A\,\delta \leq B\,\delta$. *The subcontext relation* $\leq$ *is reflexive and transitive. If* $\Gamma' \leq \Gamma$ *and* $\Gamma \vdash J$ *is a valid judgement, then so is* $\Gamma' \vdash J$.

Lemma 3 (Inversion lemma)

1. *If* $\Gamma \vdash \mathsf{Fun}\,A\,B\colon C$, *then* $\exists l : \Gamma \vdash \mathsf{U}_l \leq C$, $\Gamma \vdash A\colon \mathsf{U}_l$ *and* $\Gamma.A \vdash B\colon \mathsf{U}_l$.
2. *If* $\Gamma \vdash \mathsf{Fun}\,A\,B$, *then* $\Gamma \vdash A$ *and* $\Gamma.A \vdash B$.

Equal Terms.

$$\frac{\Gamma \vdash t\colon A}{\Gamma \vdash t = t\colon A}\ (\text{REFL-ET}) \qquad \frac{\Gamma \vdash t = t'\colon A \qquad \Gamma \vdash t = t''\colon A}{\Gamma \vdash t' = t''\colon A}\ (\text{SYM-TRAN-ET})$$

$$\frac{\Gamma \vdash t\colon A}{\Gamma \vdash t\,\mathsf{id} = t\colon A}\ (\text{NEUT-ET}) \qquad \frac{\Gamma \vdash \delta\colon \Delta}{\Gamma \vdash \mathsf{U}_l\,\delta = \mathsf{U}_l\colon \mathsf{U}_{l+1}}\ (\text{U-SUBS-ET})$$

$$\frac{\Gamma \vdash \delta\colon \Delta \qquad \Delta \vdash A \qquad \Gamma \vdash t\colon A\,\delta}{\Gamma \vdash \mathsf{q}\,(\delta, t) = t\colon A\,\delta}\ (\text{SND-ET})$$

$$\frac{\Gamma \vdash A = A'\colon \mathsf{U}_l \qquad \Gamma.A' \vdash B'\colon \mathsf{U}_l \qquad \Gamma.A \vdash B = B'\colon \mathsf{U}_l}{\Gamma \vdash \mathsf{Fun}\,A\,B = \mathsf{Fun}\,A'\,B'\colon \mathsf{U}_l}\ (\text{CONG-FUN-ET})$$

$$\frac{\Delta \vdash A\colon \mathsf{U}_l \qquad \Delta.A \vdash B\colon \mathsf{U}_l \qquad \Gamma \vdash \delta\colon \Delta}{\Gamma \vdash (\mathsf{Fun}\,A\,B)\,\delta = \mathsf{Fun}\,(A\,\delta)\,(B\,(\delta\,\mathsf{p}, \mathsf{q}))\colon \mathsf{U}_l}\ (\text{FUN-SUBS-ET})$$

$$\frac{\Gamma \vdash A \qquad \Gamma.A \vdash B \qquad \Gamma.A \vdash t = t'\colon B}{\Gamma \vdash \lambda t = \lambda t'\colon \mathsf{Fun}\,A\,B}\ (\text{XI-ET})$$

$$\frac{\Delta \vdash A \qquad \Delta.A \vdash B \qquad \Delta.A \vdash t\colon B \qquad \Gamma \vdash \delta\colon \Delta}{\Gamma \vdash (\lambda t)\,\delta = \lambda(t\,(\delta\,\mathsf{p}, \mathsf{q}))\colon (\mathsf{Fun}\,A\,B)\,\delta}\ (\text{ABS-SUBS-ET})$$

$$\frac{\Gamma \vdash A \qquad \Gamma.A \vdash B \qquad \Gamma \vdash t = t'\colon \mathsf{Fun}\,A\,B \qquad \Gamma \vdash r = r'\colon A}{\Gamma \vdash \mathsf{App}\ t\ r = \mathsf{App}\ t'\ r'\colon B\,[r]}\ (\text{CONG-APP-ET})$$

$$\frac{\begin{array}{c}\Delta \vdash A \\ \Delta.A \vdash B \qquad \Delta \vdash t\colon \mathsf{Fun}\,A\,B \qquad \Delta \vdash r\colon A \qquad \Gamma \vdash \delta\colon \Delta\end{array}}{\Gamma \vdash (\mathsf{App}\ t\ r)\,\delta = \mathsf{App}\ (t\,\delta)\ (r\,\delta)\colon (B\,[r])\,\delta}\ (\text{APP-SUBS-ET})$$

$$\frac{\Delta \vdash t = t'\colon A \qquad \Gamma \vdash \delta = \delta'\colon \Delta}{\Gamma \vdash t\,\delta = t'\,\delta'\colon A\,\delta}\ (\text{CONG-SUBS-ET})$$

$$\frac{\Delta \vdash A \qquad \Delta \vdash t\colon A \qquad \Sigma \vdash \delta\colon \Delta \qquad \Gamma \vdash \sigma\colon \Sigma}{\Gamma \vdash (t\,\delta)\,\sigma = t\,(\delta\,\sigma)\colon (A\,\delta)\,\sigma}\ (\text{ASSOC-ET})$$

$$\frac{\Gamma \vdash A \qquad \Gamma.A \vdash B \qquad \Gamma.A \vdash t\colon B \qquad \Gamma \vdash r\colon A}{\Gamma \vdash \mathsf{App}\ (\lambda t)\ r = t\,[r]\colon B\,[r]}\ (\text{BETA-ET})$$

$$\frac{\Gamma \vdash t = t'\colon A \qquad \Gamma \vdash A \leq B}{\Gamma \vdash t = t'\colon B}\ (\text{SUB-ET})$$

Fig. 3. Equality Rules for Terms

3. *If $\Gamma \vdash \lambda t\colon C$, then $\exists A, B : \Gamma \vdash \mathsf{Fun}\,A\,B \leq C$ and $\Gamma.A \vdash t\colon B$.*
4. *If $\Gamma \vdash \mathsf{App}\ t\ r\colon C$, then $\exists A, B : \Gamma \vdash B\,[r] \leq C$, $\Gamma \vdash t\colon \mathsf{Fun}\,A\,B$, $\Gamma \vdash r\colon A$.*
5. *If $\Gamma \vdash \mathsf{U}_l\colon C$, then $\Gamma \vdash \mathsf{U}_{l+1} \leq C$.*
6. *If $\Gamma \vdash \mathsf{p}^i\colon \Delta$, then $\Gamma = \Gamma'.A_i \ldots .A_1$ and $\Gamma' \leq \Delta$ for some $A_1, \ldots, A_i, \Gamma'$.*
7. *If $\Gamma \vdash \mathsf{q}\,\mathsf{p}^i\colon C$, then $\Gamma = \Gamma'.A_i \ldots .A_0$ and $\Gamma \vdash A_i\,\mathsf{p}^{i+1} \leq C$ for some $A_0, \ldots, A_i, \Gamma'$.*

Remark 1. If $\Gamma \vdash A$, then there exists $l \in \mathbb{N}$ such that $\Gamma \vdash A\colon \mathsf{U}_l$. Therefore rules (FUN-F) and (SUBS-TYPE) are redundant; but they would not be anymore even with slight changes in the system, so we keep them.

Trickier inversion results. Other inversion results can be formulated. Despite their similar nature, their proofs are more difficult, they will follow from the normalisation by evaluation construction and results:

Lemma 4 (Injectivity and subtyping reflection)

1. *If* $\Gamma \vdash \mathsf{U}_l \leq \mathsf{U}_{l'}$, *then* $l \leqslant l'$. *If* $\Gamma \vdash \mathsf{U}_l = \mathsf{U}_{l'}$ *then* $l = l'$.
2. *If* $\Gamma \vdash \mathsf{Fun}\, A\, B \leq \mathsf{Fun}\, A'\, B'$, *then* $\Gamma \vdash A' \leq A$ *and* $\Gamma.A' \vdash B \leq B'$.
3. *If* $\Gamma \vdash \mathsf{Fun}\, A\, B = \mathsf{Fun}\, A'\, B'$, *then* $\Gamma \vdash A' = A$ *and* $\Gamma.A \vdash B = B'$.

An attempt to prove them directly by induction on derivations would fail when considering transitivity. For the proof to succeed, it becomes necessary to show that a universe and a product type are not comparable. This is related to the consistency of the calculus which will only follow once we have a model that distinguishes terms headed with different constructors.

3 Semantics and Normalisation by Evaluation

In this section we set up the mathematical structures to define a model which can be used to normalise well-formed types and well-typed terms. Due to the lack of (η), we simplify the machinery of NbE introduced in [2], where η-expansion is performed at the semantic level. Here it is enough to have a function $\mathsf{R} : D \to \mathit{Terms}$ to reify values from the semantic domain D to the syntax.

This reification function helps mimicking syntactic concepts semantically: a semantic normal form will be an element of the domain that is reified to a term in normal form, and similarly for semantic neutral values. Thus, we think of semantic normal forms as canonical representations of terms (and types).

The set *Types* of semantic types is defined using induction-recursion: at the same time that we introduce an element X in *Types*, we also specify the set of semantic terms $[X]$ for that (semantic) type. A key property is that every member of *Types* or of $[X]$, for any $X \in \mathit{Types}$, is a semantic normal form; thus we know that elements in those sets will be reified as normal forms. This result, as several others, is proved by induction on *Types*.

Domain. Our model is built on top of a domain of values. Let D be the least solution of the following domain equation in the category $\mathbf{DOM}_\perp$ of pointed *ω-cpo* and continuous functions [3]:

$$D \cong \{\top\}_\perp \oplus D \times D \oplus \mathbb{N}_\perp \oplus [D \to D] \oplus D \times [D \to D] \oplus D \times D \oplus \mathbb{N}_\perp \ .$$

where $\oplus$ stands for the coalesced sum.

Elements in D are either $\perp$ or a non bottom element of one of the summands, properly tagged and mapped to D. Instead of usual tags and isomorphisms, we employ a more friendly notation for elements in D:

$\top$	(d, d')	$\mathsf{App}\ d\ d'$	for $d, d' \in D$
x_i	U_l		for $i, l \in \mathbb{N}$
$\mathsf{Lam}\ f$	$\mathsf{Fun}\ d\ f$		for $d \in D$, and $f \in [D \to D]$

Reification. When reifying an element from the semantic domain D to the syntax, one has to keep track of the free and bound variables. We keep this information in a parameter of the partial function R.[1]

Definition 2 (Reification function)

$$\begin{aligned} \mathsf{R}_j\,\mathsf{U}_l &= \mathsf{U}_l \\ \mathsf{R}_j\,(\mathsf{App}\;d\;d') &= \mathsf{App}\;(\mathsf{R}_j\,d)\;(\mathsf{R}_j\,d') \\ \mathsf{R}_j\,(\mathsf{Lam}\;f) &= \lambda(\mathsf{R}_{j+1}\,(f\;x_j)) \end{aligned} \qquad \begin{aligned} \mathsf{R}_j\,(\mathsf{Fun}\;X\;F) &= \mathsf{Fun}\,(\mathsf{R}_j\,X)\,(\mathsf{R}_{j+1}\,(F\;x_j)) \\ \mathsf{R}_j\,(x_i) &= \begin{cases} \mathsf{q} & \text{if } j \leqslant i+1 \\ \mathsf{q}\,\mathsf{p}^{j-(i+1)} & \text{if } j > i+1 \end{cases} \end{aligned}$$

Definition 3 (Semantic neutral values and normal forms)

$$Ne = \{d \mid \mathsf{R}_i\,d \textit{ is defined and is a neutral term for all } i \in \mathbb{N}\}$$
$$Nf = \{d \mid \mathsf{R}_i\,d \textit{ is defined and is a normal form for all } i \in \mathbb{N}\}$$

Remark 2. Note that $Ne \subseteq Nf$ and $\bot \notin Nf$. For all j, l, $x_j \in Ne$ and $\mathsf{U}_l \in Nf$. If $d \in Ne$ and $d' \in Nf$ then $\mathsf{App}\;d\;d' \in Ne$. Since the only elements of *Terms* that are neutral are applications or variables, i.e. $\mathsf{q}\,\mathsf{p}^i$ for some $i \in \mathbb{N}$, it is clear that $d \in Ne$ implies $d = \mathsf{App}\;d'\;e$ or $d = x_i$.

Applicative structure. We define the binary operation of application $\cdot$ in the domain D and the projections, which are denoted with $\mathsf{p} = \pi_1$ and $\mathsf{q} = \pi_2$. Notice that the application of a semantic neutral value to a semantic normal form produces a semantic neutral value.

$$d \cdot d' = \begin{cases} f\;d' & \text{if } d = \mathsf{Lam}\;f \\ \mathsf{App}\;d\;d' & \text{if } d \in Ne \\ \bot & \text{otherwise} \;. \end{cases} \qquad \pi_i\,d = \begin{cases} e_i & \text{if } d = (e_1, e_2) \\ \bot & \text{otherwise} \;. \end{cases}$$

Predomain model. Now we define the subsets of D which will be used to interpret the typing relation: we first introduce the subset of D for each universe. Semantically the set of all types is the universe with level ∞. These subsets are introduced using the schema of inductive-recursive definition of [9].

Definition 4. *For each $l \in \mathbb{N} \cup \{\infty\}$, we give the following simultaneous inductive definition of $\mathcal{U}_l \subseteq D$ and $[d]_l \subseteq D$ for each $d \in \mathcal{U}_l$. Assume $\mathcal{U}_{l'}$ and $[\_]_{l'}$ defined for all $l' < l$.*

1. *Neutrals: $Ne \subseteq \mathcal{U}_l$, and $[d]_l = Ne$ for all $d \in Ne$.*
2. *Lower universes, let $l' < l$: $\mathsf{U}_{l'} \in \mathcal{U}_l$ and $[\mathsf{U}_{l'}]_l = \mathcal{U}_{l'}$.*
3. *Function spaces, let $X \in \mathcal{U}_l$ and $F\,e \in \mathcal{U}_l$ for all $e \in [X]_l$:*
   – $\mathsf{Fun}\,X\,F \in \mathcal{U}_l$*, and* $[\mathsf{Fun}\;X\;F]_l = \{d \mid d \cdot e \in [F\;e]_l,$ *for all* $e \in [X]_l\}$.

Remark 3. As noticed in [13], there is a well-founded order on $\mathcal{U}_l$: the minimal elements are $\mathsf{U}_{l'}$ with $l' < l$, and elements in Ne; $X \sqsubset \mathsf{Fun}\;X\;F$, and for all $e \in [X]$, $F\;e \sqsubset \mathsf{Fun}\;X\;F$. This order is used to justify most of the following proofs by induction on $\mathcal{U}_l$.

[1] To be precise, the reification function R is the least fixed point of a suitable functional $F\colon (\mathbb{N} \times D \to \mathit{Terms}_\bot) \to (\mathbb{N} \times D \to \mathit{Terms}_\bot)$.

Lemma 5. *If* $l' < l \in \mathbb{N} \cup \{\infty\}$, *then* $\mathcal{U}_{l'} \subset \mathcal{U}_l$ *and* $[d]_{l'} = [d]_l$ *for each* $d \in \mathcal{U}_{l'}$.

We define $Types = \mathcal{U}_\infty$ and $[d] = [d]_\infty$; due to this lemma we can always write $[d]$ instead of $[d]_l$.

The following lemma states that every semantic type X and every element of $[X]$ is a semantic normal form; an immediate consequence of this is that neither *Types* nor any $[X]$ are subdomains of D.

Lemma 6 (Closure of semantic sub-sets)

1. $Ne \subseteq Types$ *and for all* $X \in Types$, $Ne \subseteq [X]$.
2. $Types \subseteq Nf$ *and for all* $X \in Types$, $[X] \subseteq Nf$.

Modelling subtyping. We define a binary relation $\preccurlyeq \subseteq Types \times Types$ among the denotation of types and interpret subtyping with that relation.

Definition 5. *We define inductively the relation* $\preccurlyeq \subseteq Types \times Types$ *with the following clauses:*

1. *Neutrals: if* $d \in Ne$ *then* $d \preccurlyeq d$.
2. *Universes: if* $l \leqslant l'$ *then* $\mathsf{U}_l \preccurlyeq \mathsf{U}_{l'}$.
3. *Function spaces: if* $X' \preccurlyeq X$ *and for all* $e \in [X'], F\,e \in Types$ *and* $F\,e \preccurlyeq F'\,e$ *then* $\mathsf{Fun}\ X\ F \preccurlyeq \mathsf{Fun}\ X'\ F'$.

Notice the condition $F\,e \in Types$ in the third clause, which can be dropped once we have shown that $[\_]$ is monotone (Lem. 7), since one would then have $e \in [X]$ and the condition would follow from the assumption $\mathsf{Fun}\ X\ F \in Types$.

Remark 4. In the proof of the following lemma we use $d \preccurlyeq d'$ holds only if $d, d' \in Ne$, $d = \mathsf{U}_l$ and $d' = \mathsf{U}_{l'}$, or $d = \mathsf{Fun}\ X\ F$ and $d' = \mathsf{Fun}\ X'\ F'$. By the lemma we can see that $\preccurlyeq$ is finer than set-theoretical inclusion: for instance $[d] \subseteq [\mathsf{U}_l]$ but $d \not\preccurlyeq \mathsf{U}_l$ for any neutral value d.

Lemma 7. *If* $X \preccurlyeq X'$, *then* $[X] \subseteq [X']$. *The relation* $\preccurlyeq$ *is a preorder.*

Semantics and Validity. The semantic equations using the applicative structure just defined are given by the pair of functions $[\![\_]\!]^s\_ \in Subs \to [D \to D]$ and $[\![\_]\!]^t\_ \in Terms \to [D \to D]$. In the following we omit the superscript from the semantic brackets. Notice that both semantic functions are defined by induction on the set of preterms and when partially applied to a term (or substitution) are continuous endofunctions over D.

Substitutions

$$[\![\langle\rangle]\!]^s d = \top \qquad [\![\mathsf{id}]\!]^s d = d \qquad [\![\mathsf{p}]\!]^s d = \mathsf{p}\ d$$

$$[\![(\gamma, t)]\!]^s d = ([\![\gamma]\!]^s d, [\![t]\!]^t d) \qquad [\![\gamma\,\delta]\!]^s d = [\![\gamma]\!]^s([\![\delta]\!]^s d)$$

Terms (and types)

$$\llbracket \mathsf{U}_l \rrbracket^t d = \mathsf{U}_l \qquad \llbracket \mathsf{Fun}\, A\, B \rrbracket^t d = \mathsf{Fun}\ (\llbracket A \rrbracket^t d)\ (e \mapsto \llbracket B \rrbracket^t (d, e))$$
$$\llbracket \lambda t \rrbracket^t d = \mathsf{Lam}\ (e \mapsto \llbracket t \rrbracket^t (d, e)) \qquad \llbracket \mathsf{App}\ t\ r \rrbracket^t d = \llbracket t \rrbracket^t d \cdot \llbracket r \rrbracket^t d$$
$$\llbracket t\, \gamma \rrbracket^t d = \llbracket t \rrbracket^t (\llbracket \gamma \rrbracket^s d) \qquad \llbracket \mathsf{q} \rrbracket^t d = \mathsf{q}\ d$$

In the following definition we define simultaneously the notion of validity of judgements and the interpretation of well-formed contexts. Subtyping between types is modeled by $\preccurlyeq$, while subtyping of contexts by set-theoretical inclusion.

Definition 6 (Validity)

1. *Contexts:*
   - *Validity:* $\diamond \vDash$ *always, and* $\Gamma.A \vDash$ *if and only if* $\Gamma \vDash A$.
   - *Interpretation:* $\llbracket \diamond \rrbracket = \{\top\}$; *and when* $\Gamma.A \vDash$, *then* $\llbracket \Gamma.A \rrbracket = \{(d, e) \mid d \in \llbracket \Gamma \rrbracket$ *and* $e \in [\llbracket A \rrbracket d]\}$.
2. *Types:* $\Gamma \vDash A$ *if and only if* $\Gamma \vDash$ *and for all* $d \in \llbracket \Gamma \rrbracket$, $\llbracket A \rrbracket d \in$ *Types*.
3. *Subtyping:* $\Gamma \vDash A \leq B$ *if and only if* $\Gamma \vDash A$, $\Gamma \vDash B$, *and for all* $d \in \llbracket \Gamma \rrbracket$, $\llbracket A \rrbracket d \preccurlyeq \llbracket B \rrbracket d$.
4. *Subtyping of contexts:* $\Gamma \leq \Delta \vDash$ *if and only if* $\llbracket \Gamma \rrbracket \subseteq \llbracket \Delta \rrbracket$.
5. *Terms:* $\Gamma \vDash t : A$ *if and only if* $\Gamma \vDash A$ *and for all* $d \in \llbracket \Gamma \rrbracket$, $\llbracket t \rrbracket d \in [\llbracket A \rrbracket d]$.
6. *Substitutions:* $\Gamma \vDash \sigma : \Delta$ *if and only if* $\Gamma \vDash$, $\Delta \vDash$, *and for all* $d \in \llbracket \Gamma \rrbracket$, $\llbracket \sigma \rrbracket d \in \llbracket \Delta \rrbracket$.

We extend the definition of validity to equality judgements, for example $\Gamma \vDash t = t' : A$ *if and only if* $\Gamma \vDash t : A$, $\Gamma \vDash t' : A$, *and* $\llbracket t \rrbracket d = \llbracket t' \rrbracket d$, *for all* $d \in \llbracket \Gamma \rrbracket$.

The following theorem states that this construction over the applicative structure models the calculus. We omit the proof of soundness which can be done by a tedious, but straightforward, induction on derivations.

Theorem 1 (Soundness). *If* $\Gamma \vdash J$, *then* $\Gamma \vDash J$. □

By Lem. 6 we know $x_n \in [X]$, for $X \in$ *Types* and $n \in \mathbb{N}$; by Thm. 1 we also have $\Gamma \vdash A$ implies that $\llbracket A \rrbracket d \in$ *Types*. These facts permit us to introduce a canonical environment, that can be used to normalise terms and types. Notice that it takes a (well-formed) context as a parameter, but, since we do not have (η), it would suffice to keep into account the length of the context.

Definition 7. *By induction on* $\Gamma \vdash$ *we define the environment* $\rho_\Gamma \in \llbracket \Gamma \rrbracket$:

$$\rho_\diamond = \top \qquad \text{and} \qquad \rho_{\Gamma.A} = (\rho_\Gamma, x_n) \text{ where } n = |\Gamma|$$

The normalisation function is the composition of evaluation under the canonical environment and readback:

$$\mathbf{nbe}_\Gamma(\_) \in \{t \mid \Gamma \vdash t \text{ or } \Gamma \vdash t\colon A, \text{ for some } A\} \to \mathrm{Nf}$$
$$\mathbf{nbe}_\Gamma(t) = \mathsf{R}_{|\Gamma|}\,(\llbracket t \rrbracket \rho_\Gamma)$$

Two corollaries can be drawn from Thm. 1; first the normalisation function maps well-typed terms (and well-formed types) to syntactic normal forms, remember Lem. 6; moreover if two terms are probably equal, then they are normalised to the same syntactical normal form.

Corollary 1 (Completeness of NbE). *If* $\Gamma \vdash t = t' \colon A$, *then* $\mathbf{nbe}_\Gamma(t) \equiv \mathbf{nbe}_\Gamma(t')$; *also, if* $\Gamma \vdash A = A'$, *then* $\mathbf{nbe}_\Gamma(A) \equiv \mathbf{nbe}_\Gamma(A')$.

An interesting result is that the normalisation function on well-typed normal forms is the identity function.[2] It is proved by induction on normal forms and by virtue of Rem. 1 it extends to types.

Lemma 8. *If* $v \in \mathrm{Nf}$ *and* $\Gamma \vdash v \colon A$, *then* $\mathbf{nbe}_\Gamma(v) \equiv v$.

Remark 5. As we commented before Def. 7, only the length of the context is relevant: so $\rho_\Gamma = \rho_\Delta$, when $|\Gamma| = |\Delta|$; thus $\mathbf{nbe}_\Gamma(\_) = \mathbf{nbe}_\Delta(\_)$, in particular when $\Gamma \leq \Delta$.

4 Correctness of NbE via Logical Relations

In this section we prove correctness of NbE, that is, if $\Gamma \vdash t \colon A$, then $\Gamma \vdash t = \mathbf{nbe}_\Gamma(t) \colon A$. In order to do this, we introduce three families of logical relations: one for types, one for terms and another one for substitutions.

The proof of correctness follows the same path as in [2]: first one shows that if $\Gamma \vdash t \colon A \sim d \in [X]$, then one can derive $\Gamma \vdash t = \mathsf{R}_{|\Gamma|}\, d \colon A$; then it is proved that every term is logically related with its denotation under the canonical environment. This proof proceeds by induction on derivation; it is here where one can see the benefit of modelling subtyping with the order $\preccurlyeq$. In fact, the definition of the logical relations is not affected by the presence of subtyping; there is, however, a new lemma proving that logical relations are stable under subtyping.

Definition 8 (Logical relations). *There are two families of logical relations;*

1. $\Gamma \vdash \_ \sim \_ \in \mathit{Types} \subseteq \{A \mid \Gamma \vdash A\} \times \mathit{Types}$.
2. $\Gamma \vdash \_ \colon A \sim \_ \in [X] \subseteq \{t \mid \Gamma \vdash t \colon A\} \times [X]$, *for* $\Gamma \vdash A$ *and* $X \in \mathit{Types}$.

- *Neutral types:* $X \in \mathit{Ne}$.
  - $\Gamma \vdash A \sim X \in \mathit{Types}$ *if and only if for all* $\Theta \leqslant^i \Gamma$, $\Theta \vdash A\,\mathsf{p}^i = \mathsf{R}_{|\Theta|}\, X$.
  - $\Gamma \vdash t \colon A \sim d \in [X]$ *if and only if* $\Gamma \vdash A \sim X \in \mathit{Types}$, *and for all* $\Theta \leqslant^i \Gamma$, $\Theta \vdash t\,\mathsf{p}^i = \mathsf{R}_{|\Theta|}\, d \colon A\,\mathsf{p}^i$.
- *Universes.*
  - $\Gamma \vdash A \sim \mathsf{U}_l \in \mathit{Types}$ *if and only if* $\Gamma \vdash A = \mathsf{U}_l$.
  - $\Gamma \vdash t \colon A \sim d \in [\mathsf{U}_l]$ *if and only if* $\Gamma \vdash A = \mathsf{U}_l$, $\Gamma \vdash t \sim d \in \mathit{Types}$, *and for all* $\Theta \leqslant^i \Gamma$, $\Theta \vdash t\,\mathsf{p}^i = \mathsf{R}_{|\Theta|}\, d \colon \mathsf{U}_l$.
- *Function spaces.*

[2] This result is mentioned and proved in [10]; we note that it was not proved in [2].

- $\Gamma \vdash A \sim \mathsf{Fun}\ X\ F \in \mathit{Types}$ *if and only if there are* B, C *such that* $\Gamma \vdash A = \mathsf{Fun}\, B\, C$, $\Gamma \vdash B \sim X \in \mathit{Types}$, *and for all* $\Theta \leqslant^i \Gamma$ *and all* $\Theta \vdash s \colon B\,\mathsf{p}^i \sim e \in [X]$, $\Theta \vdash C\,(\mathsf{p}^i, s) \sim F\,e \in \mathit{Types}$.
- $\Gamma \vdash t \colon A \sim d \in [\mathsf{Fun}\ X\ F]$ *if and only if there are* B, C *such that* $\Gamma \vdash A = \mathsf{Fun}\, B\, C$, $\Gamma \vdash B \sim X \in \mathit{Types}$, *and for all* $\Theta \leqslant^i \Gamma$ *and all* $\Theta \vdash s \colon B\,\mathsf{p}^i \sim e \in [X]$, $\Theta \vdash \mathsf{App}\ (t\,\mathsf{p}^i)\ s \colon C\,(\mathsf{p}^i, s) \sim d \cdot e \in [F\,e]$. *In addition: if* $d = \mathsf{Lam}\ f$, *then there is* t' *such that* $\Gamma.B \vdash t' \colon C$ *and* $\Gamma \vdash t = \lambda t' \colon A$; *and if* $d \in \mathit{Ne}$, *then* $\Theta \vdash t\,\mathsf{p}^i = \mathsf{R}_{|\Theta|}\, d \colon A\,\mathsf{p}^i$.

We call the attention to the last point of the logical relation for terms, in the case of $\mathsf{Fun}\ X\ F$: it is needed because there are neutral elements in $[\mathsf{Fun}\ X\ F]$, with (η) every member of that semantic type is of the form $\mathsf{Lam}\ f$ and for every term with type $\Gamma \vdash t \colon \mathsf{Fun}\, B\, C$, there exists t' such that $\Gamma.B \vdash t' \colon C$ and $\Gamma \vdash t = \lambda t' \colon \mathsf{Fun}\, B\, C$.

Remark 6. A consequence of the definition is that logical relations for terms implies that for types: if $\Gamma \vdash t \colon A \sim d \in [X]$, then $\Gamma \vdash A \sim X \in \mathit{Types}$. This can be proved by induction on $X \in \mathit{Types}$.

The following two lemmata prove that logical relations are stable under subtyping of contexts, cf. second point of Lem. 2, and by judgemental equality; both make easier the proofs of other results.

Lemma 9. *Let* $\Gamma' \leq \Gamma$. *If* $\Gamma \vdash A \sim X \in \mathit{Types}$, *then* $\Gamma' \vdash A \sim X \in \mathit{Types}$. *Moreover, if* $\Gamma \vdash t \colon A \sim d \in [X]$, *then* $\Gamma' \vdash t \colon A \sim d \in [X]$.

Lemma 10. *If* $\Gamma \vdash A = A'$ *and* $\Gamma \vdash A \sim X \in \mathit{Types}$, $\Gamma \vdash A' \sim X \in \mathit{Types}$. *If in addition* $\Gamma \vdash t = t' \colon A$ *and* $\Gamma \vdash t \colon A \sim d \in [X]$, *then* $\Gamma \vdash t' \colon A' \sim d \in [X]$.

Lemma 11. *Let* $\Gamma \vdash A \sim X \in \mathit{Types}$ *and* $n = |\Gamma|$.

1. $\Gamma \vdash A = \mathsf{R}_n\, X$.
2. *If* $\Gamma \vdash t \colon A \sim d \in [X]$, *then* $\Gamma \vdash t = \mathsf{R}_n\, d \colon A$.
3. *If for all* $\Theta \leqslant^i \Gamma$, $\Theta \vdash t\,\mathsf{p}^i = \mathsf{R}_{n+i}\, k \colon A\,\mathsf{p}^i$, *then* $\Gamma \vdash t \colon A \sim k \in [X]$.

Remark 7. An immediate corollary from lemmata 10 and 11 is that two types are judgementally equal if they are related to the same semantic type: if $\Gamma \vdash A \sim X \in \mathit{Types}$ and $\Gamma \vdash A' \sim X \in \mathit{Types}$, then $\Gamma \vdash A = A'$.

The following lemma shows that semantic subtyping implies judgemental subtyping. It is an important stepping stone in our way for two reasons: it allows us to prove stability of judgements under subtyping and it also permits to prove that Fun reflects subtyping.

Lemma 12. *If* $\Gamma \vdash A \sim X \in \mathit{Types}$, $\Gamma \vdash B \sim Y \in \mathit{Types}$, *and* $X \preccurlyeq Y$, *then* $\Gamma \vdash A \leq B$.

The next lemma, which could be called subsumption for logical relations, was self-evident to us as soon as we realised that subtyping should be modelled by a finer relation than set inclusion.

Lemma 13. *Let $\Gamma \vdash A \sim X \in \mathit{Types}$, $\Gamma \vdash B \sim Y \in \mathit{Types}$, and $X \preccurlyeq Y$. If $\Gamma \vdash t\colon A \sim d \in [X]$, then $\Gamma \vdash t\colon B \sim d \in [Y]$.*

To prove that each term is related to its denotation, we need the more general result that logical relations are preserved under substitutions.

Definition 9 (Logical relations for substitutions). *Given two well-formed contexts $\Gamma \vdash$ and $\Delta \vdash$, then $\Gamma \vdash \_\colon \Delta \sim \_ \in [\![\Delta]\!] \subseteq \{\sigma \mid \Gamma \vdash \sigma\colon \Delta\} \times [\![\Delta]\!]$. By induction on Δ we define:*

- $\Gamma \vdash \sigma\colon \diamond \sim d \in [\![\diamond]\!]$.
- $\Gamma \vdash \sigma\colon \Delta.A \sim (d, e) \in [\![\Delta.A]\!]$ *if and only if* $\Gamma \vdash \sigma = (\delta, t)\colon \Delta.A$, $\Gamma \vdash \delta\colon \Delta \sim d \in [\![\Delta]\!]$, $\Gamma \vdash A\,\delta \sim [\![A]\!]d \in \mathit{Types}$, *and* $\Gamma \vdash t\colon A\,\delta \sim e \in [[\![A]\!]d]$.

Notice that we cannot state, let alone prove, at this point a corresponding lemma to Lem. 13 for substitutions; instead, we will prove subsumption for substitutions in Thm. 2, when we will have all the inductive hypotheses needed. On the other hand the analogous of Lem. 10 can be proved now.

Lemma 14. *If $\Gamma \vdash \sigma\colon \Delta \sim d \in [\![\Delta]\!]$ and $\Gamma \vdash \sigma = \gamma\colon \Delta$, $\Gamma \vdash \gamma\colon \Delta \sim d \in [\![\Delta]\!]$.*

Lemma 15 (Monotonicity of logical relations). *Let $\Theta \leqslant^i \Gamma$.*

1. *If $\Gamma \vdash A \sim X \in \mathit{Types}$, then $\Theta \vdash A\,\mathsf{p}^i \sim X \in \mathit{Types}$.*
2. *If $\Gamma \vdash t\colon A \sim d \in [X]$, then $\Theta \vdash t\,\mathsf{p}^i\colon A\,\mathsf{p}^i \sim d \in [X]$.*
3. *If $\Gamma \vdash \sigma\colon \Delta \sim d \in [\![\Delta]\!]$ and then $\Theta \vdash \sigma\,\mathsf{p}^i\colon \Delta \sim d \in [\![\Delta]\!]$.*

The significance of the following theorem is that types and terms are logically related with their denotation, first and third points, respectively; the proof goes by induction on derivations and, obviously, needs an analogous case for substitutions, fourth point. The second point arises because we need its conclusion to apply the Lem. 13 on the cases (SUB) and (SUBS).

Theorem 2 (Fundamental theorem of logical relations)

1. *if $\Delta \vdash A$ and $\Gamma \vdash \delta\colon \Delta \sim d \in [\![\Delta]\!]$, then $\Gamma \vdash A\,\delta \sim [\![A]\!]d \in \mathit{Types}$;*
2. *if $\Delta \vdash A \leq B$ and $\Gamma \vdash \delta\colon \Delta \sim d \in [\![\Delta]\!]$, then $\Gamma \vdash A\,\delta \sim [\![A]\!]d \in \mathit{Types}$ and $\Gamma \vdash B\,\delta \sim [\![B]\!]d \in \mathit{Types}$.*
3. *if $\Delta \vdash t\colon A$ and $\Gamma \vdash \delta\colon \Delta \sim d \in [\![\Delta]\!]$, then $\Gamma \vdash t\,\delta\colon A\,\delta \sim [\![t]\!]d \in [[\![A]\!]d]$; and*
4. *if $\Theta \vdash \gamma\colon \Gamma$ and $\Gamma \vdash \delta\colon \Delta \sim d \in [\![\Delta]\!]$, then $\Theta \vdash \gamma\,\delta\colon \Delta \sim [\![\gamma]\!]d \in [\![\Delta]\!]$.*

Theorem 3 (Correctness of NbE). *If $\Gamma \vdash A$, then $\Gamma \vdash A = \mathbf{nbe}_\Gamma(A)$ and if $\Gamma \vdash t\colon A$, then $\Gamma \vdash t = \mathbf{nbe}_\Gamma(t)\colon A$.*

Proof. By induction on well-formed contexts one proves $\Gamma \vdash \mathsf{id}\colon \Gamma \sim \rho_\Gamma \in [\![\Gamma]\!]$; thus by Thm. 2, $\Gamma \vdash t\,\mathsf{id}\colon A\,\mathsf{id} \sim [\![t]\!]\rho_\Gamma \in [[\![A]\!]\rho_\Gamma]$; by Lem. 11, $\Gamma \vdash t\,\mathsf{id} = \mathsf{R}_{|\Gamma|}\,([\![t]\!]\rho_\Gamma)\colon A\,\mathsf{id}$, by (SYM-TRANS) and (SUB-ET), we have $\Gamma \vdash t = \mathsf{R}_{|\Gamma|}\,([\![t]\!]\rho_\Gamma)\colon A$.

An immediate consequence of correctness of NbE is that to decide subtyping it is enough to decide subtyping between normal forms.

Lemma 16. *Let $\Gamma \vdash A$ and $\Gamma \vdash B$, then $\Gamma \vdash A \leq B$ iff $\Gamma \vdash \mathbf{nbe}(A) \leq \mathbf{nbe}(B)$.*

Injectivity can be proved by rather direct means — in fact, the second part of the first of point of Lem. 4 can be deduced from Thm. 1 and Lem. 8. On the other hand, reflection of subtyping requires some more work. The first point of Lem. 4 can be shown in a similar way.

Corollary 2 (Injectivity of Fun)

1. *If $\Gamma \vdash \mathsf{Fun}\, A\, B = \mathsf{Fun}\, A'\, B'$, then $\Gamma \vdash A' = A$ and $\Gamma.A' \vdash B = B'$.*
2. *If $\Gamma \vdash \mathsf{Fun}\, A\, B \leq \mathsf{Fun}\, A'\, B'$, then $\Gamma \vdash A' \leq A$ and $\Gamma.A' \vdash B \leq B'$.*

Proof. In both cases we can invert the equality and subtyping judgements, respectively, to obtain $\Gamma \vdash A$, $\Gamma \vdash A'$, and $\Gamma.A \vdash B$, and $\Gamma.A' \vdash B'$. Let $V = \mathbf{nbe}_\Gamma(A)$, $V' = \mathbf{nbe}_\Gamma(A')$, $W = \mathbf{nbe}_{\Gamma.A}(B)$, and $W = \mathbf{nbe}_{\Gamma.A'}(B')$. By Thm. 1 $[\![\mathsf{Fun}\, A\, B]\!]\rho_\Gamma \preccurlyeq [\![\mathsf{Fun}\, A'\, B']\!]\rho_\Gamma$, thus $[\![A']\!]\rho_\Gamma \preccurlyeq [\![A]\!]\rho_\Gamma$ and $[\![B]\!]\rho_{\Gamma.A} \preccurlyeq [\![B']\!]\rho_{\Gamma.A'}$. Moreover, by Thm. 2, $\Gamma \vdash A \sim [\![A]\!]\rho_\Gamma \in \mathit{Types}$, $\Gamma \vdash A' \sim [\![A']\!]\rho_\Gamma \in \mathit{Types}$. Thus by Lem. 12 we get $\Gamma \vdash A' \leq A$. Again by Thm. 2, we have $\Gamma.A \vdash B \sim [\![B]\!]\rho_{\Gamma.A} \in \mathit{Types}$ and $\Gamma.A' \vdash B' \sim [\![B']\!]\rho_{\Gamma.A'} \in \mathit{Types}$. As in the proof of the Lem. 12 we can prove, by Lem. 9, $\Gamma.A' \vdash B \sim [\![B]\!]\rho_{\Gamma.A} \in \mathit{Types}$, so by Lem. 12, we get $\Gamma.A' \vdash B \leq B'$.

Just as we deduced from Lem. 16 that it is enough to decide subtyping among normal forms, reflection of subtyping tells us that it is sufficient to check subtyping of arguments to decide subtyping between Fun.

5 Type-Checking

In this section we define a bi-directional type-checking algorithm [7,16,1]. Since we can only infer types for neutrals, because abstractions are domain-free, the type-checking algorithm works only for normal forms.

In the following definition we give a rule system which conforms a procedure for checking subtyping between well-formed types in normal form. It is informative to think of the order $\trianglelefteq$ as the syntactical version of $\preccurlyeq$.

Definition 10 (Algorithmic subtyping). *Let $\Gamma \vdash V$ and $\Gamma \vdash V'$, then $V \trianglelefteq V'$ if and only if*

1. *$V \equiv V'$ and $V \in \mathrm{Ne}$, or*
2. *$V \equiv \mathsf{U}_l$ and $V' \equiv \mathsf{U}_{l'}$ and $l \leqslant l'$, or*
3. *$V \equiv \mathsf{Fun}\, V_0\, W_0$ and $V' \equiv \mathsf{Fun}\, V_1\, W_1$, $V_1 \trianglelefteq V_0$ and $W_0 \trianglelefteq W_1$.*

Reflexivity and transitivity of algorithmic subtyping are easy to prove. This lemma is needed to prove completeness of algorithmic subtyping.

Lemma 17 (Algorithmic subtyping is a preorder). *Let $\Gamma \vdash V$, then $V \trianglelefteq V$. Moreover let $\Gamma \vdash V'$ and $\Gamma \vdash V''$. If $V \trianglelefteq V'$ and $V' \trianglelefteq V''$, then $V \trianglelefteq V''$.*

Correctness of algorithmic subtyping follows from a weaker version, v.g. if $\mathbf{nbe}_\Gamma(A) \trianglelefteq \mathbf{nbe}_\Gamma(B)$, then $\Gamma \vdash \mathbf{nbe}_\Gamma(A) \leq \mathbf{nbe}_\Gamma(B)$.

Theorem 4 (Correctness and completeness of algorithmic subtyping). *Let $\Gamma \vdash A$ and $\Gamma \vdash B$, then $\Gamma \vdash A \leq B$ if and only if $\mathbf{nbe}_\Gamma(A) \trianglelefteq \mathbf{nbe}_\Gamma(B)$.*

The type-checking algorithm consists of two mutually recursive components: the first one checks if a normal form v can be typed under a (well-formed) context Γ and a type A, well-formed under Γ; this component is directed by the form of v; if it succeeds, we write $\Gamma \vdash v \Leftarrow A$. The second procedure tries to compute a type B given a (well-formed) context Γ and a neutral term k; its success is denoted by $\Gamma \vdash k \Rightarrow B$.

Definition 11 (Type-checking algorithm). *Let $\Gamma \vdash$, $\Gamma \vdash A$, $V, W, v \in \mathrm{Nf}$, and $k \in \mathrm{Ne}$.*

$$(\textsc{u-tc})\ \frac{l' < l}{\Gamma \vdash \mathsf{U}_{l'} \Leftarrow \mathsf{U}_l} \qquad \frac{\Gamma \vdash V \Leftarrow \mathsf{U}_l \quad \Gamma.V \vdash W \Leftarrow \mathsf{U}_l}{\Gamma \vdash \mathsf{Fun}\, V\, W \Leftarrow \mathsf{U}_l}\ (\textsc{fun-tc})$$

$$(\textsc{abs-tc})\ \frac{\Gamma.A \vdash v \Leftarrow B}{\Gamma \vdash \lambda v \Leftarrow \mathsf{Fun}\, A\, B} \qquad \frac{\Gamma \vdash k \Rightarrow B \quad \mathbf{nbe}_\Gamma(B) \trianglelefteq \mathbf{nbe}_\Gamma(A)}{\Gamma \vdash k \Leftarrow A}\ (\textsc{neut-tc})$$

$$(\textsc{var-ti})\ \frac{}{\Gamma.A \vdash \mathsf{q} \Rightarrow A\,\mathsf{p}} \qquad \frac{0 < i \leqslant n}{A_n.A_{n-1}.\dots A_0 \vdash \mathsf{q}\,\mathsf{p}^i \Rightarrow A_i\,\mathsf{p}^{i+1}}\ (\textsc{weak-ti})$$

$$\frac{\Gamma \vdash k \Rightarrow B \quad \mathbf{nbe}_\Gamma(B) = \mathsf{Fun}\, V\, W \quad \Gamma \vdash v \Leftarrow V}{\Gamma \vdash \mathsf{App}\ k\ v \Rightarrow W\,[v]}\ (\textsc{app-ti})$$

Notice that both in (NEUT-TC) and (APP-TI) we use the normalisation function; its use is justified because, by correctness of type-checking, we know that the inferred type is well-formed under the given context.

Theorem 5 (Correctness). *Let $\Gamma \vdash A$, $\Gamma' \vdash$, $v \in \mathrm{Nf}$ and $k \in \mathrm{Ne}$. If $\Gamma \vdash v \Leftarrow A$, then $\Gamma \vdash v \colon A$; if $\Gamma' \vdash k \Rightarrow B$, then $\Gamma' \vdash B$ and $\Gamma' \vdash k \colon B$.*

The following lemma is fundamental in the proof of completeness; it shows that algorithmic type-checking judgements are preserved under subtyping of contexts and super-typing of types.

Lemma 18. *Let $\Gamma \leq \Delta$. If $\Delta \vdash v \Leftarrow A$ and $\Delta \vdash A \leq B$, then $\Gamma \vdash v \Leftarrow \mathbf{nbe}_\Delta(B)$. If $\Delta \vdash k \Rightarrow D$, then exists C, such that $\Gamma \vdash k \Rightarrow C$ and $\Gamma \vdash C \leq D$.*

Completeness of type-checking proceeds by induction on the normal form being checked and for each case use Lem. 3 to analyse the possible shape of the type.

Theorem 6 (Completeness). *If $\Gamma \vdash v \colon A$, then $\Gamma \vdash v \Leftarrow \mathbf{nbe}_\Gamma(A)$.*

6 Conclusion

The main contributions of this paper are the decidability results for MLTT with subtyping and the proof of completeness and correctness of algorithmic subtyping and type-checking algorithms. To prove them we have developed the meta-theory for MLTT with subtyping. In particular, injectivity and subtyping reflection of constructors are important meta-theoretical results. We have not found previous uses of a relation finer than set-theoretical inclusion for semantic subtyping.

Related work. We briefly comment on the most relevant papers on subtyping related to our work. We refer to [12] for a discussion between coercive subtyping and subsumptive subtyping, the one considered in the present paper; and to [17] for subtyping for PTSs. Subtyping for MLTT was first presented in [6], that work introduces a subtyping relation between dependent record types; it includes a typechecker, but it lacks any meta-theoretical analysis. In [8] the meta-theory of a MLTT with dependent records, subtyping, and singletons is developed. Both [5] and [16] study logical frameworks with subtyping and their meta-theory, concluding with type-checking algorithms. Unlike our calculus, those logical frameworks lack computation at the level of types.

A cumulative hierarchy of universes together with subtyping for dependent products is considered in ECC [11], but subtyping is allowed only on the covariant argument and equality in the contravariant argument of Fun; with this restriction a minimal type property can be proved and set-theoretic inclusion is enough to model subtyping.

Upon submission of this article we learned about [14], where a similar calculus is considered: a cumulative hierarchy of universes, a contravariant rule for subtyping and typed equality. But they have typed abstraction which leads them to a principal type property. They obtain similar decidability results employing different techniques.

Further work. An interesting open problem is the equivalence of ECC, with its untyped (in)equality, and a presentation where conversion and subtyping are typed, as in our system. It is not clear if the technique of Siles and Herbelin [15] can be extended to handle subtyping. As remarked by Adams [4] having injectivity of Fun of the system with typed equality is enough to prove the equivalence; it would be interesting to try to use NbE to prove injectivity of Fun for ECC with typed (in)equality and check if the equivalence holds.

We want to study in the near future some extensions of the core calculus presented in this paper: for example, to add inductive types, singleton types, or dependent records. A first issue to address is for which of these features one needs explicit coercions and when they can be inferred. Another interesting question is if we can handle (η) combining semantic subtyping of this paper with the PER model of [2]. Another direction to advance is to prove the correctness of the normalisation function as implemented in Haskell; since it is not affected by the presence of subtyping, we aim to prove it for the minimal MLTT with one universe, dependent products and without (η).

Acknowledgements. We are thankful to Thierry Coquand for pointing out that the absence of (η) would make semantic constructions simpler. We are also grateful to Andreas Abel for letting us know about [14] and to three anonymous reviewers whose comments helped us to improve the final version of the article.

References

1. Abel, A., Coquand, T., Dybjer, P.: Verifying a semantic $\beta\eta$-conversion test for Martin-Löf type theory. In: Audebaud, P., Paulin-Mohring, C. (eds.) MPC 2008. LNCS, vol. 5133, pp. 29–56. Springer, Heidelberg (2008)
2. Abel, A., Coquand, T., Pagano, M.: A modular type-checking algorithm for type theory with singleton types and proof irrelevance. Logical Methods in Computer Science 7(2) (2011)
3. Abramsky, S., Jung, A.: Domain Theory. In: Handbook of Logic in Computer Science, pp. 1–168. Oxford University Press (1994)
4. Adams, R.: Pure type systems with judgemental equality. Journal of Functional Programming 16(2), 219–246 (2006)
5. Aspinall, D., Compagnoni, A.B.: Subtyping dependent types. Theor. Comput. Sci. 266(1-2), 273–309 (2001)
6. Betarte, G., Tasistro, A.: Extension of Martin-Löf type theory with record types and subtyping. Oxford Logic Guides, pp. 20–42. Oxford University Press, USA (1998)
7. Coquand, T.: An algorithm for type-checking dependent types. Science of Computer Programming 26, 167–177 (1996)
8. Coquand, T., Pollack, R., Takeyama, M.: A logical framework with dependently typed records. Fundamenta Informaticæ 65(1-2), 113–134 (2005)
9. Dybjer, P.: A general formulation of simultaneous inductive-recursive definitions in type theory. Journal of Symbolic Logic 65 (2000)
10. Fiore, M.: Semantic analysis of normalisation by evaluation for typed lambda calculus. In: Proceedings of the 4th ACM SIGPLAN International Conference on Principles and Practice of Declarative Programming, PPDP 2002, pp. 26–37. ACM, New York (2002)
11. Luo, Z.: Computation and reasoning: a type theory for computer science. International series of monographs on computer science. Clarendon Press (1994)
12. Luo, Z., Soloviev, S., Xue, T.: Coercive subtyping: Theory and implementation. Inf. Comput. 223, 18–42 (2013)
13. Martin-Löf, P.: An Intuitonistic Theory of Types: Predicative Part. In: Logic colloquium 1973: Proceedings of the Logic Colloquium, Bristol, pp. 73–118 (1975)
14. Scherer, G.: Universe subtyping in Martin-Löf type theory. Technical report, Ludwig-Maximilians-Universität München (2011)
15. Siles, V., Herbelin, H.: Pure type system conversion is always typable. J. Funct. Program. 22(2), 153–180 (2012)
16. Stone, C.A., Harper, R.: Extensional equivalence and singleton types. ACM Trans. Comput. Logic 7(4), 676–722 (2006)
17. Zwanenburg, J.: Pure type systems with subtyping. In: Girard, J.-Y. (ed.) TLCA 1999. LNCS, vol. 1581, pp. 381–396. Springer, Heidelberg (1999)

Small Induction Recursion

Peter Hancock[2], Conor McBride[2], Neil Ghani[2], Lorenzo Malatesta[2], and Thorsten Altenkirch[1]

[1] University of Nottingham*
[2] University of Strathclyde**

Abstract. There are several different approaches to the theory of data types. At the simplest level, polynomials and containers give a theory of data types as free standing entities. At a second level of complexity, dependent polynomials and indexed containers handle more sophisticated data types in which the data have an associated indices which can be used to store important computational information. The crucial and salient feature of dependent polynomials and indexed containers is that the index types are defined in advance of the data. At the most sophisticated level, induction-recursion allows us to define data and indices simultaneously.

This work investigates the relationship between the theory of *small* inductive recursive definitions and the theory of dependent polynomials and indexed containers. Our central result is that the expressiveness of small inductive recursive definitions is exactly the same as that of dependent polynomials and indexed containers. A second contribution of this paper is the definition of morphisms of small inductive recursive definitions. This allows us to extend our main result to an equivalence between the category of small inductive recursive definitions and the category of dependent polynomials/indexed containers. We comment on both the theoretical and practical ramifications of this result.

1 Introduction

One of the most important concepts in computer science is the notion of an inductive definition. It is difficult to trace back its origin since this concept permeates the history of proof theory and a large part of theoretical computer science. In recent years, the desire to explore, understand, and extend the concept of an inductive definition has led different researchers to different but (extensionally) equivalent notions. The theory of containers [1], and polynomial functors [18,12] are some of the outcomes of this research These theories give a comprehensive account of those data types such as `Nat` (the natural numbers), `List a` (lists containing data of a given type `a`), and `Tree a` (trees containing, once more, data of a given type `a`) which are free-standing in that their definition does not require the definition of other inter-related data types.

* Supported by a grant from the Institute of Advanced Studies, Princeton, and EPSRC grant EP/G03298X/1.
** Supported by EPSRC grant EP/G033056/1.

M. Hasegawa (Ed.): TLCA 2013, LNCS 7941, pp. 156–172, 2013.

These theories are too simple to capture more sophisticated data types possessing features such as: (*i*) variable binding as in the untyped and typed lambda calculus; (*ii*) constraints as in red black trees; and (*iii*) extra information about data having such types - e.g. vectors which record the lengths of lists. Therefore containers and polynomials have been generalised to indexed containers [2] and dependent polynomials [12,13] to capture not only free standing data types such as those mentioned above, but also data types where the data are indexed by an index storing computationally relevant information. Containers and (non-dependent) polynomials arise as specific instances of these generalised notions where the type of indices is chosen to be a singleton type.

However, even dependent polynomials and indexed containers fail to cover all data types of interest as they require the indices to be defined before the data. Induction-recursion (IR), developed in the seminal works of Peter Dybjer and Anton Setzer [9,10,11], remedies this deficiency. The key feature of an inductive-recursive definition is the simultaneous inductive definition of a small type X of indices together with the recursive definition of a function $T : X \to D$ from X into a type D which may be large or small. Since X and T can be defined at the same time, the indices need not be defined in advance of the data. Universes (introduced by Martin-Löf in the early 70's [17]) are paradigm examples of inductive recursive definitions.

It is natural to ask what is the relationship between dependent polynomials and induction recursion. Can we characterise those inductive-recursive definitions which correspond to dependent polynomials? This paper makes the following concrete contributions: i) we show that dependent polynomial and indexed containers correspond exactly to small inductive-recursive definitions, where "smallness" refers to the size of the target-type D; ii) we define morphisms of small inductive recursive definitions and use this notion to show that the resulting category of small inductive-recursive definitions is equivalent to the category of dependent polynomials and indexed containers; and iii) we extend these results to cover small indexed induction recursion.

These results have theoretical and practical importance. At the theoretical level, while it has been conjectured that the power of induction recursion lies in the case where D is large, no formal proof exists before this paper. Further, we contribute to the theory of induction recursion by giving a notion of morphism between inductive recursive definitions in the same way that containers, indexed containers and dependent polynomials have morphisms. Finally, dependent polynomials and indexed containers have a rich algebraic structure so our work shows that structure can be transported to Small IR - see the conclusion for details. Note that this structure is defined by universal properties and hence its transportation would not be possible without the work in Section 5 on morphisms. At a practical level, while systems such as Agda accept induction recursion, some systems, eg Coq, do not. This work gives a simple way to add small induction recursion to Coq by showing how to translate such definitions into indexed containers. It also allows programmers to convert definitions between the two forms, according to which works better for their own applications. To achieve this, and to make the paper more accessible and non-hermetic, and to type check our translations, we have implemented our translations in Agda and provide lots of Agda examples.

The paper is organised as follows: in Section 2 we set our notation, while Section 3 recalls indexed containers, dependent polynomials and induction recursion. In Section 4, we show an equivalence between data types definable by small IR and those data types definable using dependent polynomials and/or indexed containers. In Section 5 we introduce the category of small inductive-recursive definitions and show it equivalent to the category of dependent polynomials/indexed containers. In Section 6 we briefly recall the theory of indexed inductive-recursive definitions, and extend the previous equivalence to the case of indexed small induction recursion. We conclude in Section 7.

The sources, proofs and additional materials for this paper are available from `http://personal.cis.strath.ac.uk/~conor/pub/SmallIR`.

2 Preliminaries and Internal Languages

We follow the standard approach of using extensional Martin-Löf type theory as the internal language to formalise reasoning with the locally cartesian closed structure of the category of sets — see [19,14] for details [1]. Our notation follows Agda — indeed, this paper is a literate Agda development. We write identity types as $x \equiv y$ and assume uniqueness of identity proofs. We write ΣT or $(s : S) \times T\ s$ and ΠT or $(s : S) \to T\ s$ for the dependent sum and dependent product in Martin-Löf type theory of $T : S \to \mathsf{Set}$. The elements of $(s : S) \times T\ s$ are pairs (s, t) where $s : S$ and $t : T\ s$ may be projected by π_0 and π_1. The elements of $(s : S) \to T\ s$ are functions $\lambda\, x \to t\ x$ mapping each element $s : S$ to an element $t\ s$ of $T\ s$.

Categorically, we think of an I-indexed type as a morphism $f : X \to I$ with codomain I. These are objects of the slice category Set/I. Morphisms in Set/I from object $f : X \to I$ to object $f' : X' \to I$ are given by functions $h : X \to X'$ such that $f = f' \circ h$. Type theoretically, we can represent matters in more or less the same way – that is, an object in a slice Set/I is a pair (X, f) of a set X (the domain), and a function $f : X \to I$. However, another possibility is to model an I-indexed type by a function $F : I \to \mathsf{Set}$ where $F\ i$ represents the fibre of f above i, i.e. as $(X, f)^{-1}\ i$, defined as follows.

$$
\begin{array}{ll}
\cdot^{-1} : \mathsf{Set}/I \to (I \to \mathsf{Set}) & \exists. : (I \to \mathsf{Set}) \to \mathsf{Set}/I \\
(X, f)^{-1}\ i = (x : X) \times f\ x \equiv i & \exists.F = (\Sigma F, \pi_0)
\end{array}
$$

We write $\exists.F$ for the inverse of this operator: that these are inverse (given uniqueness of identity proofs) is at the heart of the well known equivalence between the categories Set/I and $I \to \mathsf{Set}$ which, in a sense, underlies the equivalences we describe in this paper.

[1] The correspondence between lcccs and Martin Löf type theories is affected by coherence problems related to the interpretation of substitution. We refer to [7], [14] and more recently [5] for different solutions to these problems.

Given a function $k : I \rightarrow J$, we can form three very important functors. The pullback along k of an object $f : X \rightarrow J$ of Set/J defines a *reindexing* functor $\Delta_k : \mathsf{Set}/J \rightarrow \mathsf{Set}/I$. Δ_k has both a left adjoint and a right adjoint, respectively $\Sigma_k, \Pi_k : \mathsf{Set}/I \rightarrow \mathsf{Set}/J$. In the internal language, we define these for $\cdot \rightarrow \mathsf{Set}$, as follows:

$$\begin{array}{ll} \Delta_k : (J \rightarrow \mathsf{Set}) \rightarrow (I \rightarrow \mathsf{Set}) & \Sigma_k : (I \rightarrow \mathsf{Set}) \rightarrow (J \rightarrow \mathsf{Set}) \\ \Delta_k\, F\, i = F\,(k\, i) & \Sigma_k\, F\, j = (i\!:\!I) \times k\, i \equiv j \times F\, i \\ & \Pi_k : (I \rightarrow \mathsf{Set}) \rightarrow (J \rightarrow \mathsf{Set}) \\ & \Pi_k\, F\, j = (i\!:\!I) \rightarrow k\, i \equiv j \rightarrow F\, i \end{array}$$

3 Three Theories of Data Types

The foundation of our understanding of data types is initial algebra semantics. Thus, formally our theories of data types are in fact theories of functors which have initial algebras. In this section we recall the notions of dependent polynomials, indexed containers and induction recursion, each of which define certain classes of functors and hence data types.

Definition 1. *The collection of* dependent polynomials *with input indices I and output indices O is written* $\mathsf{Poly}\, I\, O$ *and consists of triples (r, t, q) where $I \leftarrow^r P \rightarrow^t S \rightarrow^q O$. A* dependent polynomial functor *is any functor isomorphic to some $[\![(r,t,q)]\!]_{\mathsf{Poly}} = \Sigma_q \circ \Pi_t \circ \Delta_r : \mathsf{Set}/I \rightarrow \mathsf{Set}/O$, illustrated as follows:*

$$\mathsf{Set}/I \xrightarrow{\Delta_r} \mathsf{Set}/P \xrightarrow{\Pi_t} \mathsf{Set}/S \xrightarrow{\Sigma_q} \mathsf{Set}/O\,.$$

While the definition above is concise, some readers may prefer a more concrete presentation. So we turn to the representation of dependent polynomials in the internal language. This leads us to the notion of an *indexed container.*

Definition 2. Indexed containers *with input indices I and output indices O is written* $\mathsf{IC}\, I\, O$ *and consists of triples (S, P, n) where $S : O \rightarrow \mathsf{Set}$, $P : (o\!:\!O) \rightarrow S\, o \rightarrow \mathsf{Set}$ and $n : (o\!:\!O) \rightarrow (s\!:\!S\, o) \rightarrow P\, o\, s \rightarrow I$. Its extension is the functor*

$$\begin{array}{l} [\![\cdot]\!]_{\mathsf{IC}} : \mathsf{IC}\, I\, O \rightarrow (I \rightarrow \mathsf{Set}) \rightarrow (O \rightarrow \mathsf{Set}) \\ [\![(S,P,n)]\!]_{\mathsf{IC}}\, X\, o = (s\!:\!S\, o) \times (p\!:\!P\, o\, s) \rightarrow X\,(n\, o\, s\, p) \end{array}$$

Every dependent polynomial functor (r, t, q) gives rise to an indexed container $(\hat{S}, \hat{P}, n)$.

$$\begin{array}{ll} \hat{S}\, o & = (S, q)^{-1}\, o \\ \hat{P}\, o\,(s, \_) & = (P, t)^{-1}\, s \\ n\ o\,(s,\_)\,(p,\_) & = r\, p \end{array}$$

We may readily check that

$$\begin{array}{rl} [\![(\hat{S}, \hat{P}, n)]\!]_{\mathsf{IC}}\, F\, o = & (sq\!:\!((S,q)^{-1}\, o)) \times (pq\!:\!((P,t)^{-1}\,(\pi_0\, sq))) \rightarrow F\,(r\,(\pi_0\, pq)) \\ \cong & (s\!:\!S) \times (q\, s \equiv o) \times (p\!:\!P) \rightarrow (t\, s \equiv p) \rightarrow F\,(r\, p) \\ = & (\Sigma_q \circ \Pi_t \circ \Delta_r)\, F\, o \end{array}$$

confirming the equivalence between indexed containers and dependent polynomials.

Polynomials (resp. containers) arise as a special case of dependent polynomials (indexed containers) by choosing $I = O = 1$. Notice the salient feature of both dependent polynomials and indexed containers — that the input and output indices I and O are fixed and must be defined in advance. This restriction means that neither dependent polynomials nor indexed containers suffice to define all the data types in which we are interested. Paradigmatic undefinable data types are universes of types. These are pairs (U, T) consisting of a set U, thought as a set of names or codes, and of a function $T : U \rightarrow Set$, thought as a "decoding function" which assigns a set $T\,u$ to every element u of U. For example, consider a universe containing the type of natural numbers $\mathbb{N}$ and closed under Σ-types. Such a universe will be the least solution of the

$$\begin{aligned} U &= 1 + (u\colon U) \times T\ u \rightarrow\ U \\ T\,(\mathsf{inl}\ \star) &= \mathbb{N} \\ T\,(\mathsf{inr}\,(u, f)) &= (x\colon T\ u) \times T\ (f\ x) \end{aligned}$$

Note how, in this example, the set of codes U must be defined simultaneously with the decoding function T - something not possible with dependent polynomials or indexed containers which require that U be defined before T. Dybjer and Setzer developed the theory of induction recursion to cover exactly such inductive definitions where the indices and the data must be defined simultaneously. The first presentation of induction-recursion [8] was as an external schema. In later presentations [9,10], inductive recursive definitions are given via a type of codes $\mathsf{IR}\ I\ O$. [2]

Definition 3. *Let I, O be types. The type of* IR *I O-codes has the following constructors*

$$\begin{array}{llr} \textbf{data}\ \mathsf{IR}\ (I\ O\ :\ \mathsf{Set})\ :\ \mathsf{Set1}\ \textbf{where} & & \\ \quad \iota\ :\ (o\ :\ O) & & \rightarrow\ \mathsf{IR}\,I\ O \\ \quad \sigma\ :\ (S\ :\ \mathsf{Set})\ (K\ :\ S & \rightarrow\ \mathsf{IR}\,I\ O) & \rightarrow\ \mathsf{IR}\,I\ O \\ \quad \delta\ :\ (P\ :\ \mathsf{Set})\ (K\ :\ (P\ \rightarrow\ I)\ \rightarrow & \mathsf{IR}\,I\ O) & \rightarrow\ \mathsf{IR}\,I\ O \end{array}$$

In general I and O may be large types such as Set *or* Set $\rightarrow$ Set *etc. Above, we encode* small *induction recursion (small* IR*) we mean the cases where I and O are sets.*

Dybjer and Setzer [9,10] prove that every IR code defines a functor. In the case of *small* IR, this functorial semantics can be given in terms of slice categories. Before giving this semantics, we note that slice categories have set-indexed coproducts. That is, given a set A, and an A-indexed collection of objects $f_a : X_a \rightarrow I$ of Set/I, the cotuple $[f_a]_{a:A} : \coprod_{a:A} X_a \rightarrow I$ is the coproduct of the objects f_a in Set/I. We use $\mathsf{in}_a : X_a \rightarrow \coprod_{a:A} X_a$ for the a-th injection. In the internal

[2] Dybjer and Setzer treated only the case where I and O are the same. Our mild generalization allows the construction of partial fixed points.

language, the coproduct of an A-indexed family $X_a : I \to \mathsf{Set}$ is the function mapping i to $(a{:}A) \times X_a\ i$. We use these coproducts to give a definition of the functor denoted by an IR code more compact than - but of course equivalent to - that originally provided by Dybjer and Setzer.

Definition 4. *Let I, O be sets, $\gamma : \mathsf{IR}\ I\ O$. The action of the functor $[\![\gamma]\!] : \mathsf{Set}/I \to \mathsf{Set}/O$ on an object $f : X \to I$ of Set/I is defined by recursion on γ as follows*

– *if $\gamma = \iota\ o$ for some $o : O$*

$$[\![\iota\ o]\!]\,(f : X \to I) = (\lambda\_.o) : 1 \to O$$

– *if $\gamma = \sigma\ S\ K$ for some $S : \mathsf{Set}$, $K : S \to \mathsf{IR}\ I\ O$*

$$[\![\sigma\ S\ K]\!]\,(f : X \to I) = \textstyle\coprod_{s:S} [\![K\ s]\!]\ f$$

– *if $\gamma = \delta\ P\ K$ for some $P : \mathsf{Set}$, $K : (P \to I) \to \mathsf{IR}\ I\ O$*

$$[\![\delta\ P\ K]\!]\,(f : X \to I) = \textstyle\coprod_{x:P \to X} [\![K(f \circ x)]\!]\ f$$

An IR *functor is any functor isomorphic to one of the form $[\![\gamma]\!]$ for some $\gamma : \mathsf{IR} I O$.*

We can give the above construction in type theory, using the *direct* translation of slices, closed under dependent sum, yielding an interpretation in the style of Dybjer and Setzer:

$$
\begin{array}{lcl}
[\![\cdot]\!]_{\mathsf{DS}} : \mathsf{IR}\ I\ O \to \mathsf{Set}/I \to \mathsf{Set}/O & & \\
[\![\iota\ o]\!]_{\mathsf{DS}}\ \ (X, f) & = & (1, \lambda\_ \to o) \\
[\![\sigma\ S\ K]\!]_{\mathsf{DS}}\ (X, f) & = & (s{:}S) \times \qquad\quad\ [\![K\ s]\!]_{\mathsf{DS}}\ (X, f) \\
[\![\delta\ P\ K]\!]_{\mathsf{DS}}\ (X, f) & = & (x{:}P \to X) \times [\![K\ (f \circ x)]\!]_{\mathsf{DS}}\ (X, f)
\end{array}
$$

For any $\gamma : \mathsf{IR}\ I\ I$, we can construct of an inductive datatype simultaneously with its recursive decoder as the initial algebra, $((\mu\,\gamma, \mathsf{decode}\ \gamma), \mathsf{in})$, of $[\![\gamma]\!]_{\mathsf{DS}}$.

```
data μ (γ : IR I I) : Set where
  in : dom ([[γ]]DS (μ γ, decode γ)) → μ γ

decode : (γ : IR I I) → μ γ → I
decode γ (in t) = fun ([[γ]]DS (μ γ, decode γ)) t
```

Here dom computes the domain of a universe and fun the decoder of a universe. As an example, we show that all containers [1] can be defined by induction recursion:

Example 1 (containers and W*-types).* Given a simple container (S, P), where $S : \mathsf{Set}$ and $P : S \to \mathsf{Set}$, we can represent it by an $\mathsf{IR}\ 1\ 1$ code as follows:

```
cont : (S : Set) → (P : S → Set) → IR 1 1
cont S P = (σ S λ s → δ (P s) λ p → ι ⋆)
```

We note that $\mathsf{dom}\ [\![\mathsf{cont}\ S\ P]\!]_{\mathsf{DS}}\ (X, \_) \ = \ (s : S) \times (P\ s \ \to\ X) \ \times\ 1$ and that $\mu\ (\mathsf{cont}\ S\ P)$ thus amounts to Martin-Löf's well-ordering type $\mathsf{W}\ S\ P$. As a corollary of our main result we shall see that $\mathsf{IR}\ 1\ 1$ codes describe exactly the category of containers and their morphisms.

Example 2 (A Language of Sums and Products). If $\mathsf{Fin} : \mathbb{N} \to \mathsf{Set}$ maps n to a set with n elements, we can implement finitary summation and product with types:

$$\mathsf{sum}\ \mathsf{prod} : (n{:}\mathbb{N}) \to (\mathsf{Fin}\ n \to \mathbb{N}) \to \mathbb{N}$$

Next we can encode a datatype of numerical expressions closed under constants, sums and products, where each expression decodes to its numerical value — we need to know these values to compute the correct domains for the sums and the products.

$$\begin{array}{l}
\mathbf{data}\ \mathsf{Tag} : \mathsf{Set}\ \mathbf{where}\ \mathsf{fin}'\ \mathsf{sum}'\ \mathsf{prod}' : \mathsf{Tag} \\
\mathsf{lang} : \mathsf{IR}\ \mathbb{N}\ \mathbb{N} \\
\mathsf{lang} = \sigma\ \mathsf{Tag}\ \lambda\ \{\mathsf{fin}' \to \sigma\ \mathbb{N}\ \lambda\ n \to \iota\ n \\
\qquad\qquad\qquad ; \mathsf{sum}' \to \delta\ 1\ \lambda\ n \to \delta\ (\mathsf{Fin}\ (n\ \star))\ \lambda f \to \iota\ (\mathsf{sum}\ (n\ \star)\ f) \\
\qquad\qquad\qquad ; \mathsf{prod}' \to \delta\ 1\ \lambda\ n \to \delta\ (\mathsf{Fin}\ (n\ \star))\ \lambda f \to \iota\ (\mathsf{prod}\ (n\ \star)\ f)\} \\
\mathsf{example} : \mu\ \mathsf{lang} \\
\mathsf{example} = \mathsf{in}\ (\mathsf{sum}', (\lambda\ _ \to \mathsf{in}\ (\mathsf{fin}', 5, \star)), (\lambda\ n \to \mathsf{in}\ (\mathsf{fin}', n, \star)), \star)
\end{array}$$

The example expression denotes $\sum_{n<5} n$, and indeed, $\mathsf{decode}\ \mathsf{lang}\ \mathsf{example} = \mathit{10}$.

Having introduced dependent polynomials, indexed containers and small induction recursion, we can now turn to the main focus of the paper, namely showing that they define the same class of functors and hence define the same class of data types. The key to the construction is observing that we may just as well interpret $\mathsf{IR}\ I\ O$ with our $I \to \mathsf{Set}$ presentation of slices.

$$\begin{array}{l}
[\![\cdot]\!]_{\mathsf{IR}} : \mathsf{IR}\ I\ O \to (I \to \mathsf{Set}) \to (O \to \mathsf{Set}) \\
[\![\iota\ o']\!]_{\mathsf{IR}}\quad F\ o = o' \equiv o \\
[\![\sigma\ S\ K]\!]_{\mathsf{IR}}\ F\ o = (s{:}S) \times ([\![K\ s]\!]_{\mathsf{IR}}\ F\ o) \\
[\![\delta\ P\ K]\!]_{\mathsf{IR}}\ F\ o = (\mathit{if}{:}P \to \Sigma F) \times ([\![K\ (\pi_0 \circ \mathit{if})]\!]_{\mathsf{IR}}\ F\ o)
\end{array}$$

The correspondence up to trivial isomorphism between $[\![\cdot]\!]_{\mathsf{IR}}$ and $[\![\cdot]\!]_{\mathsf{DS}}$ is readily observed by considering F here to be an arbitrary $(X, f)^{-1}$.

4 From Poly to Small IR and Back

We divide this section into two: (*i*) we first show how to translate dependent polynomials, and hence indexed containers, into IR codes; and (*ii*) we then show how every small IR code can be translated into a dependent polynomial. Crucially, we show that these translations preserve the functorial semantics of dependent polynomials and IR codes.

From Poly to Small IR. We have already seen (example 1) that the extension of a container is an IR functor. We now extend this result to indexed containers and dependent polynomials.

Lemma 1. *Every dependent polynomial functor is an* IR *functor.*

It is enough to show that, for every dependent polynomial (r, t, q) : Poly I O, there is an IR I O-code, whose interpretation is isomorphic to the dependent polynomial functor $[\![(r, t, q)]\!]_{\mathsf{Poly}}$. Our candidate for this IR-code is given and interpreted as follows

$$[\![\sigma\ S\ \lambda\ s \rightarrow \delta\ ((P, t)^{-1}\ s)\ \lambda\ i \rightarrow \sigma\ (i \equiv r \circ \pi_0)\ \lambda\ \_ \rightarrow \iota\ (q\ s)]\!]_{\mathsf{IR}}\ F\ o =$$
$$(s\!:\!S) \times (if : ((P, t)^{-1}\ s \rightarrow \Sigma F)) \times (\pi_0 \circ if \equiv r \circ \pi_0) \times (q\ s \equiv o)$$

which is readily seen to be isomorphic to $\Sigma_q\ (\Pi_t\ (\Delta_r\ F))\ o$

$$(s\!:\!S) \times (q\ s \equiv o) \times (p\!:\!P) \rightarrow (t\ p \equiv s) \rightarrow F\ (r\ p)$$

as the former effectively constrains the function if to choose $r\ p$ as the index of its F, for each position $(p, \_) : (P, t)^{-1}\ s$.

From Small IR to Poly. The essence of our embedding of IR I O into Poly I O consists of showing how three constructors for IR I O-codes can be interpreted in Poly I O.

Definition 5. *To each code* γ : IR I O *we use structural recursion on* γ *to define a dependent polynomial* $I \leftarrow^{t_\gamma} P_\gamma \rightarrow^{r_\gamma} S_\gamma \rightarrow^{q_\gamma} O$*:*

– *if* γ *is* $\iota\ o$*, then we define* $\mathsf{S}\ \gamma = 1$, $\mathsf{P}\ \gamma = 0$, $\mathsf{r}\ \gamma = !_I$, $\mathsf{t}\ \gamma = !_1$, *and* $\mathsf{q}\ \gamma\ \star = o$. *As a diagram, this is as follows.* $I \leftarrow^{!} 0 \rightarrow^{!} 1 \rightarrow^{o} O$*:*
– *if* γ *is* $\sigma\ S\ K$ *then the diagram is as follows.*

$$I \xleftarrow{[\mathsf{r}\ (K\ s)]_{s:S}} \textstyle\coprod_{s:S} \mathsf{P}\ (K\ s) \xrightarrow{\coprod_{s:S} \mathsf{t}\ (K\ s)} \coprod_{s:S} \mathsf{S}\ (K\ s) \xrightarrow{[\mathsf{q}\ (K\ s)]_{s:S}} O$$

Here (and in the next clause) we use $\coprod_{s:S} m\ s$ *to abbreviate the cotuple* $[\mathsf{in}_s \circ m\ s]_{s:S}$.
– *if* γ *is* $\delta\ P\ K$*, the diagram is as follows.*

$$\begin{array}{ccc} \coprod_{i:P\rightarrow I} (P \times \mathsf{S}\ (K\ i)) + \mathsf{P}\ (K\ i) & \xrightarrow{\coprod_{i:P\rightarrow I} [\pi_0, \mathsf{t}\ (K\ i)]} & \coprod_{i:P\rightarrow I} \mathsf{S}\ (K\ i) \\ \downarrow {\scriptstyle [[i \circ \pi_0, \mathsf{r}\ (K\ i)]]_{i:P\rightarrow I}} & & \downarrow {\scriptstyle [\mathsf{q}\ (K\ i)]_{i:P\rightarrow I}} \\ I & & O \end{array}$$

Note that in the last clause, it is crucial that we are dealing with small IR so that I *is a set, hence* $P \rightarrow I$ *is a set and hence the coproducts used are also small.*

We can now state the result concerning the second half of our isomorphism.

Lemma 2. *Every small* IR *functor is a dependent polynomial functor.*

To prove the lemma we define a function ϕ : IR I O $\rightarrow$ Poly I O by recursion on the structure of IR codes and then we prove by induction that the functorial semantics is preserved. Details of the proof can be found in the online in the expanded version of the paper at the url given in the introduction.

5 Equivalence between Small IR and Poly

We now extend our previous results to cover not just functors but also natural transformations. We will do this by (*i*) recalling the notion of morphism between dependent polynomials/indexed containers; (*ii*) introducing morphisms of IR codes, showing that the interpretation function, $[\![\_]\!]_{\mathsf{IR}} : \mathsf{IR}\ I\ O \to [\mathsf{Set}/I, \mathsf{Set}/O]$ can be extended to a functor which is full and faithful; and (*iii*) finally we prove the equivalence between the two resulting categories $\mathsf{IR}\ I\ O$ and $\mathsf{Poly}\ I\ O$. Note that our definition of morphisms for IR codes also covers the cases where I and O are large.

The Categories $\mathsf{Poly}\ I\ O$ **and** $\mathsf{IC}\ I\ O$**.** Dependent polynomials/indexed containers with fixed input and output index sets, I and O, form a category. In this section we recall the definition of the morphisms between dependent polynomials and their interpretation as natural transformations. We conclude by stating some properties of the categories of dependent polynomials/indexed containers which allows us to recast in elementary terms the dependent polynomials introduced in definition 5.

Definition 6. *A morphism between dependent polynomials* (r, t, q) *and* (r', t', q') *is given by a diagram of the form (where the bottom square is a pullback of* u *and* t'*).*

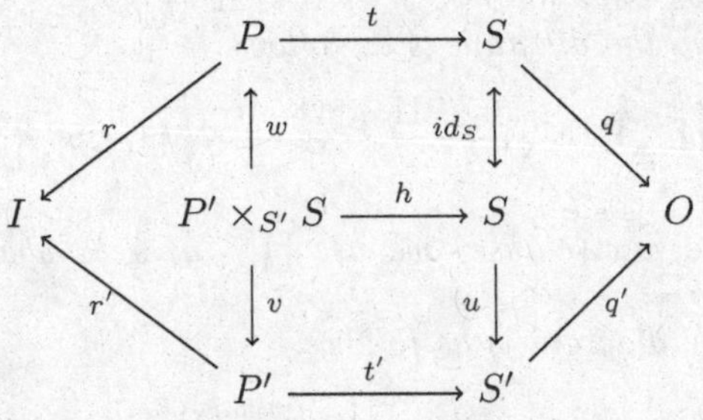

From now on, $\mathsf{Poly}\ I\ O$ will indicate the category of dependent polynomials with fixed input and output index sets I, O and their morphisms. In a similar manner we can define morphism between indexed containers.

Definition 7. *A morphism between* (S, P, n) *and* (S', P', n') *consists of*

- *a function* $u : (o : O) \to S\ o \to S'\ o$;
- *a function* $f : (o : O) \to S\ o \to P'\ o\ (u\ o\ s) \to P\ o\ s$;

such that for every $o : O$, $s : S\ o$ *and* $p' : P\ o\ (u\ o\ s)$ *we have* $n\ o\ s\ (f\ o\ s\ p') = n'\ o\ (u\ o\ s)\ p'$*.*

We will indicate with $\mathsf{IC}\ I\ O$ the category of indexed containers and their morphisms. The main result concerning these morphisms is the following (Theorem 2.12 in [13]). We state the result for dependent polynomials but clearly an analogue result holds also for indexed containers.

Theorem 1 ([13] Theorem 2.12). *Given dependent polynomials* (r, t, q) *and* (r', t', q')*, every natural transformation* $[\![(r, t, q)]\!] \to [\![(r', t', q')]\!]$ *is represented in an essentially unique way by a commuting diagram as in definition 6.*

This theorem ensures that the assignment to each dependent polynomial of its extension is a functor, and moreover this functor is full and faithful. In the following we indicate with *PolyFun I O* the full subcategory of $[\mathsf{Set}/I, \mathsf{Set}/O]$ whose objects are dependent polynomial functors and whose morphisms are natural transformation between them[3].

Corollary 1 (Representation). *For any pair of sets* I, O *the functor*

$$[\![\ ]\!] : \mathsf{Poly}\ I\ O \to \mathit{PolyFun}(I, O)$$

is an equivalence of categories.

Dependent polynomials and indexed containers have several interesting closure properties. Here we only need closure under set-indexed coproducts and binary product. Note that we had to define morphisms before introducing these closure properties to ensure that they have the required categorical universal properties. The sum of a K-indexed family of dependent polynomials $\{Q_k = (r_k, t_k, q_k) \mid k : K\}$, for an arbitrary set K, is the dependent polynomial $\coprod_{k:K} Q_k$ given by the following diagram

$$I \xleftarrow{[r_k]_{k:K}} \coprod_{k:K} P_k \xrightarrow{\coprod_{k:K} t_k} \coprod_{k:K} S_k \xrightarrow{[q_k]_{k:K}} O$$

where $\coprod_{k:K} t_k = [\mathsf{in}_k \circ t_k]_{k:K}$. Note that the dependent polynomial associated to $\sigma\ S\ K : \mathsf{IR}\ I\ O$ is of exactly this form. The product of two dependent polynomials (r, t, q) and (r', t', q') is the evident dependent polynomial

$$I \longleftarrow (P' \times_O S) + (P \times_O S') \longrightarrow S \times_O S' \longrightarrow O.$$

We can now describe the dependent polynomial associated to a code $\delta\ P\ K : \mathsf{IR}\ I\ O$ as the sum of products of a family of dependent polynomials. We start with a family of dependent polynomials $\{(\mathsf{r}\ (K\ i), \mathsf{t}\ (K\ i), \mathsf{q}\ (K\ i) \mid i : P \to I\}$. For each element of this family we take the product of it with the dependent polynomial

$$I \xleftarrow{i \circ \pi_0} P \times O \xrightarrow{\pi_1} O \xrightarrow{id_O} O$$

and then we take the sum of these products over the set $P \to I$.

[3] The original result for polynomial functors (Theorem 2.12 in [13]) is stated in terms of strong natural transformations. We can avoid mention of strength since natural transformations between functors on slices of Set are automatically strong.

The Category of Small IR Codes. We know how to define small IR codes and interpret them as functors between slices of Set. In this section we introduce morphisms between small IR I O-codes. Our definition will ensure that every such morphism gives rise to a natural transformation between the corresponding IR functors – and *vice versa.* We start this section developing the appropriate categorical description of the semantics of IR constructors. The constructor ι simply represents constant functors while the constructor σ takes coproducts of functors. The following lemma tells us more about the semantics of δ.

Lemma 3. *Given an object $k : X \to I$, there is a natural isomorphism*

$$[\![\delta\ P\ K]\!]\ k \cong \coprod_{i:P\to I} Hom_{\mathsf{Set}/I}(i,k) \otimes [\![K\ i]\!]_{\mathsf{IR}}\ k$$

Here $\otimes$ indicates the tensor product. Given a set X and $i : Y \to I$, the object $X \otimes i$ is nothing but the copower $\coprod_{x:X} i$, i.e the X-fold coproduct of the object i.

Proof. We have a natural isomorphism

$$\begin{aligned} [\![\delta\ P\ K]\!]\ k &= \coprod_{x:P\to X} [\![K(k\circ x)]\!]_{\mathsf{IR}}\ k \\ &\cong \coprod_{i:P\to I} \coprod_{x:P\to X} (i \equiv k\circ x) \otimes [\![K\ i]\!]_{\mathsf{IR}}\ k. \end{aligned}$$

Then observe that $\coprod_{x:P\to X}(i \equiv k\circ x) \cong Hom_{\mathsf{Set}/I}(i,k)$.

Thanks to this lemma, we are able to characterise the semantics of δ-codes through a well-known universal construction in category theory: the left Kan extension.

If $i : X \to I$ is an object in Set/I we use $(+\ i)$, in the following lemma, to indicate the functor

$$\begin{aligned} (+i) : \mathsf{Set}/I &\longrightarrow \mathsf{Set}/I \\ k &\longmapsto [i,k]. \end{aligned}$$

Theorem 2. *There is a natural isomorphism*

$$[\![\delta\,P\,F]\!] \cong \coprod_{i:P\to I} Lan_{(+i)}[\![F\ i]\!]$$

Our definition of IR I O-morphisms is based on this isomorphism. First, we recall the universal property characterising the left Kan extension $Lan_G F : \mathbb{B} \to \mathbb{C}$ of a functor $F : \mathbb{A} \to \mathbb{C}$ along $G : \mathbb{A} \to \mathbb{B}$; for every functor $H : \mathbb{B} \to \mathbb{C}$ there is a bijection

$$Nat(Lan_G F, H) \cong Nat(F, H\circ G)$$

natural in H. We also need to check that IR I O-functors are closed by precomposition with functors of the form $(+i)$. Fortunately, this can be easily checked by structural induction on codes. We just state the result.

Lemma 4. *Given $\gamma :$ IR I O, and a function $i : P \to I$ there exists $\gamma^i :$ IR I O-code such that*

$$[\![\gamma]\!]_{\mathsf{IR}} \circ (+i) = [\![\gamma^i]\!]_{\mathsf{IR}}$$

We can now define IR morphisms by structural induction on codes as follows.

Definition 8. *Let $\gamma, \gamma' : \mathsf{IR}\ I\ O$ we define the homset $\mathsf{IR}(\gamma, \gamma')$ as follows.*
Morphisms from ι-codes:

1A. $IR(\iota\ o, \iota\ o') = o \equiv o'$
1B. $IR(\iota\ o, \sigma\ S\ K) = \coprod_{s:S} IR(\iota\ o, K\ s)$
1C. $IR(\iota\ o, \delta\ P\ K) = \coprod_{e:P\to\emptyset} IR(\iota\ o, K\ (!\circ e))$

Morphisms from σ-codes:

2. $IR(\sigma\ S\ K, \gamma) = \prod_{s:S} IR(K\ s, \gamma)$

Morphisms from δ-codes:

3. $IR(\delta\ P\ K, \gamma) = \prod_{i:P\to I} IR(K\ i, \gamma^i)$

The following theorem shows we have the right notion of morphism for IR codes.

Theorem 3. *The interpretation $[\![\text{-}]\!]_{\mathsf{IR}}$ of $\mathsf{IR}\ I\ O$-codes can be extended to morphisms: we can associate to each $\mathsf{IR}\ I\ O$-morphism $f : \gamma \to \gamma'$ a natural transformation $[\![f]\!]_{\mathsf{IR}} : [\![\gamma]\!]_{\mathsf{IR}} \to [\![\gamma']\!]_{\mathsf{IR}}$. Moreover the following assignment is full and faithful.*

$$[\![\text{-}]\!]_{\mathsf{IR}} : \mathsf{IR}\ I\ O \ \to\ [\mathsf{Set}/I, \mathsf{Set}/O]$$

The theorem is proved by induction on the structure of IR morphisms. Full and faithfulness allows us to reflect functor composition to the composition of small IR codes and hence we have the following important result.

Corollary 2. IR *I O-codes and their morphisms define a category.*

An Equivalence. In the previous sections we have seen how to represent $\mathsf{IR}\ I\ O$-codes as dependent polynomials in $\mathsf{Poly}\ I\ O$ and vice versa. To sum up:

- We saw how to define a function $\psi : \mathsf{Poly}\ I\ O \to \mathsf{IR}\ I\ O$ such that $[\![\text{-}]\!]_{\mathsf{IC}} \cong [\![\text{-}]\!]_{\mathsf{IR}} \circ \psi$
- We saw how to define a function $\phi : \mathsf{IR}\ I\ O \to \mathsf{Poly}\ I\ O$ such that $[\![\text{-}]\!]_{\mathsf{IR}} \cong [\![\text{-}]\!]_{\mathsf{IC}} \circ \phi$.

We sum up these results in the following corollary.

Corollary 3. *For every $\gamma : \mathsf{IR}\,I\,O$ and, for every $(r, t, q) : \mathsf{Poly}\ I\ O$*

1) $[\![\psi \circ \phi\,(\gamma)]\!]_{\mathsf{IR}} \cong [\![\gamma]\!]_{\mathsf{IR}}$,
2) $[\![\phi \circ \psi\,(r, t, q)]\!]_{\mathsf{Poly}} \cong [\![(r, t, q)]\!]_{\mathsf{Poly}}$

These isomorphisms deal just with objects of the two categories $\mathsf{IR}\,I\,O$ and $\mathsf{Poly}\ I\ O$. But what can we say about morphisms? As we show in the next theorem the equivalence of these two categories is an immediate consequence of the previous results combined with full and faithfulness of the respective interpretation functions:

Theorem 4. *The two categories $\mathsf{IR}\ I\ O$ and $\mathsf{Poly}\ I\ O$ are equivalent.*

It is immediate to show full and faithfulness of ϕ (or, equivalently of ψ):

$$\begin{aligned}
\mathsf{IR}\, I\, O(\gamma, \gamma') &\cong Nat([\![\gamma]\!]_{\mathsf{IR}}, [\![\gamma']\!]_{\mathsf{IR}}) && \text{(Corollary 2)}\\
&\cong Nat([\![\phi(\gamma)]\!]_{\mathsf{Poly}}, [\![\phi(\gamma')]\!]_{\mathsf{Poly}}) && \text{(Lemma 2)}\\
&\cong \mathsf{Poly}\; I\;\, O(\phi(\gamma), \phi(\gamma')) && \text{(Corollary 1)}
\end{aligned}$$

Now, since we have already showed that each dependent polynomial, (r, t, q) is isomorphic to $\phi(\gamma)$ for some $\gamma : \mathsf{IR}\, I\, O$ (namely $\gamma = \psi(r, t, q)$), this is enough to conclude the stated equivalence (see theorem 1, par. 4, ch. IV in [16]). Here is a commutative diagram which represents the statement of theorem 4:

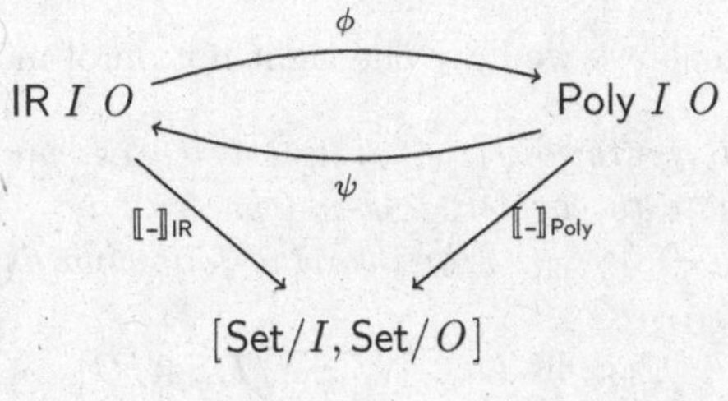

6 Small Indexed Induction Recursion

The theory of induction recursion has been extended by Dybjer and Setzer in [11] in order to capture more sophisticated inductive-recursive definitions. As indexed container and dependent polynomials generalise polynomials and containers respectively, the theory of indexed induction-recursion (IIR) generalises the theory inductive-recursive definitions in order to capture not only ordinary inductive-recursive definition, but also families of inductive-recursive definitions which admit extra indexing. IR then appears as the fragment of IIR given by those definitions indexed over a singleton.

We will briefly recall the axiomatic presentation of IIR which closely follows that of IR. We then show how the theory of small indexed inductive-recursive definitions (*small* IIR) can be reduced to small IR. This simple fact will automatically transfer the results of the previous sections to small IIR, allowing to conclude a generalisation of the equivalence stated in theorem 4. We now give the coding for small IIR.

```
data IIR (D : I → Set) (E : J → Set) : Set1 where
  ι : (je : ΣE)                                        → IIR D E
  σ : (S : Set)          (K : S             → IIR D E) → IIR D E
  δ : (P : Set) (i : P → I) (K : ((p : P) → D (i p)) → IIR D E) → IIR D E
```

Note that δ carries an extra argument i, selecting the index for each position in P. One way to interpret these codes is by translation to the codes for $\mathsf{IR}\ \Sigma D\ \Sigma E$, as follows:

$$
\begin{aligned}
&\lfloor \cdot \rfloor : \mathsf{IIR}\ D\ E \rightarrow \mathsf{IR}\ \Sigma D\ \Sigma E\\
&\lfloor \iota\ je \rfloor \quad = \iota\ je\\
&\lfloor \sigma\ S\ K \rfloor \quad = \sigma\ S\ \lambda\ s\ \rightarrow\ \lfloor K\ s \rfloor\\
&\lfloor \delta\ P\ i\ K \rfloor = \delta\ P\ \lambda\ iD\ \rightarrow\ \sigma\ (i \equiv (\pi_0\ \circ\ iD))\ \lambda\ q\ \rightarrow\ \lfloor K\ (\pi_1\ \circ\ iD) \rfloor
\end{aligned}
$$

In the δ case, the generated IR code yields a ΣD for each position in P, so we constrain its first component to coincide with the index required by the i in the IIR code. Given this embedding, we can endow small IIR with the categorical machinery developed for small IR. We therefore can straightforwardly define a category of IIR D E -codes and their morphisms. Theorem 4 in Section 5 immediately give us the following corollary.

Corollary 4. *The category* IIR D E *and the category* Poly ΣD ΣE *are equivalent.*

We can also follow Dybjer and Setzer by giving a direct interpretation of an IIR code as a functor between families of slice categories.

$$
\begin{aligned}
&[\![\cdot]\!]_{\mathsf{IIR}} : \mathsf{IIR}\ D\ E \rightarrow ((i : I) \rightarrow \mathsf{Set}/(D\ i)) \rightarrow ((j : J) \rightarrow \mathsf{Set}/(E\ j))\\
&[\![\iota\ (j', e)]\!]_{\mathsf{IIR}}\ G\ j = ((j' \equiv j), \lambda\ q\ \rightarrow\ \cdot\ q\ e)\\
&[\![\sigma\ S\ K]\!]_{\mathsf{IIR}}\quad G\ j = (s : S) \times \qquad ([\![K\ s]\!]_{\mathsf{IIR}}\ G\ j)\\
&[\![\delta\ P\ i\ K]\!]_{\mathsf{IIR}}\ G\ j = (ig : (p\ :\ P) \rightarrow \mathsf{dom}\ (G\ (i\ p))) \times ([\![K\ (\lambda\ p \rightarrow \mathsf{fun}\ (G\ (i\ p))\ (ig\ p))]\!]_{\mathsf{IIR}}\ G\ j)
\end{aligned}
$$

We note that keeping I and D small ensures the following:

$$(i : I) \rightarrow \mathsf{Set}/(D\ i)\ \cong\ (i : I) \rightarrow D\ i \rightarrow \mathsf{Set}\ \cong\ \Sigma D \rightarrow \mathsf{Set}\ \cong\ \mathsf{Set}/\Sigma D$$

Consider $G\ i\ =\ (\exists.(F\ \circ\ (i,)))$ for some $F\ :\ \Sigma D\ \rightarrow\ \mathsf{Set}$ to see that $[\![\gamma]\!]_{\mathsf{IIR}}\ G$ corresponds to $[\![\lfloor\gamma\rfloor]\!]_{\mathsf{IR}}\ F$, up to bureaucratic isomorphism.

Once again, we construct simultaneously an indexed family of data types $\mu\ \gamma\ i$ and their decoders decode i as the initial algebra for $[\![\gamma]\!]_{\mathsf{IIR}}$.

```
μd : (γ : IIR D D) → (i : I) → Set/(D i)
μd γ i = (μ γ i, decode γ i)
data μ (γ : IIR D D) (i : I) : Set where
  in : dom ([[γ]]IIR (μd γ) i) → μ γ i
decode : (γ : IIR D D) → (i : I) → μ γ i → D i
decode γ i (in t) = fun ([[γ]]IIR (μd γ) i) t
```

The corresponding fixpoint of $[\![\lfloor\gamma\rfloor]\!]_{\mathsf{IR}}$ gives the inductive family indexed by pairs in ΣD.

Example 3. **The Bove-Capretta method, applied to call-by-value computation.** Bove and Capretta [4] make use of indexed induction-recursion to model the *domains* of partial functions. A partial function $d\ :\ (i\ :\ I)\ \rightharpoonup\ D\ i$ has a domain given by a code $\gamma\ :\ \mathsf{IIR}\ D\ D$. If $h\ :\ \mu\ \gamma\ i$ gives evidence that the domain is inhabited at argument i, then decode $\gamma\ i\ h$ is sure to compute the result.

Let us take a concrete example. One might define a type of λ-terms and seek to give a call-by-value evaluator for them, as follows.

```
data Tm : Set where                  cbv : Tm ⇀ Tm
  var : ℕ            → Tm            cbv (var x) = var x
  app : Tm → Tm → Tm                 cbv (lam t) = lam t
  lam : Tm           → Tm            cbv (app f s) with cbv f
                                     ...     | lam t = cbv (subst0 (cbv s) t)
                                     ...     | f'    = app f (cbv s)
```

where, say, we adopt a de Bruijn indexing convention and define subst0 s t to substitute s for variable *0* in t. Of course, cbv is not everywhere defined. When it is defined? It is hard to define the domain *inductively*, because the app f s case will require that subst0 (cbv s) t is in the domain whenever f is in the domain *and evaluates to* lam t. We need to define the domain simultaneously with evaluation — a job for IR.

It will prove convenient to define the special case of δ when $P = 1$.

$$\delta_1 : (i : I) \to (K : D\, i \to \mathsf{IIR}\, D\, E) \to \mathsf{IIR}\, D\, E$$
$$\delta_1\, i\, K = \delta\, 1\, (\lambda\, \_ \to i)\, \lambda\, d \to K\, (d\, \star)$$

In the code for a domain predicate, a recursive call at i gives rise to a δ_1 i K code, where K explains how to carry on if the call returns. Let us give the domain of cbv.

```
cbvD : IIR {Tm} {Tm} (λ _ → Tm) (λ _ → Tm)
cbvD = σ Tm λ
  { (var x)   → ι (var x, var x)
  ; (lam t)   → ι (lam t, lam t)
  ; (app f s) → δ₁ f λ { (lam t) → δ₁ s λ s' → δ₁ (subst0 s' t) λ t' → ι (app f s, t')
                       ; f'      → δ₁ s λ s'                        → ι (app f s, app f' s')
  }                    }
```

Note the way the application case makes key use of the delivered values in subsequent recursive calls, and in every case, the final ι delivers an input-output pair. The type μ cbvD t thus contains the evidence that cbv t terminates without presupposing a particular value — decoding that evidence will yield t's value. The equivalence we have demonstrated in this paper ensures that the corresponding inductive *family* indexed over Tm $\times$ Tm is exactly the big-step evaluation relation for cbv.

7 Conclusion and Further Work

Despite its evident potential, the theory of induction recursion has not become as widely understood and used as it should be. In this paper we seek to broaden appreciation of Dybjer and Setzer's work by comparing it with better-known theories of data types based on dependent polynomials, and more practically with indexed containers. In the case of *small* IR, these three analyses coincide. We can now pick up the fruits of our central result (theorem 4).

Initial Algebras. When interpreting codes in $\mathsf{IR}\ I\ I$ we get endofunctors on Set/I. Theorem 4 ensures that initial algebras for these functors always exist, since they are initial algebras for dependent polynomial endofunctors. Altenkirch and Morris have [2] given parametrized initial algebras of indexed containers of type $\mathsf{IC}\ (I + O)\ O$: as a result of this work, the same construction carries over into $\mathsf{IR}\ (I + O)\ O$ functors. We also now know that functors definable by Small IR also have final coalgebras - these are just the final coalgebras for dependent polynomial functors/indexed containers. They have recently investigated by Capretta in unpublished work under the name of Wander types.

Closure Properties. The axiomatization of small IR and its semantics provides a new (but equivalent) grammar to work with the categories Poly and IC. It is known that these categories have very rich closure properties such as sums, products, composition, as well as linear and differential structure. Clearly we can transport these properties along the equivalence of theorem 4.

Compositions. A difficult open question in the theory of induction-recursion is whether the Dybjer-Setzer functors are closed under composition: given codes $\gamma : \mathsf{IR}\ I\ J$ and $\gamma' : \mathsf{IR}\ J\ O$ is it always possible to find a code ξ in $\mathsf{IR}\ I\ O$ such that $[\![\gamma']\!] \circ [\![\gamma]\!] \cong [\![\xi]\!]$? Theorem 4 ensures that we can transport composition in Poly or IC to obtain closure under composition of small IR functors.

Further Work. We have proved that $\mathsf{Poly}\ I\ O$, $\mathsf{IC}\ I\ O$ and $\mathsf{IR}\ I\ O$ are equivalent categories which define the same class of functors. It is easy to generalize this result to a biequivalence of bicategories. Since it is possible to define reindexing of IR codes and IR functors, in future work we would like to explore this extra-structure of small IR and compare it with the double category of Poly. Abstracting from the category of sets we also aim to investigate to which extent this result applies to arbitrary LCCCs.

References

1. Abbott, M., Altenkirch, T., Ghani, N.: Containers. Constructing Strictly Positive Types. TCS 342, 3–27 (2005)
2. Altenkirch, T., Morris, P.: Indexed containers. In: Procs. of the 24th Annual IEEE Symposium on Logic in Computer Science (LICS 2009). IEEE Computer Society (2009)
3. Aczel, P.: An introduction to inductive definition. In: Barwise, J. (ed.) Handbook of Mathematical Logic, pp. 739–782. North-Holland, Amsterdam (1977)
4. Bove, A., Capretta, V.: Nested General Recursion and Partiality in Type Theory. In: Boulton, R.J., Jackson, P.B. (eds.) TPHOLs 2001. LNCS, vol. 2152, pp. 121–135. Springer, Heidelberg (2001)
5. Clairambault, P., Dybjer, P.: The Biequivalence of Locally Cartesian Closed Category and Martin Löf Type Theories, arXiv:1112.3456v1 [cs.LO], December 15 (2011)
6. Coquand, T., Dybjer, P.: Inductive Definitions and Type Theory an Introduction. In: Thiagarajan, P.S. (ed.) FSTTCS 1994. LNCS, vol. 880, pp. 60–76. Springer, Heidelberg (1994)

7. Curien, P.-L.: Substitution up to isomorphism. Fundamenta Informaticae 19(1-2), 51–86 (1993)
8. Dybjer, P.: A general formulation of simultaneous inductive-recursive definitions in type theory. Journal of Symbolic Logic 65(2), 525–549 (2000)
9. Dybjer, P., Setzer, A.: A Finite Axiomatization of Inductive-Recursive Definitions. In: Girard, J.-Y. (ed.) TLCA 1999. LNCS, vol. 1581, pp. 129–146. Springer, Heidelberg (1999)
10. Dybjer, P., Setzer, A.: Induction-recursion and initial algebras. Annales of Pure and Applied Logic 124, 1–47 (2003)
11. Dybjer, P., Setzer, A.: Indexed Induction-Recursion. Journal of Logic and Algebraic Programming 66(1), 1–49 (2006)
12. Gambino, N., Hyland, M.: Wellfounded trees and dependent polynomial functors. In: Berardi, S., Coppo, M., Damiani, F. (eds.) TYPES 2003. LNCS, vol. 3085, pp. 210–225. Springer, Heidelberg (2004)
13. Gambino, N., Kock, J.: Polynomial functors and polynomial monads arXiv:0906.4931v2 [math.CT], March 6 (2010)
14. Hofmann, M.: On the interpretation of type theory in locally cartesian closed categories. In: Pacholski, L., Tiuryn, J. (eds.) CSL 1994. LNCS, vol. 933, pp. 427–441. Springer, Heidelberg (1995)
15. Kock, J.: Notes on Polynomial functors, `http://www.mat.uab.es/~kock/cat/polynomial.html`
16. Mac Lane, S.: Categories for the working mathematician, 2nd edn. Springer, New York (1998)
17. Martin-Löf, P.: An intuitionistic theory of types: Predicative part. In: Logic Colloquium 1973, pp. 73–118. North-Holland, Amsterdam (1973)
18. Moerdijk, I., Palmgren, E.: Wellfounded trees in categories. Annals of Pure and Applied Logic 104, 189–218 (2000)
19. Seely, R.A.G.: Locally cartesian closed categories and type theory. Math. Proc. Cambridge Philos. Soc. (95), 33–48 (1984)

Generalizations of Hedberg's Theorem

Nicolai Kraus[1], Martín Escardó[2], Thierry Coquand[3,*], and Thorsten Altenkirch[1,**]

[1] University of Nottingham
[2] University of Birmingham
[3] University of Gothenburg

Abstract. As the groupoid interpretation by Hofmann and Streicher shows, *uniqueness of identity proofs* (UIP) is not provable. Generalizing a theorem by Hedberg, we give new characterizations of types that satisfy UIP. It turns out to be natural in this context to consider constant endofunctions. For such a function, we can look at the type of its fixed points. We show that this type has at most one element, which is a nontrivial lemma in the absence of UIP. As an application, a new notion of anonymous existence can be defined. One further main result is that, if every type has a constant endofunction, then all equalities are decidable. All the proofs have been formalized in Agda.

Keywords: Hedberg's Theorem, homotopy type theory, propositional equality, truncation, squash types, bracket types, anonymous existence, constant endofunctions.

1 Introduction

Although the identity types in Martin-Löf type theory (MLTT) are defined by one constructor refl and by one eliminator J that matches the constructor, the statement that every identity type has at most one inhabitant is not provable [9]. Thus, *uniqueness of identity proofs* (UIP), or, equivalently, *Streicher's axiom K* are principles that have to be assumed, and have often been assumed, as additional rules of MLTT. In recent years, there is a growing interest in type theory without these assumptions, in particular with the development of *Homotopy Type Theory* (HoTT) and *Univalent Foundations* (UF) - see [4] for a brief and [13] for a detailed introduction. While we do not use any axioms of HoTT or UF (other than those of standard MLTT), we make use of their notation and intuition. For a better understanding of our arguments, is useful to think of a type as a space, and a propositional equality proof as a path. Notation and some basic definitions are listed in Section 2.

---

* Supported by the ERC project 247219, and grants of The Ellentuck and The Simonyi Fund.

** Supported by the EPSRC grant EP/G03298X/1 and by a grant of the Institute for Advanced Study.

M. Hasegawa (Ed.): TLCA 2013, LNCS 7941, pp. 173–188, 2013.

As said above, we do not assume the principle of unique identity proofs. However, certain types do satisfy it naturally, and such types are often called *h-sets*. A sufficient condition for a type to be an h-set, given by Hedberg [8], is that it has decidable propositional equality. In Section 3, we analyze Hedberg's original argument, which consists of two steps:

1. A type X is an h-set iff for all $x, y : X$ there is a constant map $x = y \to x = y$.
2. If X has decidable equality then such constant endomaps exist.

Here, we write $x = y$ for the identity type $\mathrm{Id}_X(x, y)$ of an implicitly given type X.

Decidable equality means that, for all x and y, we have $(x = y) + (x \neq y)$. Thus, a natural weakening is $\neg\neg$-*separated* equality,

$$\neg\neg(x = y) \to x = y,$$

which occurs often in constructive mathematics. In this case we say that the type X is *separated*. For example, going beyond MLTT, the reals and the Cantor space in Bishop mathematics and topos theory are separated. In MLTT, the Cantor type of functions from natural numbers to booleans is separated under the assumption of functional extensionality,

$$\forall f\, g : X \to Y, (\forall x : X, f\, x = g\, x) \to f = g.$$

We observe that under functional extensionality, a separated type X is an h-set, because there is always a constant map $x = y \to x = y$.

In order to obtain a further characterization of the notion of h-set, we consider *truncations* (also known as *bracket* or *squash* types), written $\|X\|$ in accordance with recent HoTT notation. The idea is to collapse all inhabitants of X so that $\|X\|$ has at most one inhabitant. We refer the reader to the technical development for a precise definition. We observe that

1′. A type X is an h-set iff $\|x = y\| \to x = y$ for all $x, y : X$,

and we mention a couple of other simple, but noteworthy, connections.

While Section 3 gives properties and arguments involving path spaces (i. e. equality types), we go beyond that in Section 4. Dealing with a path space opens up many possibilities that are not available for a general type. For that reason, we find it somewhat surprising that the equivalence of two of the above mentioned properties can be translated to general spaces, though that requires a nontrivial argument. This is done in Section 4:

A type X satisfies $\|X\| \to X$ iff it has a constant endomap.

We find this interesting, as it says that from the anonymous existence of a point of X, that is, from the inhabitedness of $\|X\|$, one can get an inhabitant of X, provided a constant endomap is available. It is important here (and above) that our definition of constant function does not require X to be inhabited: we say that a function is constant if any two of its values are equal, and this may happen vacuously. The main technical lemma to prove this, which is noteworthy on its own right, is our Fixed Point Lemma:

For any type X and any constant map $f : X \to X$, the type of fixed points of f is an h-proposition.

Here, an h-proposition is defined to be a type with at most one element. The proof of this lemma would be trivial if UIP was assumed, but in its absence, it is not.

Section 5 can, together with the just described results, be seen as the highlight of this paper. The assumption that every type has a constant endomap has an interesting status. It is not a constructive principle, but at the same time, it is seemingly weaker that typical classical statements. But this is only partially true: While we cannot make a strong conclusion for arbitrary types, such as excluded middle, we prove that the assumption implies that all equalities are decidable.

The just discussed section depends crucially on the Fixed Point Lemma, and so does Section 6: We describe how the lemma gives rise to another notion of anonymous existence, which we call *populatedness*. We say that X is populated, written $\langle\langle X \rangle\rangle$, if every constant endofunction on X has a fixed point. Unlike $\|X\|$, this new notion is thus defined internally, instead of using a postulate.

In our final Section 7, we discuss the relationship between the different notions of existence, starting with a chain of implications:

$$X \longrightarrow \|X\| \longrightarrow \langle\langle X \rangle\rangle \longrightarrow \neg\neg X.$$

We have formalized and proved all our statements in the dependently typed programming language Agda [3] and presented parts on the HoTT blog [1].

2 Preliminaries

We work in a standard version of Martin-Löf Type Theory with dependent sums, dependent function types and identity types. For the latter, we assume the eliminator J and, as it is standard, its computational β-rule, but not the definitional η-law. We further do not assume the eliminator K, or the principle of unique identity proofs. Summarized, our setting is very minimalistic. Sometimes, additional principles (*function extensionality* and *truncation*, as introduced later) are assumed, but this will be stated clearly.

We use standard notation whenever it is available. Regarding the identity types, we write, for two elements $a, b : A$, the expression $a = b$ for the type of *equality proofs*, or *paths* from a to b, keeping A implicit. Other common notations for the same thing are $a =_A b$, as well as $\mathrm{Id}(a, b)$ and $\mathrm{Id}_A(a, b)$. If $a = b$ is inhabited, it is standard to say that a and b are *propositionally equal*. In contrast, *definitional equality* is a meta-level concept, referring to two terms, rather than two (hypothetical) elements, with the same β (and, sometimes, η in a restricted sense) normal form. Recently, it has become standard to use the symbol $\equiv$ for definitional equality.

Propositional equality satisfies the *Groupoid Laws*: If we have $p : a = b$ and $q : b = c$, there is a canonical path $p \bullet q : a = c$ (the *composition* of p and q). Further, we have $p^{-1} : b = a$. There always is $\mathsf{refl}_a : a = a$, which behaves as a neutral element when composed with another path. Pairs of inverses cancel each other out when composed, and the obvious associativity law holds. In general, these statements are valid only up to propositional equality.

An important special case of the J eliminator is *substitution*, for which the name *transport* has been established in HoTT: If P is a family of types over A, and there are two elements (or *points*) $a, a' : A$, together with some $p : a = a'$, then a point $x : P(a)$ can be "transported along the path p" to get an element of $P(a')$:

$$\text{transport } p\, x : P(a').$$

Another useful function, easily derived from the J eliminator, is the follwing: If we have a function $f : A \to B$ and a path $p : a = a'$ in A, we get a path of type $f(a) = f(a')$ in B:

$$\text{ap}_f\ p : f(a) = f(a')$$

Our hope is that all of the notions in the following definition are as intuitive as possible, if not already known. The only notions that are not standard are *collapsible*, meaning that a type has a constant endomap, and *path-collapsible*, saying that every path space over the type is collapsible.

Definition 1. *We say that a type X is an* h-proposition *if all its inhabitants are equal:*

$$\text{hprop}\, X \equiv \forall\, x\, y : X,\ x = y.$$

Further, X satisfies UIP *(uniqueness of identity proofs), or is an* h-set, *if its path spaces are all h-propositional:*

$$\text{h-set}\, X \equiv \forall\, x\, y : X,\ \text{hprop}(x = y).$$

The property of being h-propositional or an h-set are all h-propositional themselves, which the following properties are not.

X is decidable *if it is either inhabited or empty:*

$$\text{decidable}\, X \equiv X + \neg X.$$

We therefore say that X has decidable equality, *if the equality type of any two inhabitants of X is decidable:*

$$\text{discrete}\, X \equiv \forall\, x\, y : X,\ \text{decidable}(x = y).$$

Based on the terminology in [11], we also call a type with decidable equality discrete.

A function (synonymously, map) $f : X \to Z$ is constant *if it maps any two elements to the same inhabitant of Y:*

$$\text{const}\, f \equiv \forall\, x\, y : X,\ f(x) = f(y).$$

We call a type X collapsible *if it has a constant endomap:*

$$\text{coll}\, X \equiv \Sigma_{f : X \to X}\ \text{const}\, f.$$

Finally, X is called path-collapsible *if any two points x, y of X have a collapsible path space:*

$$\text{path-coll}\, X \equiv \forall\, x\, y : X,\ \text{coll}\,(x = y).$$

For some statements, but only if clearly indicated, we use *functional extensionality.* This principle says that two functions f, g of the same type are equal as soon as they are pointwise equal:

$$(\forall x\,,\ f\, x = g\, x) \to f = g.$$

An important equivalent formulation (see Voevodsky [14]) is that the set of h-propositions is closed under $\forall$. More precisely,

$$(\forall a : A, \text{hprop}\, B) \to \text{hprop}\,(\forall a : A, B).$$

In the case of non-dependent function types, this can be read as follows: If B is h-propositional, then so is $A \to B$.

3 Hedberg's Theorem

Before discussing possible generalizations, we discuss Hedberg's Theorem.

Theorem 1 (Hedberg). *Every discrete type has unique identity proofs,*

$$\text{discrete}\, X \to \text{h-set}\, X.$$

We shortly state Hedberg's original proof [8], consisting of two steps.

Lemma 1. *If a type has decidable equality, it is path-collapsible:*

$$\text{discrete}\, X \to \text{path-coll}\, X.$$

Proof. Given inhabitants x and y of X, the assumptions provide an inhabitant of $\text{decidable}(x = y) \equiv (x = y) + \neg(x = y)$. If it is an inhabitant of $x = y$, we construct the required constant map $(x = y) \to (x = y)$ by mapping everything to this path. If it is an inhabitant of $\neg(x = y)$, there is only a unique such map which is constant automatically. □

Lemma 2. *If a type is path-collapsible, it has unique identity proofs:*

$$\text{path-coll}\, X \to \text{h-set}\, X.$$

Proof. Assume f is a parametrized constant endofunction on the path spaces. Let p be a path from x to y. We claim that $p = (f\, p) \bullet (f\, \text{refl}_x)^{-1}$. Using the equality eliminator on (x, y, p), we only have to give a proof for the triple (x, x, refl_x), which is one of the groupoid laws that equality satisfies. Using the fact f is constant on every path space, the right-hand side expression is independent of p, and in particular, equal to any other path of the same type. □

Hedberg's proof [8] is just the concatenation of the two lemmas. A slightly more direct proof can be found in a post on the HoTT blog [10], and in the HoTT Coq repository [12]. The first of the two lemmas uses the rather strong assumption of decidable equality. In contrast, the assumption of the second lemma is equivalent its conclusion, which means that we cannot do much there. We include a proof of this simple claim in Theorem 2 below and concentrate on weakening the assumption of the first lemma. Let us first introduce the notions of *stability* and *separatedness*.

Definition 2. *For any type X, define*

$$\text{stable}\, X \equiv \neg\neg X \to X,$$
$$\text{separated}\, X \equiv \forall x\, y : X, \text{stable}(x = y).$$

We can see stable X as a *classical* condition, similar to decidable $X \equiv X + \neg X$, but strictly weaker. Indeed, we get a first strengthening of Hedberg's Theorem as follows:

Lemma 3. *If functional extensionality holds, any separated type has unique identity proofs,*

$$\text{separated}\, X \to \text{h-set}\, X.$$

Proof. There is, for any $x, y : X$, a canonical map $(x = y) \to \neg\neg(x = y)$. Composing this map with the proof that X is separated yields an endofunction on the path spaces. With functional extensionality, the first map has an h-propositional codomain, which implies that the endofunction is constant, fulfilling the requirements of lemma 2. □

We remark that full functional extensionality is actually not needed here. Instead, a weaker version that only works with the empty type is sufficient. Similar statements hold true for all further applications of extensionality in this paper. Details can be found in the Agda file [3].

In a constructive setting, the question how to express that "there exists something" in a type X is very subtle. One possibility is to ask for an inhabitant of X, but in many cases, this is stronger than one can hope. A second possibility, which corresponds to our above definition of *separated*, is to ask for a proof of $\neg\neg X$. Then again, this is very weak, and often too weak, as one can in general only prove negative statements from double-negated assumptions.

This fact has inspired the introduction of *squash types* (the Nuprl book [6]), and similar, *bracket types* (Awodey and Bauer [5]). These lie in between of the two extremes mentioned above. In our intensional setting, we talk of *h-propositional truncations*: For any type X, we postulate that there is a type $\|X\|$ that is an *h-proposition*, representing the statement that X is inhabited. The rules are that if we have a proof of X, we can, of course, get a proof of $\|X\|$, and from $\|X\|$, we can conclude the same statements as we can conclude from X, but only if the actual representative of X does not matter:

Definition 3. *For a given type X : **Type**, we postulate the existence of a type $\|X\|$: **Type**, satisfying the following properties:*

1. $\eta : X \to \|X\|$
2. $\text{hprop}(\|X\|)$
3. $\forall P : \textbf{\textit{Type}}, \text{hprop}\, P \to (X \to P) \to \|X\| \to P.$

We say that X is h-inhabited *if $\|X\|$ is inhabited.*

Note that this amounts to saying that the operator $\|\cdot\|$ is left adjoint to the inclusion of the subcategory of h-propositions into the category of all types. Therefore, it can be seen as the *h-propositional reflection*.

There is a type expression that is equivalent to h-inhabtedness:

Proposition 1. *For any given X : **Type**, we have*

$$\|X\| \longleftrightarrow \forall P : \textbf{\textit{Type}}, \text{hprop}\, P \to (X \to P) \to P.$$

The trouble with the expression on the right-hand side is that it is not living in universe **Type**. This size issue is really the only thing that is disturbing here, as the expression satisfies all the properties of the above definition, at least under the assumption of functional extensionality. Voevodsky [14] uses *resizing rules* to get rid of the problem.

Proof. The direction "$\rightarrow$" of the statement is not more than a rearrangement of the assumptions of property (3). For the other direction, we only need to instantiate P with $\|X\|$ and observe that the properties (1) and (2) in the definition of $\|X\|$ are exactly what is needed. □

With this definition at hand, we can provide an even stronger variant of Hedberg's Theorem. Completely analogous to the notions of stability and separatedness, we define *h-stable* and *h-separated:*

Definition 4. *For any type X, define*

$$\text{h-stable}\, X \equiv \|X\| \rightarrow X,$$
$$\text{h-separated}\, X \equiv \forall\, x\, y : X,\ \|x = y\| \rightarrow (x = y).$$

In fact, h-separated X is a strictly weaker condition than separated X. Not only can we conclude h-set X from h-separated X, but even the converse. We also include the simple, but until here unmentioned fact that path-collapsibility is also equivalent to these statements:

Theorem 2. *For a type X in MLTT with h-propositional truncation, the following are equivalent:*

(i) X is an h-set.
(ii) X is path-collapsible.
(iii) X is h-separated.

Proof. *(ii)* $\Rightarrow$ *(i)* is just Lemma 2.

(i) $\Rightarrow$ *(iii)* uses simply the the definition of the h-propositional truncation: Given $x, y : X$, the fact that X is an h-set tells us exactly that $x = y$ is h-propositional, implying that we have a map $\|x = y\| \rightarrow (x = y)$.

Concerning *(iii)* $\Rightarrow$ *(ii)*, it is enough to observe that the composition of η : $(x = y) \rightarrow \|x = y\|$ and the map $\|x = y\| \rightarrow (x = y)$, provided by the fact that X is h-separated, is a parametrized constant endofunction. □

As a conclusion of this part of the paper, we observe that h-propositional truncation has some kind of extensionality built-in: In Lemma 3, we have given a proof for the simple statement that separated types are h-sets in the context of functional extensionality. This is not true in pure MLTT. Let us now drop functional extensionality and assume instead that h-propositional truncation is available. Every separated type is h-separated - more generally, we have

$$(\neg\neg A \rightarrow A) \rightarrow \|A\| \rightarrow A$$

for any type A -, and every h-separated space is an h-set. Notice that the mere availability of h-propositional truncation suffices to solve a gap that functional extensionality would usually fill.

4 Collapsibility Implies H-Stability

If we unfold the definitions in the statements of Theorem 2, they all involve the path spaces over some type X:

(i) $\forall x\, y : X$, hprop$(x = y)$
(ii) $\forall x\, y : X$, coll$\,(x = y)$
(iii) $\forall x\, y : X$, h-stable$(x = y)$.

We have proved that these statements are logically equivalent. It is a natural question to ask whether the properties of path spaces are required. The possibilities that path spaces offer are very powerful and we have used them heavily. Indeed, if we formulate the above properties for an arbitrary type A instead of path types

(i') hprop(A)
(ii') coll$\,(A)$
(iii') h-stable$\,A$,

we notice immediately that (i') is significantly and strictly stronger than the other two properties. (i') says that A has at most one inhabitant, (ii') says that there is a constant endofunction on A, and (iii') gives us a possibility to get an explicit inhabitant of A from the proposition that A has an anonymous inhabitant. An h-propositional type has the other two properties trivially, while the converse is not true. In fact, as soon as we know an inhabitant $a : A$, we can very easily construct proofs of (ii') and (iii'), while it does not help at all with (i').

The implication $(iii') \Rightarrow (ii')$ is also simple: If we have $h : \|A\| \to A$, the composition $h \circ \eta : A \to A$ is constant, as for any $a, b : A$, we have $\eta(a) = \eta(b)$ and therefore $h(\eta(a)) = h(\eta(b))$.

In summary, we have $(i') \Rightarrow (iii') \Rightarrow (ii')$ and we know that the first implication cannot be reversed. What is less clear is the reversibility of the second implication: If we have a constant endofunction on A, can we get a map $\|A\| \to A$? Put differently, what does it take to get out of $\|A\|$? Of course, a proof that A is h-stable is fine for that, but does a constant endomap on A also suffice? Surprisingly, the answer is positive, and there are interesting applications (Section 6). The main ingredient of our proof, and of much of the rest of the paper, is the following crucial lemma about fixed points:

Lemma 4 (Fixed Point Lemma). *Given a constant endomap f on a type X, the type of fixed points is h-propositional, where this type is defined by*

$$\mathrm{fix}\, f \equiv \Sigma_{x:X}\, x = f(x).$$

Before we can give the proof, we first need to formulate two observations. Both of them are simple on their own, but important insights for the Fixed Point Lemma. Let X and Y be two types.

Proposition 2. *Assume $h, k : X \to Y$ are two functions and $t : x = y$ as well as $p : h(x) = k(x)$ are paths. Then, substituting along t into p can be expressed as a composition of paths:*

$$(\text{transport}\, t\, p) = \left((\text{ap}_h\ t)^{-1} \bullet p \bullet (\text{ap}_k\ t) \right).$$

Proof. This is immediate if t is the trivial reflexivity path, i.e. if (x, y, t) is just (x, x, refl_x), and for all other cases, it follows as a direct application of the equality eliminator J. □

Even if the latter proof is trivial, the statement is essential. In the proof of Lemma 4, we need a special case, were x and y are the same. However, this special version cannot be proved directly. We consider the second observation the key insight for the Fixed Point Lemma:

Proposition 3. *If $f : X \to Y$ is constant and $x : X$ some point, then ap_f maps every path between x and x to $\text{refl}_{f(x)}$, up to propositional equality.*

Proof. It is not possible to prove this directly. Instead, we state a slight generalization: If c is the proof of $\text{const}\, f$, then ap_f maps a path $p : x = y$ to $(c\, x\, x)^{-1} \bullet c\, x\, y$. This is easily seen to be correct for (x, x, refl_x), which is enough to apply the eliminator. As the expression is independent of p, but only depends on its endpoints, it is for $p : x = x$ equal to $\text{refl}_{f(x)}$, as claimed. Note that the proposition can also be stated as: For all x and y, the function $\text{ap}_f\ x\, y : (x = y) \to (f\, x = f\, y)$ is constant. □

With these lemmas at hand, the rest is fairly simple:

Proof (of the Fixed Point Lemma). Assume $f : X \to X$ is a function and $c : \text{const}\, f$ is a proof that it is constant. For any two pairs (x, p) and $(x', p') : \text{fix}\, f$, we need to construct a path connection them.

First, we simplify the situation by showing that we can assume that x and x' are the same: By composing $p : x = f\, x$ with $c\, x\, x' : f\, x = f\, x'$ and $(p')^{-1} : f\, x' = x'$, we get a path $p'' : x = x'$. A path between two pairs corresponds to two paths: One path between the first components, and one between the second, where a substitution along the first path is needed. We therefore now get that $(x, \text{transport}\, (p'')^{-1}\, p')$ and (x', p') are propositionally equal: p'' is a path between the first components, which makes the second component trivial. Write q for the term $\text{transport}\, (p'')^{-1}\, p'$.

We are now in the (nicer) situation that we have to construct a path between (x, p) and $(x, q) : \text{fix}\, f$. Again, such a path has to consist of two paths, for the two components. Let us assume that we use some path $t : x = x$ for the first component. We then have to show that $\text{transport}\, t\, p$ equals q. In the situation with (x, p) and (x', p'), it might have been tempting to use p'' as a path between the first components, and that would correspond to choosing refl_x for t. However, one quickly convinces oneself that this cannot work in the general case.

By Proposition 2, with the identity for h and f for k, the first of the two terms, i. e. $\text{transport}\, t\, p$, corresponds to $t^{-1} \bullet p \bullet \text{ap}_f\ t$. With Proposition 3, that term can be further simplified to $t^{-1} \bullet p$. What we have to prove is now just $(t^{-1} \bullet p) = q$, so let us just choose $q \bullet p^{-1}$ for t, thereby making it into a straight-forward application of the standard lemmas. □

We are now finally in the position to prove the statement that is announced in Section 4:

Theorem 3. *A type A is collapsible, i.e. has a constant endomap, iff it is h-stable in the sense that $\|A\| \to A$.*

Proof. As already mentioned in Section earlier, the "if-part" is simple: If there is a map $\|A\| \to A$, we just need to compose it with $\eta : A \to \|A\|$ to get a constant endomap on A.

For the other direction, let c be the proof that f is constant, just as before. Observe that we have $A \to \mathrm{fix}\, f$ by mapping a on $(f\,a, c\,a\,(f\,a))$. As $\mathrm{fix}\, f$ is an h-proposition by the previous lemma, we get a map $\|A\| \to \mathrm{fix}\, f$ by the elimination rule for h-propositional truncation. That map can be composed with the first projection of type $\mathrm{fix}\, f \to A$, yielding a function $\|A\| \to A$ as required. □

Looking at the just proved theorem, it makes sense to ask the following question: Given a constant function $f : A \to B$, is it possible to construct a function $\overline{f} : \|A\| \to B$? We can do that if B is an h-set. For the general case, we have evidence that the answer is likely to be negative.

5 Global Collapsibility Implies Decidable Equality

If X is some type, having a proof of $\|X\|$ is, intuitively, much weaker than a proof of X. While the latter consists of a concrete element of X, the first is given by an *anonymous* inhabitant of X. This is actually nothing more than the intention of the truncation: $\|X\|$ allows us to make the statement that "there exists something in X", without giving away a concrete element. It is therefore unreasonable to suppose that

$$\forall X : \mathbf{Type}, \|X\| \to X,$$

can be proved, but it is interesting to consider what it would imply. Using Theorem 3, the above type is logically equivalent to the statement

Every type has a constant endomap.

From a constructive type of view, this is an interesting statement. It clearly follows from the *Principle of Excluded Middle*, $\forall X : \mathbf{Type}, X + \neg X$: If we know an inhabitant of a type, we can immediately construct a constant endomap, and for the empty type, considering the identity function is sufficient. Thus, we understand *"Every type has a constant endomap"* as a weak form of the excluded middle: It seems to use that every type is either empty or inhabited, but there is no way of knowing in which case we are. We are unable to show that it implies excluded middle.

However, what we can conclude is excluded middle for all path spaces. We can prove the following statement in basic MLTT, without h-propositional truncation, without extensionality, and even without a universe:

Lemma 5. *Let A be a type and $a_0, a_1 : A$ two points. If for all $x : A$ the type $(a_0 = x) + (a_1 = x)$ is collapsible, then $a_0 = a_1$ is decidable.*

Before giving the proof, we state an immediate corollary:

Theorem 4. *If every type has a constant endomap (equivalently, is h-stable), then every type has decidable equality.*

Proof (of Lemma 5). Let us define $E_x \equiv (x = a_0) + (x = a_1)$. The assumption says that we have a family of endomaps $f_x : E_x \to E_x$, together with proofs of their constancy $c_x : \mathrm{const}\, f_x$. We show that the identity map on $\Sigma_{x:A}\, \mathrm{fix}\, f_x$ factorizes pointwise through **Bool**. Note that an element of $\Sigma_{x:A}\, \mathrm{fix}\, f_x$ is a pair of an $x : A$ and a point in $\mathrm{fix}\, f_x$; and such a point consists itself of a pair (c, p), where $c : E_x$ and $p : c = f_x(c)$. There is a canonical inhabitant of $\mathrm{fix}\, f_{a_0}$, given by $f_{a_0}(\mathrm{inl}\, \mathrm{refl}_{a_0})$ for the first component, and $c_{a_0}\, (\mathrm{inl}(\mathrm{refl}_{a_0}))\, (f_{a_0}(\mathrm{inl}(\mathrm{refl}_{a_0}))$ for the second. We call it k_0, and analogously, we write k_1 for the canonical inhabitant of $\mathrm{fix}\, f_{a_1}$.

$$\begin{aligned} r : \Sigma_{x:A}\, \mathrm{fix}\, f_x &\to \mathbf{Bool} \\ (x, (\mathrm{inl}\, q, p)) &\mapsto \mathrm{true}, \\ (x, (\mathrm{inr}\, q, p)) &\mapsto \mathrm{false}, \end{aligned} \qquad \begin{aligned} s : \mathbf{Bool} &\to \Sigma_{x:A}\, \mathrm{fix}\, f_x \\ \mathrm{true} &\mapsto (a_0, k_0), \\ \mathrm{false} &\mapsto (a_1, k_1). \end{aligned}$$

We claim that any pair (x, k) is equal to $s \circ r(x, k)$. An equality of pairs corresponds to a pair of equalities. As the second component is, by the Fixed Point Lemma, an equality over an h-propositional type, it is enough to show that x equals the first component of $s \circ r(x, k)$. Let k be (c, p). We can now perform case analysis on c: If c is of the form $\mathrm{inl}\, q$, we need to prove $x = a_0$; but this is shown by q. If c is $\mathrm{inr}\, q$, we proceed analogously. Therefore, equality of any two such pairs is decidable, as we just have to check whether r maps them to the same value in **Bool**.

Again because $\mathrm{fix}\, f_x$ is an h-proposition, the pairs (a_0, k_0) and (a_1, k_1) are equal iff $a_0 = a_1$, and, therefore, $a_0 = a_1$ is decidable. □

6 Populatedness

In this section we discuss a notion of *anonymous existence*, similar, but weaker (see Section 7.2) than h-propositional truncation. It crucially depends on the Fixed Point Lemma 4. Let us start by discussing another perspective of what we have explained in the previous section.

Trivially, for any type X, we can prove the statement

$$\|X\| \to (\|X\| \to X) \to X. \tag{1}$$

By Lemma 3, this is equivalent to

$$\|X\| \to \mathrm{coll}\, X \to X, \tag{2}$$

which can be read as: If we have a constant endomap on X and we wish to get an inhabitant of X (or, equivalently, a fixed point of the endomap), then $\|X\|$ is sufficient to do so. Now, we can ask whether it is also necessary: Can we replace the first assumption $\|X\|$ by something weaker? Looking at formula 1, it would be natural to conjecture that this is not the case, but it is. In this section, we discuss by what it can be replaced, and in Section 7.2, we give a proof that it is indeed weaker.

For answering the question what is needed to get from h-stable A to A, let us define the following notion:

Definition 5 (populatedness). *For a given type X, we say that X is populated, written $\langle\!\langle X \rangle\!\rangle$, if every constant endomap on X has a fixed point:*

$$\langle\!\langle X \rangle\!\rangle \equiv \forall f : X \to X, \operatorname{const} f \to \operatorname{fix} f,$$

where fix f *is the type of fixed points, defined as in Lemma 4.*

This definition allows us to comment on the question risen above. If $\langle\!\langle X \rangle\!\rangle$ is inhabited and X is collapsible, then X has an inhabitant, as such an inhabitant can be extracted from the type of fixed points by projection. Hence, $\langle\!\langle X \rangle\!\rangle$ instead of $\|X\|$ in 2 would be sufficient as well (we discuss in Section 7 whether it is weaker). Therefore,

$$\langle\!\langle X \rangle\!\rangle \to (\|X\| \to X) \to X.$$

Next we draw a parallel between populatedness and h-inhabitedness.

Theorem 5. *For any given X :* ***Type****, the following holds:*

$$\langle\!\langle X \rangle\!\rangle \longleftrightarrow \forall P : \textbf{\textit{Type}}, \operatorname{hprop} P \to (P \to X) \to (X \to P) \to P.$$

This statement can be read as "X is populated iff every h-proposition logically equivalent to X is inhabited." Note that the only difference to the type expression in Proposition 1 is that we only quantify over *sub-propositions* of X, i. e. over those that satisfy $P \to X$, while we quantify over all propositions in the case of $\|X\|$. Therefore, $\|X\|$ is clearly at least as strong as $\langle\!\langle X \rangle\!\rangle$.

Proof. Let us first prove the direction "$\to$". Assume an h-propositional P is given, together with functions $X \to P$ and $P \to X$. Composition of these gives us a constant endomap on X, exactly as in the proof of Theorem 2. But then $\langle\!\langle X \rangle\!\rangle$ makes sure that this constant endomap has a fixed point, which is (or allows us to extract) an inhabitant of X. Using $X \to P$ again, we get P.

For the direction "$\leftarrow$", assume we have a constant endomap f. We need to construct an inhabitant of fix f. In the expression on the right-hand side, choose P to be fix f. By the Fixed Point Lemma, this is an h-proposition. Further, P and X are logically equivalent (i.e. there are maps in both directions), where the non-trivial direction makes use of Theorem 3. Then, the right-handed expression shows P, which is just the required fix f. □

This proof uses the Fixed Point Lemma twice: Once, as we needed P to be an h-proposition, and once hidden, as we used Theorem 3.

The similarities between $\|X\|$ and $\langle\langle X\rangle\rangle$ do not stop here. The following statement, together with the direction "$\rightarrow$" of the theorem that we have just proved, is worth to be compared to the definition of $\|X\|$ (that is, Definition 3):

Proposition 4. *For any type X, the type $\langle\langle X\rangle\rangle$ has the following properties:*
(1) $X \rightarrow \langle\langle X\rangle\rangle$
(2) $\text{hprop}(\langle\langle X\rangle\rangle)$ *(if functional extensionality holds).*

The proof is fairly simple, and, of course, again an application of the Fixed Point Lemma.

Proof. Regarding (1), given $x : X$ and a constant endomap f, we need to prove that f has a fixed point. We just take $f\,x$ and use the fact that $f\,x$ is propositionally equal to $f(f\,x)$, by constancy of f.

For (2), we need to use that $\text{fix} f$ is an h-proposition, by Lemma 4. By functional extensionality, a (dependent) function type is h-propositional if the codomain is (see Section 2) and we are done. □

7 Taboos and Counter-Models

In this final section we look at the differences between the various notions of (anonymous) inhabitedness we have encountered. We have, for any type X, the following chain of implications:

$$X \longrightarrow \|X\| \longrightarrow \langle\langle X\rangle\rangle \longrightarrow \neg\neg X.$$

The first implication is trivial and the second has already been mentioned after Theorem 5. Maybe somewhat surprisingly, the last implication does not require functional extensionality, as we do not need to prove that $\neg\neg X$ is h-propositional: To show

$$\langle\langle X\rangle\rangle \rightarrow \neg\neg X\,,$$

let us assume $f : \neg X$. But then, f can be composed with the unique function from the empty type into X, yielding a constant endomap on X, and obviously, this function does not have a fixed point. Therefore, the assumption of $\langle\langle X\rangle\rangle$ would lead to a contradiction, as required.

Intuitively, none of the implications should be reversible. To make that precise, we use two techniques: Taboos, showing that the provability of a statement would imply the provability of another, better understood statement, that is known to be not provable. As the second technique, we use HoTT models.

1. Theorem 4 shows that, if the first implication can be reversed, then all types have decidable equality. Using Hedberg's Theorem, this immediately implies that every type is an h-set, and thus, it is inconsistent with the Univalence Axiom of HoTT. But the conclusion that every type is an h-set can be derived much more directly: If we assume $\|X\| \rightarrow X$ for all types X, we have this in particular for all path spaces. Then, by Theorem 2, every type is an h-set. As an alternative argument, if every type is h-stable, a form of choice that does not belong to type theory is implied.

2. It would be wonderful if the second implication could be reversed, as this would imply that h-propositional truncation is definable in MLTT. However, this is equivalent to a certain h-propositional axiom of choice discussed below, which is not provable but holds under excluded middle.
3. If the last implication can be reversed, excluded middle for h-propositions holds (a constructive taboo, which is not valid in recursive models).

7.1 Inhabited and H-Inhabited

The question whether the first implication in the chain above can be reversed has already been analyzed in Section 5. This cannot be possible as long as equality is not globally decidable. Here, we want to state another noteworthy consequence of

$$\forall X : \mathbf{Type},\ \|X\| \to X.$$

In [2], we show that this assumption allows us to show that any relation has a functional subrelation with the same domain. This is a form of the axiom of choice that does not pertain to intuitionistic type theory. Here, we only sketch the proof. Given a binary relation A on the type X. Define

$$A_x \equiv \Sigma_{y:X}\, A(x,y), \qquad F(x,y) \equiv \Sigma_{a:A(x,y)}\,(y,a) = k_x(y,a),$$

where $k_x : A_x \to A_x$ is the constant map induced by the hypothesis $\|A_x\| \to A_x$. By the Fixed Point Lemma, $F(x,y)$ is an h-proposition. If $(a,p) : F(x,y)$ and $(a',p') : F(x,y')$, then

$$(y,a) = k_x(y,a) = k_x(y',a') = (y',a')$$

because k_x is constant and hence $y = y'$, and so F is single-valued. But in fact, with a subtler argument, it is single-valued in the stronger sense that F_x is an h-proposition. Moreover, F has the same domain as A in the sense that F_x is inhabited iff A_x is inhabited.

7.2 H-Inhabited and Populated

Assume that the second implication can be reversed, meaning that we have

$$\forall X : \mathbf{Type},\ \langle\!\langle X \rangle\!\rangle \to \|X\|.$$

Repeated use of the Fixed Point Lemma leads to a couple of interesting equivalent statements. We discuss one that is particularly interesting: Every populated type is h-inhabited iff for every type, the statement that it is h-stable is h-inhabited.

In the previous subsection, we have discussed that we cannot prove the statement that every type is h-stable. However, we can always populate it:

Lemma 6. $\forall X$: ***Type***, $\langle\!\langle \|X\| \to X \rangle\!\rangle$.

Proof. Assume we are given a constant endomap f on h-stable X. We need to construct a fixed point of that endomap, which amounts to construction an inhabitant of h-stable X. By the Fixed Point Lemma, a constant endomap $g : X \to X$ is enough for this. From f, we can construct g easily: Given $x : X$, we get a canonical inhabitant of h-stable X. We apply f on this inhabitant, and we apply the result on $\eta(x)$, yielding an inhabitant of X. We define $g\,x$ to be this inhabitant. It is easy to see that g is constant. □

An alternative proof is available in the Agda file.

Theorem 6. *The implication $\|X\| \to \langle\!\langle X \rangle\!\rangle$ can always be reversed iff the statement that that a type is h-stable can always be h-inhabited:*

$$(\forall X : \textbf{\textit{Type}}, \langle\!\langle X \rangle\!\rangle \to \|X\|) \longleftrightarrow (\forall X : \textbf{\textit{Type}}, \|\|X\| \to X\|).$$

Proof. The direction "$\to$" is an immediate application of Lemma 6 above. The other direction is slightly trickier: If we knew h-stable X, we would have a constant endomap on X, and with the assumption $\langle\!\langle X \rangle\!\rangle$, this constant endomap would have a fixed point. Hence, we would have an inhabitant of X, and therefore and inhabitant of $\|X\|$. We observe that $\|X\|$ is h-propositional, so, by definition, we do not necessarily need h-stable X, but $\|$ h-stable $X\|$ is enough, and that completes the proof. □

It is also easy to see (cf. our Agda file [3]) that

$$\langle\!\langle X \rangle\!\rangle \quad \longleftrightarrow \quad \|\|X\| \to X\| \to \|X\|,$$

which gives an alternative route to the above theorem. Moreover, the statement $\forall X : \textbf{Type}, \|\|X\| \to X\|$ is equivalent to the *h-propositional axiom of choice*: For every h-proposition P and any family $Y : P \to \textbf{Type}$,

$$(\forall p : P, \|Y p\|) \to \|\forall p : P, Y p\|,$$

which clearly holds under h-propositional excluded middle. When $Y p$ is a set with exactly two elements for every $p : P$, this amounts to *the world's simplest axiom of choice* [7], which fails in some toposes. Thus, by the above theorem, $\forall X : \textbf{Type}, \langle\!\langle X \rangle\!\rangle \to \|X\|$ is not provable.

7.3 Populated and Non-empty

If we can reverse the last implication of the chain, we have

$$\forall X : \textbf{Type}, \neg\neg X \to \langle\!\langle X \rangle\!\rangle.$$

To show that this is not provable, we prove that it is a taboo from the point of view of constructive mathematics, in the sense that it implies Excluded Middle for h-propositions,

$$\text{hprop-EM} \equiv \forall P\,, \text{hprop}\, P \to P + \neg P.$$

Lemma 7. *With functional extensionality, the following implication holds:*

$$(\forall X : \textit{\textbf{Type}}, \neg\neg X \to \langle\!\langle X \rangle\!\rangle) \to \text{hprop-EM}.$$

Proof. Assume P is an h-proposition. Then so is the type $P + \neg P$ (where we require functional extensionality to show that $\neg P$ is an h-proposition). Hence, the identity function on $P + \neg P$ is constant.

On the other hand, it is straightforward to construct a proof of $\neg\neg\,(P + \neg P)$. By the assumption, this means that $P + \neg P$ is populated, i.e. every constant endomap on it has a fixed point. Therefore, we can construct a fixed point of the identity function, which is equivalent to proving $P + \neg P$. □

Acknowledgments. The first-named author would like to thank Paolo Capriotti, Ambrus Kaposi, Nuo Li and especially Christian Sattler for interesting discussions and technical assistance.

References

1. Altenkirch, T., Coquand, T., Escardó, M., Kraus, N.: On h-propositional reflection and hedbergs theorem (November 2012), `http://homotopytypetheory.org/`
2. Altenkirch, T., Coquand, T., Escardó, M., Kraus, N.: Constant choice (Agda file), 2012/2013. Available at the Third-named Author's Institutional Webpage
3. Altenkirch, T., Coquand, T., Escardó, M., Kraus, N.: Generalizations of Hedberg's theorem (Agda file), 2012/2013. Available at the Third-named Author's Institutional Webpage
4. Awodey, S.: Type theory and homotopy. Technical report (2010)
5. Awodey, S., Bauer, A.: Propositions as (types). Journal of Logic and Computation 14(4), 447–471 (2004)
6. Constable, R.L., Allen, S.F., Bromley, H.M., Cleaveland, W.R., Cremer, J.F., Harper, R.W., Howe, D.J., Knoblock, T.B., Mendler, N.P., Panangaden, P., Sasaki, J.T., Smith, S.F.: Implementing Mathematics with the Nurpl Proof Development System. Prentice-Hall, NJ (1986)
7. Fourman, M.P., Ščedrov, A.: The "world's simplest axiom of choice" fails. Manuscripta Math. 38(3), 325–332 (1982)
8. Hedberg, M.: A coherence theorem for Martin-Löf's type theory. J. Functional Programming, 413–436 (1998)
9. Hofmann, M., Streicher, T.: The groupoid interpretation of type theory. In: Venice Festschrift, pp. 83–111. Oxford University Press (1996)
10. Kraus, N.: A direct proof of Hedberg's theorem. Blog post at homotopytypetheory.org (March 2012)
11. Mines, R., Richman, F., Ruitenberg, W.: A Course in constructive algebra. Universitext. Springer, New York (1988)
12. The HoTT and UF community. HoTT github repository
13. Univalent Foundations Program, IAS. Homotopy Type Theory: Univalent Foundations of Mathematics (2013)
14. Voevodsky, V.: Coq library. Availabe at the Author's Institutional Webpage

Using Models to Model-Check Recursive Schemes

Sylvain Salvati and Igor Walukiewicz*

Université de Bordeaux, INRIA, CNRS, LaBRI UMR5800

Abstract. We propose a model-based approach to the model checking problem for recursive schemes. Since simply typed lambda calculus with the fixpoint operator, λY-calculus, is equivalent to schemes, we propose the use a model of λY to discriminate the terms that satisfy a given property. If a model is finite in every type, this gives a decision procedure. We provide a construction of such a model for every property expressed by automata with trivial acceptance conditions and divergence testing. Such properties pose already interesting challenges for model construction. Moreover, we argue that having models capturing some class of properties has several other virtues in addition to providing decidability of the model-checking problem. As an illustration, we show a very simple construction transforming a scheme to a scheme reflecting a property captured by a given model.

1 Introduction

In this paper we are interested in the relation between the effective denotational semantics of the simply typed λY-calculus and the logical properties of Böhm trees. By *effective denotational* semantics we mean semantic spaces in which the denotation of a term can be computed; in this paper, these effective denotational semantics will simply be finite models of the λY-calculus, but Y will often be interpreted neither as the least nor as the greatest fixpoint.

Understanding properties of Böhm trees from a logical point of view is a problem that arises naturally in the model checking of higher-order programs. Often this problem is presented in the context of higher-order recursive schemes that generate a possibly infinite tree. Nevertheless, higher-order recursive schemes can be represented faithfully by λY-terms, in the sense that the infinite trees they generate are precisely Böhm trees of the λY-terms.

The technical question we address is whether the Böhm tree of a given term is accepted by a given tree automaton. We consider only automata with trivial acceptance conditions which we call *TAC automata*. The principal technical challenge we address here is that we allow automata to detect if a term has a head normal form. We call such automata *insightful* as opposed to *Ω-blind* automata that are insensitive to divergence. For example, the models studied by Aehlig or Kobayashi [1,10] are Ω-blind. Considering safety properties and divergence at the same time poses serious challenges to representing with denotational semantics what it means for an automaton to accept a Böhm tree. Indeed, this

* This work has been supported by ANR 2010 BLAN 0202 01 FREC.

M. Hasegawa (Ed.): TLCA 2013, LNCS 7941, pp. 189–204, 2013.

requires one to give to non-convergence a non-standard interpretation that can influence the meaning of a term in a stronger way than the usual semantics does. As we show here, Y combinator cannot be interpreted as an extremal fixpoint in this case, so known algorithms for verification of safety properties cannot take non-convergence into account in a non-trivial way.

Let us explain the difference between insightful and Ω-blind conditions. The definition of a Böhm tree says that if the head reduction of a term does not terminate then in the resulting tree we get a special symbol Ω. Yet this is not how this issue is treated in all known solutions to the model-checking problem. There, instead of reading Ω the automaton is let to run on the infinite sequence of unproductive reductions. In the case of automata with trivial conditions, this has as an immediate consequence that such an infinite computation is accepted by the automaton. From a denotational semantics perspective, this amounts to interpreting the fixpoint combinator Y as a greatest fixpoint on some finite monotonous model. So, for example, with this approach to semantics, the language of schemes that produce at least one head symbol is not definable by automata with trivial conditions. Let us note that this problem disappears once we consider Büchi conditions as they permit one to detect an infinite unproductive execution. So here we look at a particular class of properties expressible by Büchi conditions. Thus, the problem we address is a non-trivial extension of what is usually understood as the safety property for recursive schemes.

Our starting point is the proof that the usual methods for treating the safety properties of higher-order schemes cannot capture the properties described with insightful automata. The first result of the paper shows that extremal fixpoint models can only capture boolean combinations of Ω-blind TAC automata. Our main result is the construction of a model capturing insightful automata. This construction is based on an interpretation of the fixpoint operator which is neither the greatest nor the least one. The main difficulty is to obtain a definition that guaranties the existence and uniqueness of the fixpoint at every type.

In our opinion providing models capturing certain classes of properties is an important problem both from foundational and practical points of view. On the theoretical side, models need to handle all the constructions of the λ-calculus while, for example, the type systems proposed so far by Kobayashi [10], and by Kobayashi and Ong [13] do not cater for λ-abstraction. In consequence the model-based approach gives more insight into the solution. On the practical side, models capturing classes of properties set the stage to define algorithms to decide these properties in terms of evaluating λ-terms in them. One can remark that models offer most of the algorithmic advantages as other approaches, as illustrated by [16] which shows that the typing discipline of [10] can be completely rephrased in terms of simple models. This practical interest of models has been made into a slogan by Terui [20]: *better semantics, faster computation.* To substantiate further the interest of models we also present a straightforward transformation of a scheme to a scheme reflecting a given property [4]. From a larger perspective, the model based approach opens a new bridge between λ-calculus and model-checking communities. In particular the model we construct for insightful automata brings into the front stage particular non-extremal fixpoints. To our knowledge these were not much studied in the λ-calculus literature.

Related Work. The model checking problem has been solved by Ong [14] and subsequently revisited in a number of ways [8,13,17]. A much simpler proof for the same problem in the case of Ω-blind TAC automata has been given by Aehlig [1]. In his influential work, Kobayashi [10,9,11] has shown that many interesting properties of higher-order recursive programs can be analyzed with recursive schemes and Ω-blind TAC automata. He has also proposed an intersection type system for the model-checking problem. The method has been applied to the verification of higher-order programs [12,5]. Let us note that at present all algorithmic effort concentrates on Ω-blind TAC automata. In a recent work Ong and Tsukada [15] provide a game semantics model corresponding to Kobayashi's style type system. Their model can handle only Ω-blind automata, but then it is fully complete. We cannot hope to have full completeness in our approach using simple models. In turn, as we mention in [21] and show here, handling Ω-blind automata with simple models is straightforward. The reflection property for schemes has been proved by Broadbent et. al. [4]. Haddad gives a direct transformation of a scheme to an equivalent scheme without divergent computations [7].

Organization of the Paper. The next section introduces the objects of our study: λY-calculus and automata with trivial acceptance conditions (TAC automata). In the following section we briefly present the correspondence between models of λY with greatest fixpoints and boolean combinations of Ω-blind TAC automata. In Section 4 we give the construction of the model for insightful TAC automata. The last section presents a transformation of a term into a term reflecting a given property. All the missing proofs can be found in the long version of the paper [19].

2 Preliminaries

We introduce two basic objects of our study: λY-calculus and TAC automata. We will look at λY-terms as mechanisms for generating infinite trees that then are accepted or rejected by a TAC automaton. The definitions we adopt are standard ones in the λ-calculus and automata theory. The only exceptions are the notions of a tree signature used to simplify the presentation, and of Ω-blind/insightful automata that are specific to this paper.

2.1 λY-Calculus and Models

The *set of types* $\mathcal{T}$ is constructed from a unique *basic type* 0 using a binary operation $\to$. Thus 0 is a type and if α, β are types, so is $(\alpha \to \beta)$. The order of a type is defined by: $order(0) = 1$, and $order(\alpha \to \beta) = max(1 + order(\alpha), order(\beta))$.

A *signature*, denoted Σ, is a set of typed constants, that is symbols with associated types from $\mathcal{T}$. We will assume that for every type $\alpha \in \mathcal{T}$ there are constants ω^α, Ω^α and $Y^{(\alpha\to\alpha)\to\alpha}$. A constant $Y^{(\alpha\to\alpha)\to\alpha}$ will stand for a fixpoint operator. Both ω^α and Ω^α will stand for undefined, but we will need two such constants in Section 4. Of special interest to us will be *tree signatures* where all constants other than Y, ω and Ω have order at most 2. Observe that types of order 2 have the form $0 \to 0 \to \cdots \to 0 \to 0$.

Proviso. To simplify the notation we will suppose that all the constants in a tree signature are either of type 0 or of type $0 \to 0 \to 0$. So they are either a constant of the base type or a function of two arguments over the base type. This assumption does not influence the results of the paper.

The set of *simply typed* λ*-terms* is defined inductively as follows. A constant of type α is a term of type α. For each type α there is a countable set of variables $x^\alpha, y^\alpha, \dots$ that are also terms of type α. If M is a term of type β and x^α a variable of type α then $\lambda x^\alpha.M$ is a term of type $\alpha \to \beta$. Finally, if M is of type $\alpha \to \beta$ and N is a term of type α then (MN) is a term of type β. We shall use the usual convention about dropping parentheses in writing λ-terms and we shall write sequences of λ-abstractions $\lambda x_1.\dots\lambda x_n.M$ with only one λ: $\lambda x_1 \dots x_n.M$; moreover when the sequence of abstracted variables is irrelevant we shall write $\lambda \boldsymbol{x}.M$ for a sequence of variables $\boldsymbol{x}$.

The usual operational semantics of the λ-calculus is given by β-contraction. To give the meaning to fixpoint constants we use δ-contraction ($\to_\delta$).

$$(\lambda x.M)N \to_\beta M[N/x] \qquad YM \to_\delta M(YM).$$

We write $\to^*_{\beta\delta}$ for the $\beta\delta$-reduction, the reflexive and transitive closure of the sum of the two relations. Given a term $M = \lambda x_1 \dots x_n.N_0 N_1 \dots N_p$ where N_0 is of the form $(\lambda x.P)Q$ or YP, then N_0 is called the *head redex* of M. We write $M \to_{\beta\delta h} M'$ when M' is obtained by $\beta\delta$-contracting the head redex of M (when it has one). We write $\to^*_{\beta\delta h}$ and $\to^+_{\beta\delta h}$ respectively for the reflexive and transitive closure and the transitive closure of $\to_{\beta\delta h}$. The relation $\to^*_{\beta\delta h}$ is called *head reduction.* A term with no head redex is said to be in *head normal form.*

It is well known that every term has at most one normal form, but due to δ-reduction there are terms without a normal form. A term is *unsolvable* if it does not have a head normal form; otherwise the term is *solvable.* Observe that even if all the subterms of a term are solvable the reduction may generate an infinitely growing term. It is thus classical in the λ-calculus to consider a kind of infinite normal form that by itself is an infinite tree, and in consequence it is not a term of λY [3,2].

A *Böhm tree* is an unranked, ordered, and potentially infinite tree with nodes labelled by terms of the form $\lambda x_1.\dots x_n.N$; where N is a variable or a constant, and the sequence of λ-abstractions is optional. So for example x^0, Ω^0, $\lambda x^0.\omega^0$ are labels, but $\lambda y^0.\ x^{0\to 0}y^0$ is not.

Definition 1. *A* Böhm tree *of a term M is obtained in the following way.*

– *If* $M \to^*_{\beta\delta} \lambda \boldsymbol{x}.N_0 N_1 \ldots N_k$ *with* N_0 *a variable or a constant then* $BT(M)$ *is a tree having the root labelled* $\lambda \boldsymbol{x}.N_0$ *and having* $BT(N_1)$, ..., $BT(N_k)$ *as its subtrees.*
– *Otherwise* $BT(M) = \Omega^\alpha$, *where* α *is the type of* M.

Observe that a term M without the constants Ω and ω has a $\beta\delta$-normal form if and only if $BT(M)$ is a finite tree without the constants Ω and ω. In this case the Böhm tree is just another representation of the normal form.

Recall that in a tree signature all constants except of Y, Ω, and ω are of type 0 or $0 \to 0 \to 0$. A closed term without λ-abstraction and Y over such a signature is just a finite binary tree: constants of type 0 occur at leaves and those of type $0 \to 0 \to 0$ occur at internal nodes. The same holds for Böhm trees:

Lemma 1. *If* M *is a closed term of type* 0 *over a tree signature then* $BT(M)$ *is a potentially infinite binary tree.*

We will consider finitary models of λY-calculus. In the first part of the paper we will concentrate on those where Y is interpreted as the greatest fixpoint.

Definition 2. *A* GFP-model *of a signature* Σ *is a tuple* $\mathcal{S} = \langle \{\mathcal{S}_\alpha\}_{\alpha \in \mathcal{T}}, \rho \rangle$ *where* $\mathcal{S}_0$ *is a finite lattice, and for every type* $\alpha \to \beta \in \mathcal{T}$, $\mathcal{S}_{\alpha \to \beta}$ *is the lattice* mon$[\mathcal{S}_\alpha \to \mathcal{S}_\beta]$ *of monotone functions from* $\mathcal{S}_\alpha$ *to* $\mathcal{S}_\beta$ *ordered coordinatewise. The valuation function* ρ *is required to satisfy certain conditions:*

– *If* $c \in \Sigma$ *is a constant of type* α *then* $\rho(c)$ *is an element of* $\mathcal{S}_\alpha$.
– *For every* $\alpha \in \mathcal{T}$, *both* $\rho(\omega^\alpha)$ *and* $\rho(\Omega^\alpha)$ *are the greatest elements of* $\mathcal{S}_\alpha$.
– *Moreover,* $\rho(Y^{(\alpha \to \alpha) \to \alpha})$ *is the function assigning to every function* $f \in \mathcal{S}_{\alpha \to \alpha}$ *its greatest fixpoint.*

Observe that every $\mathcal{S}_\alpha$ is finite, hence all the greatest fixpoints exist without any additional assumptions on the lattice.

A *variable assignment* is a function v associating to a variable of type α an element of $\mathcal{S}_\alpha$. If s is an element of $\mathcal{S}_\alpha$ and x^α is a variable of type α then $v[s/x^\alpha]$ denotes the valuation that assigns s to x^α and that is identical to v otherwise.

The *interpretation of a term* M of type α in the model $\mathcal{S}$ under the valuation v is an element of $\mathcal{S}_\alpha$ denoted $[\![M]\!]^v_\mathcal{S}$. The meaning is defined in the standard way: for constants it is given by ρ; for variables by v; the application is interpreted as function application, and finally for abstraction $[\![\lambda x^\alpha.M]\!]^v_\mathcal{S}$ is a function mapping an element $s \in \mathcal{S}_\alpha$ to $[\![M]\!]^{v[s/x^\alpha]}_\mathcal{S}$. As usual, we will omit subscripts or superscripts in the notation of the semantic function if they are clear from the context.

It is known that Böhm trees are a kind of initial semantics for λ-terms. In particular if two terms have the same Böhm trees then they have the same semantics in every GFP model. To look at it more closely we need to formally define the semantics of a Böhm tree.

The semantics of a Böhm tree is defined in terms of its truncations. For every $n \in \mathbb{N}$, we denote by $BT(M){\downarrow_n}$ the finite term that is the result of replacing in the tree $BT(M)$ every subtree at depth n by the constant ω^α of the appropriate type. Observe that if M is closed and of type 0 then α will always be the base type 0. This is because we work with a tree signature. We define:

$$[\![BT(M)]\!]^v_{\mathcal{S}} = \bigwedge\{[\![BT(M)\downarrow_n]\!]^v_{\mathcal{S}} \mid n \in \mathbb{N}\}.$$

The above definitions are standard for λY-calculus, or more generally for PCF [2]. In particular the following proposition, in a more general form, can be found as Exercise 6.1.8 in op. cit.[1]

Proposition 1. *If $\mathcal{S}$ is a finite GFP-model and M is a closed term then:* $[\![M]\!]_{\mathcal{S}} = [\![BT(M)]\!]_{\mathcal{S}}$.

2.2 TAC Automata

Let us fix a tree signature Σ. This means that apart from ω, Ω and Y all constants have order at most 2. Let Σ_0 be the set of constants of type 0, and Σ_2 the set of constants of type $0 \to 0 \to 0$. By Lemma 1, in this case Böhm trees are potentially infinite binary trees.

Definition 3. *A* finite tree automaton with trivial acceptance condition *(TAC automaton) over the signature $\Sigma = \Sigma_0 \cup \Sigma_2$ is*

$$\mathcal{A} = \langle Q, \Sigma, q^0 \in Q, \delta_0 : Q \times (\Sigma_0 \cup \{\Omega\}) \to \{\mathit{ff}, \mathit{tt}\}, \delta_2 : Q \times \Sigma_2 \to \mathcal{P}(Q^2)\rangle$$

where Q is a finite set of states and $q^0 \in Q$ is the initial state. The transition function of TAC automaton may be the subject to the additional restriction:

$$\textbf{\textit{Ω-blind:}} \quad \delta_0(q, \Omega) = \mathit{tt} \text{ for all } q \in Q.$$

Automata satisfying this restriction are called Ω-blind. *For clarity, we use the term* insightful *to refer to automata without this restriction.*

Automata will run on Σ-labelled binary trees that are partial functions $t : \{1,2\}^* \to \Sigma \cup \{\Omega\}$ such that their domain is a binary tree, and $t(u) \in \Sigma_0 \cup \{\Omega\}$ if u is a leaf, and $t(u) \in \Sigma_2$ otherwise.

A *run of $\mathcal{A}$ on t* is a labelling $r : \{1,2\}^* \to Q$ of t such that the root is labeled by q^0 and the labelling of the successors of a node respects the transition function δ. A run is *accepting* if $\delta_0(r(u), t(u)) = \mathit{tt}$ for every leaf u of t . A tree is *accepted by $\mathcal{A}$* if there is an accepting run on the tree. The *language* of $\mathcal{A}$, denoted $L(\mathcal{A})$, is the set of trees accepted by $\mathcal{A}$.

Observe that TAC automata have acceptance conditions on leaves, expressed with δ_0, but do not have acceptance conditions on infinite paths.

As underlined in the introduction all the work on automata with trivial conditions relies on the Ω-blind restriction. Let us give some examples of properties that can be expressed with insightful automata but not with Ω-blind automata.

[1] In this paper we work with finite monotone models which are a particular case of the directed complete partial orders used in [2].

- The set of terms not having Ω in their Böhm tree. To recognize this set we take the automaton with a unique state q. This state has transitions on all the letters from Σ_2. It also can end a run in every constant of type 0 except for Ω: this means $\delta_0(q, \Omega) = \mathit{ff}$ and $\delta_0(q, c) = \mathit{tt}$ for all other c.
- The set of terms having a head normal form. We take an automaton with two states q and $q_\top$. From $q_\top$ automaton accepts every tree. From q it has transitions to $q_\top$ on all the letters from Σ_2, on letters from Σ_0 it behaves as the automaton above.
- Building on these two examples one can easily construct an automaton for a property like "every occurrence of Ω is preceded by a constant *err*".

It is immediate to see that none of these languages can be recognized by a Ω-blind automaton since if such an automaton accepts a tree t then it accepts also every tree obtained by replacing a subtree of t by Ω.

3 GFP Models and Ω-Blind TAC Automata

In this short section we summarize the relation between GFP models and Ω-blind TAC automata. We start with the expected formal definition of the set of λY-terms recognized by a model.

Definition 4. *For a GFP model $\mathcal{S}$ over the base set $\mathcal{S}_0$. The* language recognized *by a subset $F \subseteq \mathcal{S}_0$ is the set of closed λY-terms $\{M \mid [\![M]\!]_{\mathcal{S}} \in F\}$.*

Proposition 2. *For every Ω-blind TAC automaton $\mathcal{A}$, the language of $\mathcal{A}$ is recognized by a GFP model.*

Let $\mathcal{A}$ be an automaton as in Definition 3. For the model $\mathcal{S}$ in question we take a GFP model with $\mathcal{S}_0 = \mathcal{P}(Q)$. This defines $\mathcal{S}_\alpha$ for every type α. It remains to define the interpretation of constants other than ω, Ω, or Y. The meaning of a constant c of type 0 is $\{q \mid \delta_0(q, c) = tt\}$; and the meaning of a of type $0 \to 0 \to 0$ is a function whose value on $(S_0, S_1) \in \mathcal{P}(Q)^2$ is $\{q \mid \delta_2(q, a) \cap S_0 \times S_1 \neq \emptyset\}$. Finally, for the set $F_{\mathcal{A}}$ used to recognize $L(\mathcal{A})$ we will take $\{S \mid q^0 \in S\}$; recall that q^0 is the initial state of $\mathcal{A}$. With these definitions it is possible to show that for every closed term M of type 0: $BT(M) \in L(\mathcal{A})$ iff $[\![M]\!] \in F_{\mathcal{A}}$.

Next theorem shows that the recognizing power of GFP models is actually characterized by Ω-blind TAC automata. The right-to-left implication of this theorem has been stated in [21].

Theorem 1. *A language L of λ-terms is recognized by a GFP-model iff it is a boolean combination of languages of Ω-blind TAC automata.*

Using the results in [16], it can be shown that typings in Kobayashi's type systems [10] give precisely values in GFP models.

4 A Model for Insightful TAC Automata

The goal of this section is to present a model capable of recognizing languages of insightful TAC automata. Theorem 1 implies that the fixpoint operator in such a model can be neither the greatest nor the least fixpoint. In the first subsection we will construct a model containing at the same time a model with the least fixpoint and a model with the greatest fixpoint. We cannot just take the model generated by the product of the base sets of the two models as we will need that the value of a term in the least fixpoint component influences the value in the greatest fixpoint component. In the second part of this section we will show how to interpret insightful TAC automata in such a model.

4.1 Model Construction and Basic Properties

We are going to construct a model $\mathcal{K}$ intended to recognize the language of a given insightful TAC automaton. This model is built on top of the standard model $\mathcal{D}$ for detecting if a term has a head-normal form.

Consider a family of sets $\{\mathcal{D}_\alpha\}_{\alpha\in\mathcal{T}}$; where $\mathcal{D}_0 = \{\bot, \top\}$ is the two element lattice, and $\mathcal{D}_{\alpha\to\beta}$ is mon$[\mathcal{D}_\alpha \to \mathcal{D}_\beta]$. So for every α, $\mathcal{D}_\alpha$ is a finite lattice. We shall refer to the minimal and maximal element of $\mathcal{D}_\alpha$ respectively with the notations $\bot_\alpha$ and $\top_\alpha$.

Consider the model $\mathcal{D} = \langle\{\mathcal{D}_\alpha\}_{\alpha\in\mathcal{T}}, \rho\rangle$ where ω and Ω are interpreted as the least elements, and Y is interpreted as the least fixpoint operator. So $\mathcal{D}$ is a dual of a GFP model as presented in Definition 2. The reason for not taking a GFP model here is that we would prefer to use the greatest fixpoint later in the construction. To all constants other than Y, ω, and Ω the interpretation ρ assigns the greatest element of the appropriate type. The following theorem is well-known (cf [2] page 130).

Theorem 2. *For every closed term M of type 0 without ω we have:*

$$BT(M) = \Omega \quad \textit{iff} \quad [\![M]\!]_{\mathcal{D}} = \bot.$$

We fix a finite set Q and $Q_\Omega \subseteq Q$. Later these will be the set of states of a TAC automaton, and the set of states from which the automaton accepts Ω, respectively. To capture the power of such an automaton, we are going to define a model $\mathcal{K}(Q, Q_\Omega, \rho)$ of the λY-calculus with a non-standard interpretation of the fixpoint. Roughly, this model will live inside the product of $\mathcal{D}$ and the GFP model $\mathcal{S}$ for an Ω-blind automaton. The idea is that every set $\mathcal{K}_\alpha$ will have a projection on $\mathcal{D}$ but not necessarily on $\mathcal{S}$. This allows to observe whether a term converges or not, and at the same time to use this information in computing in the second component.

Definition 5. *For a given finite set Q and $Q_\Omega \subseteq Q$ we define a family of sets $\mathcal{K}_{Q,Q_\Omega} = (\mathcal{K}_\alpha)_{\alpha\in\mathcal{T}}$ by mutual recursion together with a family of relations $\mathcal{L} = (\mathcal{L}_\alpha)_{\alpha\in\mathcal{T}}$ such that $\mathcal{L}_\alpha \subseteq \mathcal{K}_\alpha \times \mathcal{D}_\alpha$:*

1. *We let* $\mathcal{K}_0 = \{(\top, P) \mid P \subseteq Q\} \cup \{(\bot, Q_\Omega)\}$ *with the order:* $(d_1, P_1) \leq (d_2, P_2)$ *iff* $d_1 \leq d_2$ *in* $\mathcal{D}_0$ *and* $P_1 \subseteq P_2$. *(cf. Figure 1)*
2. $\mathcal{L}_0 = \{((d, P), d) \mid (d, P) \in \mathcal{K}_0\}$,
3. $\mathcal{K}_{\alpha\to\beta} = \{f \in \text{mon}[\mathcal{K}_\alpha \to \mathcal{K}_\beta] \mid \exists d \in \mathcal{D}_{\alpha\to\beta}.\forall (g, e) \in \mathcal{L}_\alpha.(f(g), d(e)) \in \mathcal{L}_\beta\}$,
4. $\mathcal{L}_{\alpha\to\beta} = \{(f, d) \in \mathcal{K}_{\alpha\to\beta} \times \mathcal{D}_{\alpha\to\beta} \mid \forall (g, e) \in \mathcal{L}_\alpha.(f(g), d(e)) \in \mathcal{L}_\beta\}$.

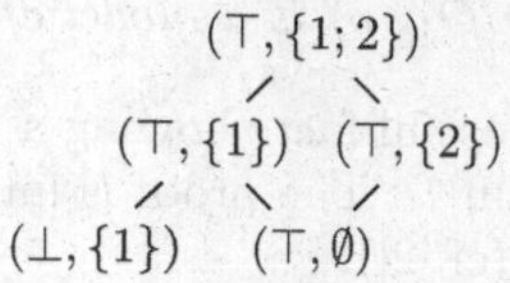

Fig. 1. The order $\mathcal{K}_0$ for $Q = \{1, 2\}$ and $Q_\Omega = \{1\}$

Note that every $\mathcal{K}_\alpha$ is finite since it lives inside the standard model constructed from $\mathcal{D}_0 \times \mathcal{P}(Q)$ as the base set. Moreover for every α, $\mathcal{K}_\alpha$ is a join semilattice and thus has a greatest element. Recall that a TAC automaton is supposed to accept unsolvable terms from states Q_Ω. So the unsolvable terms of type 0 should have Q_Ω as a part of their meaning. This is why $\bot$ of $\mathcal{D}_0$ is associated to $(\bot, Q_\Omega)$ in $\mathcal{K}_0$ via the relation $\mathcal{L}_0$. This also explains why we needed to take the least fixpoint in $\mathcal{D}$. If we had taken the greatest fixpoint then the unsolvable terms would have evaluated to $\top$ and the solvable ones to $\bot$. In consequence we would have needed to relate $\top$ with $(\top, Q_\Omega)$, and we would have been forced to relate $\bot$ with $(\bot, Q)$. But since $(\top, Q_\Omega)$ and $(\bot, Q)$ are incomparable in $\mathcal{K}_0$ we would not have been able to obtain the order preserving injection $(\cdot)^\uparrow$ from $\mathcal{D}_0$ to $\mathcal{K}_0$ that is defined below at every type:

Definition 6. *For every* $h \in \mathcal{D}_\alpha$ *we define the element* $h^\uparrow$ *of* $\mathcal{K}_\alpha$*:*

$$h^\uparrow = \bigvee \{f \mid (f, h) \in \mathcal{L}_\alpha\} .$$

It can be shown that this element always exists, and that $(\cdot)^\uparrow$ is a monotone embedding of $\mathcal{D}$ into $\mathcal{K}$. Moreover $(d^\uparrow, d)$ is in $\mathcal{L}_\alpha$ for very $d \in \mathcal{D}_\alpha$. One can also verify that the relation $\mathcal{L}_\alpha$ is functional, so we get the projection operation.

Definition 7. *For every type* α *and* $f \in \mathcal{K}_\alpha$ *we let* $\overline{f}$ *to be the unique element of* $\mathcal{D}_\alpha$ *such that* $(f, \overline{f}) \in \mathcal{L}_\alpha$.

We are now going to give the definition of the interpretation of the fixpoint combinator in $\mathcal{K}$. This definition is based on the fixpoint operator in $\mathcal{D}$. As a shorthand, we write fix_α for the operation in $\mathcal{D}_{(\alpha\to\alpha)\to\alpha}$ mapping a function of $\mathcal{D}_{\alpha\to\alpha}$ to its least fixpoint. It can be shown that for every $f \in \mathcal{K}_{\alpha\to\alpha}$ the sequence $f^n(\text{fix}_\alpha(\overline{f})^\uparrow)$ is decreasing in $\mathcal{K}_\alpha$.

Definition 8. *For every type* α *and* $f \in \mathcal{K}_\alpha$ *define*

$$\text{Fix}_\alpha(f) = \bigwedge_{n\in\mathbb{N}} (f^n(\text{fix}_\alpha(\overline{f})^\uparrow))$$

We are ready to define the model we were looking for.

Definition 9. *For a finite set Q and $Q_\Omega \subseteq Q$ consider a tuple $\mathcal{K}(Q, Q_\Omega, \rho) = (\mathcal{K}_{Q,Q_\Omega}, \rho)$ where $\mathcal{K}_{Q,Q_\Omega}$ is as in Definition 5 and ρ is a valuation such that for every type α: ω^α is interpreted as the greatest element of $\mathcal{K}_\alpha$, $Y^{(\alpha\to\alpha)\to\alpha}$ is interpreted as* Fix_α*, and Ω^0 is interpreted as $(\bot, Q_\Omega)$.*

Theorem 3. *The model $\mathcal{K}(Q, Q_\Omega, \rho)$ is a model of the λY-calculus.*

Let us mention the following useful fact showing a correspondence between the meanings of a term in $\mathcal{K}$ and in $\mathcal{D}$. The proof is immediate since, by definition, $\{\mathcal{L}_\alpha\}_{\alpha\in\mathcal{T}}$ is a logical relation (cf [2]).

Lemma 2. *For every type α and closed term M of type α:*

$$([\![M]\!]_{\mathcal{K}}, [\![M]\!]_{\mathcal{D}}) \in \mathcal{L}_\alpha\,.$$

4.2 Correctness and Completeness of the Model

It remains to show that the model we have constructed can recognize languages of TAC automata. We fix a tree signature Σ and a TAC automaton $\mathcal{A}$ as in Definition 3. So Q is the set of states of $\mathcal{A}$ and Q_Ω is the set of states q such that $\delta(q, \Omega) = tt$. Consider a model $\mathcal{K}$ based on $\mathcal{K}(Q, Q_\Omega, \rho)$ as in Definition 9. We need to specify the meaning of constants like $c : 0$ or $a : 0^2 \to 0$ in Σ:

$$\begin{aligned} \rho(c) &= (\top, \{q : \delta(q, c) = tt\}) \\ \rho(a)(d_1, R_1)(d_2, R_2) &= (\top, R) \qquad \text{where } d_1, d_2 \in \{\bot, \top\} \text{ and} \\ & \qquad\qquad R = \{q \in Q \mid \delta(q, a) \cap R_1 \times R_2 \neq \emptyset\} \end{aligned}$$

It is easy to verify that the meanings of constants are indeed in the model.

Proposition 3. *Given a closed term M of type 0: $BT(M) = \Omega^0$ iff $[\![M]\!]_{\mathcal{K}} = (\bot, Q_\Omega)$.*

As in the case of GFP-models the semantics of a Böhm tree is defined in terms of its truncations: $[\![BT(M)]\!]_{\mathcal{K}} = \bigwedge\{[\![BT(M){\downarrow_n}]\!]_{\mathcal{K}} : n \in \mathbb{N}\}$. The subtle, but crucial, difference is that now Ω^0 and ω^0 do not have the same meaning. Nevertheless the analog of Proposition 1 still holds in $\mathcal{K}$.

Theorem 4. *For very closed term M of type 0: $[\![M]\!]_{\mathcal{K}} = [\![BT(M)]\!]_{\mathcal{K}}$.*

Proof (Sketch). First we show that $[\![M]\!]_{\mathcal{K}} \leq [\![BT(M)]\!]_{\mathcal{K}}$. For this we define a finite approximation of the Böhm tree. *The Abstract Böhm tree up to depth l* of a term M, denoted $ABT_l(M)$, will be a term obtained by reducing M till it resembles $BT(M)$ up to depth l as much as possible. We define $ABT_0(M) = M$, and also $ABT_{l+1}(M) = M$ if M is unsolvable, or otherwise $ABT_{l+1}(M) = \lambda x.N_0 ABT_l(N_1)\ldots ABT_l(N_k)$, where $\lambda x.N_0 N_1 \ldots N_k$ is the head normal form of M.

Since $ABT_l(M)$ is obtained from M by a sequence of $\beta\delta$-reductions, $[\![M]\!]_{\mathcal{K}} = [\![ABT_l(M)]\!]_{\mathcal{K}}$ for every l. It remains to show that for every term M and every l:

$$[\![M]\!]_{\mathcal{K}} = [\![ABT_l(M)]\!]_{\mathcal{K}} \leq [\![BT(M){\downarrow_l}]\!]_{\mathcal{K}}.$$

Up to depth l, the two terms have the same tree structure. We check that the meaning of every leaf in $ABT_l(M)$ is not bigger than the meaning of the corresponding leaf of $BT(M){\downarrow_l}$. For leaves of depth l this is trivial since on the one hand we have a term and on the other the constant ω. For other leaves, the terms are either identical or on one side we have an unsolvable term, and on the other Ω^0. By Proposition 3 the two have the same meaning in $\mathcal{S}$.

For the inequality in the other direction observe that if a term M does not have Y combinators, then it is strongly normalizing and the theorem is trivial. So we need be able to deal with Y combinators in M. We introduce new constants c_N for every subterm YN of M. The type of c_N is $\boldsymbol{\alpha} \to \beta$ if β is the type of YN and $\boldsymbol{\alpha} = \alpha_1 \dots \alpha_k$ is the sequence of types of the sequence of free variables $\boldsymbol{x} = x_1 \dots x_k$ occurring in YN. We let the semantics of a constant c_N be

$$[\![c_N]\!]_{\mathcal{K}} = \lambda \boldsymbol{p}. \left(\mathrm{fix}_\beta(\overline{[\![N]\!]_{\mathcal{D}}^{[\boldsymbol{p}/\boldsymbol{x}]}})\right)^{\uparrow}.$$

In the full version of the paper we show that $[\![c_N]\!]$ is in $\mathcal{K}$. Moreover for every $p_1, \dots, p_k, q_1, \dots, q_l$:

$$[\![c_N]\!]_{\mathcal{K}}(p_1, \dots, p_k)(q_1, \dots, q_l) = \begin{cases} (\bot, Q_\Omega) & \text{if } [\![c_N]\!]_{\mathcal{D}}(\overline{p}_1, \dots, \overline{p}_k)(\overline{q}_1, \dots, \overline{q}_l) = \bot \\ (\top, Q) & \text{if } [\![c_N]\!]_{\mathcal{D}}(\overline{p}_1, \dots, \overline{p}_k)(\overline{q}_1, \dots, \overline{q}_l) = \top \end{cases} \tag{1}$$

We now define term $\mathit{iterate}^n(N)$ for very $n \in \mathbb{N}$.

$$\mathit{iterate}^0(N) = c_N \qquad \mathit{iterate}^{n+1}(N) = \lambda \boldsymbol{x}.N(\mathit{iterate}^n(N)\boldsymbol{x}).$$

From the definition of the fixpoint operator in $\mathcal{K}$ and the fact that $\mathcal{K}_\beta$ is finite it follows that $[\![\mathit{iterate}^n(N)]\!] = [\![\lambda \boldsymbol{x}.YN]\!]$ for some n. Now we can apply this identity to all fixpoint subterms in M starting from the innermost subterms. So the term $\mathit{expand}^i(M)$ is obtained by repeatedly replacing occurrences of subterms of the form YN in M by $\mathit{iterate}^i(N)\boldsymbol{x}$ starting from the innermost occurrences. We get that for n chosen as above $[\![M]\!]_{\mathcal{K}} = [\![\mathit{expand}^n(M)]\!]_{\mathcal{K}}$.

We come back to the proof. The missing inequality will be obtained from

$$[\![M]\!]_{\mathcal{K}} = [\![\mathit{expand}^n(M)]\!]_{\mathcal{K}} = [\![BT(\mathit{expand}^n(M))]\!]_{\mathcal{K}} \geq [\![BT(M)]\!]_{\mathcal{K}}.$$

The first equality we have discussed above. The second is trivial since $\mathit{expand}^n(M)$ does not have fixpoints. It remains to show $[\![BT(\mathit{expand}^n(M))]\!]_{\mathcal{K}} \geq [\![BT(M)]\!]_{\mathcal{K}}$.

Let us denote $BT(\mathit{expand}^n(M))$ by P. So P is a term of type 0 in a normal form without occurrences of Y. For a term K let $\tilde{K}$ be a term obtained from K by simultaneously replacing c_N by $\lambda \boldsymbol{x}.YN$. By definition of the fixpoint we have

$[\![c_N]\!]_{\mathcal{K}} \geq [\![\lambda \boldsymbol{x}.YN]\!]_{\mathcal{K}}$ which also implies that $[\![K]\!]_{\mathcal{K}} \geq [\![\tilde{K}]\!]_{\mathcal{K}}$. Moreover, as $\tilde{P} =_{\beta\delta} M$, we have that $BT(\tilde{P}) = BT(M)$. We need to show that $[\![P]\!]_{\mathcal{K}} \geq [\![BT(\tilde{P})]\!]_{\mathcal{K}}$.

Let us compare the trees $BT(P)$ and $BT(\tilde{P})$ by looking on every path starting from the root. The first difference appears when a node v of $BT(P)$ is labelled with c_N for some N. Say that the subterm of P rooted in v is $c_N K_1 \dots K_i$. Then at the same position in $BT(P')$ we have the Böhm tree of the term $(\lambda \boldsymbol{x}.YN)\tilde{K}_1 \dots \tilde{K}_i$. We will be done if we show that $[\![c_N K_1 \dots K_i]\!]_{\mathcal{K}} \geq [\![BT((\lambda \boldsymbol{x}.YN)\tilde{K}_1 \dots \tilde{K}_i)]\!]_{\mathcal{K}}$.

We reason by cases. If $[\![c_N K_1 \dots K_i]\!]_{\mathcal{D}} = \top$ then equation (1) gives us $[\![c_N K_1 \dots K_i]\!]_{\mathcal{K}} = (\top, Q)$. So the desired inequality holds since $(\top, Q)$ is the greatest element of $\mathcal{K}_0$.

If $[\![c_N K_1 \dots K_i]\!]_{\mathcal{D}} = \bot$ then $[\![c_N \tilde{K}_1 \dots \tilde{K}_i]\!]_{\mathcal{D}} = \bot$ since $[\![K_i]\!]_{\mathcal{K}} \geq [\![\tilde{K}_i]\!]_{\mathcal{K}}$. By equation (1) we get $[\![c_N \tilde{K}_1 \dots \tilde{K}_i]\!]_{\mathcal{D}} = (\bot, Q_\Omega)$. Since, by the definition of the fixpoint operator, $[\![c_N]\!]_{\mathcal{K}} \geq [\![\lambda \boldsymbol{x}.\, YN]\!]_{\mathcal{K}}$ we get $[\![YN\tilde{K}_1 \dots \tilde{K}_i]\!]_{\mathcal{K}} = (\bot, Q_\Omega)$. But then Proposition 3 implies that $YNK_1 \dots K_i$ is unsolvable. Thus we get $[\![BT((\lambda \boldsymbol{x}.YN)\tilde{K}_1 \dots \tilde{K}_i)]\!]_{\mathcal{K}} = [\![\Omega]\!]_{\mathcal{K}} = (\bot, Q_\Omega)$. □

Once we know that the semantics of the Böhm tree of a term in the model is the same as the semantics of a term, the proof of the correctness of the model is quite straightforward and very similar to the case of GFP models.

Theorem 5. *Let $\mathcal{A}$ be an insightful TAC automaton and $\mathcal{K}$ a model as at the beginning of the subsection. For every closed term M of type* 0*:*

$$BT(M) \in L(\mathcal{A}) \quad \textit{iff} \quad q^0 \textit{ is in the second component of } [\![M]\!]_{\mathcal{K}}.$$

5 Reflection

The idea behind the notion of a reflecting term is that at every moment of its evaluation every subterm should know its meaning. Knowing the meaning amounts to extra labelling of constants. Formally, we express this by the notion of a reflective Böhm tree defined below. The definition can be made more general but we will be interested only in the case of terms of type 0. In this section we will show that reflective Böhm trees can be generated by λY-terms.

As usual we suppose that we are working with a tree signature Σ. We will also need a signature where constants are annotated with elements of the model. If $\mathcal{S} = \langle \{\mathcal{S}_\alpha\}_{\alpha \in \mathcal{T}}, \rho \rangle$ is a finitary model then the extended signature $\Sigma^{\mathcal{S}}$ contains constants a^s where a is a constant in $\mathcal{S}$ and $s \in \mathcal{S}_0$; so superscripts are possible interpretations of terms of type 0 in $\mathcal{S}$.

Definition 10. *Let $\mathcal{S}$ be a finitary model and M a closed term of type* 0*. A* reflective Böhm tree *with respect to $\mathcal{S}$ is obtained in the following way:*

- *If $M \to^*_{\beta\delta} bN_1N_2$ for some constant $b : 0 \to 0 \to 0$ then $rBT_{\mathcal{S}}(M)$ is a tree having the root labelled by $b^{[\![bN_1N_2]\!]_{\mathcal{S}}}$ and having $rBT_{\mathcal{S}}(N_1)$ and $rBT_{\mathcal{S}}(N_2)$ as subtrees.*

– *If* $M \to_{\beta\delta}^{*} c$ *for some constant* $c : 0$ *then* $rBT_{\mathcal{S}}(M) = c^{[\![c]\!]_{\mathcal{S}}}$.
– *Otherwise,* M *is unsolvable and* $BT(M) = \Omega^0$.

Observe that when $\mathcal{S}$ satisfies $[\![N]\!]_{\mathcal{S}} = [\![BT(N)]\!]_{\mathcal{S}}$ for every term N then the superscripts in $rBT(M)$ are the meanings of respective subtrees in the Böhm tree. When, moreover, $\mathcal{S}$ recognizes a given property then these superscripts determine if the tree satisfies the property. These two conditions are fulfilled by the models we have considered in this paper.

We will use terms to construct reflective Böhm trees.

Definition 11. *Let* Σ *be a tree signature, and* $\mathcal{S}$ *a finitary model. Let* M *be a closed term of type* 0 *over the signature* Σ. *We say that a term* M' *over the signature* $\Sigma^{\mathcal{S}}$ *is* a reflection of M in $\mathcal{S}$ *if* $BT(M') = rBT(M)$.

The objective of this section is to construct reflections of terms. Since λY-terms can be translated to schemes and vice versa, the construction would work for schemes too. (Translations between schemes and λY-terms that do not increase the type order are presented in [18]).

Let us fix a tree signature Σ and a finitary model $\mathcal{S}$. For the construction of reflective terms we enrich λY-calculus with some syntactic sugar. Consider a type α. The set $\mathcal{S}_\alpha$ is finite for every type α; say $\mathcal{S}_\alpha = \{d_1, \dots, d_k\}$. We will introduce a new atomic type $[\alpha]$ and constants $d_1 \dots, d_k$ of this type; there will be no harm in using the same names for constants and elements of the model. We do this for every type α and consider terms over this extended type discipline. Notice that in the result there are no other closed normal terms than $d_1, \dots, d_k$ of type $[\alpha]$.

Given a term M of type $[\alpha]$ and $M_1, \dots M_n$ that are all terms of type β, we introduce the construct

$$\text{case } M\{d_i \to M_i\}_{d_i \in \mathcal{S}_\alpha}$$

which is a term of type β and which reduces to M_i when $M = d_i$. This construct is a simple syntactic sugar. We could as well represent $[\alpha]$ as the type $\beta^k \to \beta$, and a constant d_i by the ith projection $\lambda x_1 \dots x_n.x_i$. We would then get that the term $MM_1 \dots M_k$ reduces to M_i and thus behaves exactly as the *case* construct.

We define a transformation on types $\alpha^\bullet$ by induction on their structure:

$$(\alpha \to \beta)^\bullet = \alpha^\bullet \to [\alpha] \to \beta^\bullet \quad \text{and} \quad \alpha^\bullet = \alpha \text{ when } \alpha \text{ is atomic.}$$

The translation we are looking for will be an instance of a more general translation $[M, \upsilon]$ of a term M of type α into a term of type $\alpha^\bullet$ where υ is a valuation over $\mathcal{S}$.

$$[\lambda x^\alpha.M, \upsilon] = \lambda x^{\alpha^\bullet} \lambda y^{[\alpha]}.\ \text{case } y^{[\alpha]}\{d \to [M, \upsilon[d/x^\alpha]]\}_{d \in \mathcal{S}_\alpha}$$
$$[MN, \upsilon] = [M, \upsilon]\,[N, \upsilon]\,[\![N]\!]^\upsilon$$

$$[a, v] = \lambda x_1^0 \lambda y_1^{[0]} \lambda x_2^0 \lambda y_2^{[0]}.$$
$$\text{case } y_1^{[0]} \{d_1 \to \text{case } y_2^{[0]} \{d_2 \to a^{\rho(a) d_1 d_2} x_1 x_2\}_{d_2 \in \mathcal{S}_0}\}_{d_1 \in \mathcal{S}_0}$$
$$[x^\alpha, v] = x^{\alpha^\bullet}$$
$$\left[Y^{(\alpha\to\alpha)\to\alpha} M, v\right] = Y^{(\alpha^\bullet\to\alpha^\bullet)\to\alpha^\bullet} (\lambda x^{\alpha^\bullet}. [M, v]\, x^{\alpha^\bullet} [\![YM]\!]^v)$$

To prove correctness of this translation, we show that a head reduction of the original term can be simulated by a sequence of head reductions.

Lemma 3. *If $M \to_{\beta\delta h} M'$, then $[M, v] \to^+_{\beta\delta h} [M', v]$.*

Theorem 6. *For every finitary model $\mathcal{S}$ and a closed term M of type 0:*

$$BT([M, \emptyset]) = rBT_{\mathcal{S}}(M)\ .$$

Remark. If in a model $\mathcal{S}$ the divergence can be observed (as it is the case for GFP models and for the model $\mathcal{K}$, cf. Proposition 3) then in the translation above we could add the rule $[M, v] = \Omega$ whenever $[\![M]\!]^v_{\mathcal{S}}$ denotes a diverging term. We would obtain a term which would always converge. A different construction for achieving the same goal is proposed in [7].

Remark. Even though the presented translation preserves the structure of a term, it makes the term much bigger due to *case* construction in the clause for λ-abstraction. The blow-up is unavoidable due to complexity lower-bounds on the model-checking problem. Nevertheless, one can try to limit the use of *case* construct. We present a slightly more efficient translation that takes the value of the known arguments into account. For this, the translation also depends on a stack of values from $\mathcal{S}$ in order to recall the values taken by the arguments. For the sake of simplicity, we also assume that the constants always have all their arguments (this can be achieved by using terms in η-long form).

$$[\lambda x^\alpha.M, v, d :: S] = \lambda x^{\alpha^\bullet} y^{[\alpha]}. [M, v[d/x^\alpha], S]$$
$$[\lambda x^\alpha.M, v, \varepsilon] = \lambda x^{\alpha^\bullet} y^{[\alpha]}.\text{case } y^{[\alpha]} \{d \to [M, v[d/x^\alpha], \varepsilon]\}_{d \in \mathcal{S}_\alpha}$$
$$[MN, v, S] = [M, v, [\![N]\!]^v :: S]\, [N, v, \varepsilon]\, [\![N]\!]^v$$
$$[a, v, d_1 :: d_2 :: \varepsilon] = \lambda x_1^0 \lambda y_1^{[0]} \lambda x_2^0 \lambda y_2^{[0]}.\ a^{[\![a]\!] d_1 d_2} x_1 x_2$$
$$[x^\alpha, v, S] = x^{\alpha^\bullet}$$
$$[YM, v, S] = Y\, [M, v, [\![YM]\!]^v :: S]$$

6 Conclusions

We have extended the scope of the model-based approach to a larger class of properties. While a priori it is more difficult to construct a finitary model than to come up with a decision procedure, in our opinion this additional effort is justified. It allows, as we show here, to use the techniques of the theory of the λ-calculus. It opens new ways of looking at the algorithmics of the model-checking problem. Since typing in intersection type systems [10] and step functions in models are in direct correspondence [16], model-based approach can also benefit from all the developments in algorithms based on typing. Finally, this approach allows to get new constructions as demonstrated by our transformation of a scheme to a scheme reflecting a given property. Observe that this transformation is general and does not depend on our particular model.

Let us note that the model-based approach is particularly straightforward for Ω-blind TAC automata. It uses standard observations on models of the λY-calculus and Proposition 2 with a simple inductive proof. The model we propose for insightful automata may seem involved; nevertheless, the construction is based on simple and standard techniques. Moreover, this model implements an interesting interaction between components. It succeeds in mixing a GFP model for Ω-blind automaton with the model $\mathcal{D}$ for detecting solvability.

The approach using models opens several new perspectives. One can try to characterize what kinds of fixpoints correspond to what class of automata conditions. More generally, models hint a possibility to have an Eilenberg like variety theory for lambda-terms [6]. This theory would cover infinite regular words and trees too as they can be represented by λY-terms. Finally, considering model-checking algorithms, the model-based approach puts a focus on computing fixpoints in finite partial orders. This means that a number of techniques, ranging from under/over-approximations, to program optimization can be applied.

References

1. Aehlig, K.: A finite semantics of simply-typed lambda terms for infinite runs of automata. Logical Methods in Computer Science 3(3) (2007)
2. Amadio, R.M., Curien, P.-L.: Domains and Lambda-Calculi. Cambridge Tracts in Theoretical Computer Science. Cambridge University Press (1998)
3. Barendregt, H.: The Lambda Calculus, Its Syntax and Semantics. Studies in Logic and the Foundations of Mathematics, vol. 103. North-Holland (1984)
4. Broadbent, C., Carayol, A., Ong, L., Serre, O.: Recursion schemes and logical reflection. In: LICS, pp. 120–129 (2010)
5. Broadbent, C., Carayol, A., Hague, M., Serre, O.: A saturation method for collapsible pushdown systems. In: Czumaj, A., Mehlhorn, K., Pitts, A., Wattenhofer, R. (eds.) ICALP 2012, Part II. LNCS, vol. 7392, pp. 165–176. Springer, Heidelberg (2012)
6. Eilenberg, S.: Automata, Languages and Machines. Academic Press, New York (1974)
7. Haddad, A.: IO vs OI in higher-order recursion schemes. In: FICS. EPTCS, vol. 77, pp. 23–30 (2012)
8. Hague, M., Murawski, A.S., Ong, C.-H.L., Serre, O.: Collapsible pushdown automata and recursion schemes. In: LICS, pp. 452–461 (2008)
9. Kobayashi, N.: Higher-order program verification and language-based security. In: Datta, A. (ed.) ASIAN 2009. LNCS, vol. 5913, pp. 17–23. Springer, Heidelberg (2009)

10. Kobayashi, N.: Types and higher-order recursion schemes for verification of higher-order programs. In: POPL, pp. 416–428. ACM (2009)
11. Kobayashi, N.: Types and recursion schemes for higher-order program verification. In: Hu, Z. (ed.) APLAS 2009. LNCS, vol. 5904, pp. 2–3. Springer, Heidelberg (2009)
12. Kobayashi, N.: A practical linear time algorithm for trivial automata model checking of higher-order recursion schemes. In: Hofmann, M. (ed.) FOSSACS 2011. LNCS, vol. 6604, pp. 260–274. Springer, Heidelberg (2011)
13. Kobayashi, N., Ong, L.: A type system equivalent to modal mu-calculus model checking of recursion schemes. In: LICS, pp. 179–188 (2009)
14. Ong, C.-H.L.: On model-checking trees generated by higher-order recursion schemes. In: LICS, pp. 81–90 (2006)
15. Ong, C.-H.L., Tsukada, T.: Two-level game semantics, intersection types, and recursion schemes. In: Czumaj, A., Mehlhorn, K., Pitts, A., Wattenhofer, R. (eds.) ICALP 2012, Part II. LNCS, vol. 7392, pp. 325–336. Springer, Heidelberg (2012)
16. Salvati, S., Manzonetto, G., Gehrke, M., Barendregt, H.: Loader and Urzyczyn are logically related. In: Czumaj, A., Mehlhorn, K., Pitts, A., Wattenhofer, R. (eds.) ICALP 2012, Part II. LNCS, vol. 7392, pp. 364–376. Springer, Heidelberg (2012)
17. Salvati, S., Walukiewicz, I.: Krivine machines and higher-order schemes. In: Aceto, L., Henzinger, M., Sgall, J. (eds.) ICALP 2011, Part II. LNCS, vol. 6756, pp. 162–173. Springer, Heidelberg (2011)
18. Salvati, S., Walukiewicz, I.: Recursive schemes, Krivine machines, and collapsible pushdown automata. In: Finkel, A., Leroux, J., Potapov, I. (eds.) RP 2012. LNCS, vol. 7550, pp. 6–20. Springer, Heidelberg (2012)
19. Salvati, S., Walukiewicz, I.: Using models to model-check recursive schemes. Technical report, LaBRI (2012), `http://hal.inria.fr/hal-00741077`
20. Terui, K.: Semantic evaluation, intersection types and complexity of simply typed lambda calculus. In: RTA. LIPIcs, vol. 15, pp. 323–338. Schloss Dagstuhl - Leibniz-Zentrum fuer Informatik (2012)
21. Walukiewicz, I.: Simple models for recursive schemes. In: Rovan, B., Sassone, V., Widmayer, P. (eds.) MFCS 2012. LNCS, vol. 7464, pp. 49–60. Springer, Heidelberg (2012)

On Interaction, Continuations and Defunctionalization*

Ulrich Schöpp

Ludwig-Maximilians-Universität München, Germany
Ulrich.Schoepp@ifi.lmu.de

Abstract. In game semantics and related approaches to programming language semantics, programs are modelled by interaction dialogues. Such models have recently been used in the design of new compilation methods, e.g. in Ghica's approach to hardware synthesis, or in joint work with Dal Lago on programming with sublinear space. This paper relates such semantically motivated non-standard compilation methods to more standard techniques in the compilation of functional programming languages, such as continuation passing and defunctionalization. We first show for the linear λ-calculus that interpretation in a model of computation by interaction can be described as a call-by-name CPS transformation followed by a defunctionalization procedure that takes into account control-flow information. We then use the interactive model to guide the extension of the compositional translation to source languages with full contraction and recursion.

1 Introduction

A successful approach in the semantics of programming languages is to model programs by interaction dialogues [12,1]. Although dialogues are usually considered as abstract mathematical objects, it has also been argued that they are useful for *implementing* actual computation. Dialogues have been found useful especially for resource bounded computation, where they have given rise to nonstandard compilation methods for functional programming languages. For example Ghica et al. have developed methods for hardware synthesis based on game semantics [8]. A related semantic approach based on the *Int construction* [13] has been the used to design a functional programming language for sublinear space computation [4].

The aim of this paper is to relate such compilation methods based on interactive semantics to standard techniques in the compilation of functional programming languages. We consider the compilation of higher-order languages, such as PCF. A compiler would transform such a language to machine code by way of a number of intermediate languages. Typically, the higher-order source code would first be translated to first-order intermediate code, e.g. [3], from which the machine code is then generated. This paper is concerned with the first step, the translation from higher-order to first-order code. We consider two particular instances of well-known transformations that find application in compilers, CPS translation [18] and defunctionalization [19], and show that their composition is closely related to an interpretation of the source language in an interactive model of computation.

* I thank the anonymous referees for their helpful feedback and suggestions.

M. Hasegawa (Ed.): TLCA 2013, LNCS 7941, pp. 205–220, 2013.

This represents one step towards a general goal of capturing program transformations that find use in compilers by means of universal mathematical constructions. The interactive model that corresponds to CPS translation and defunctionalization is constructed using the Int construction [13], which captures a canonical way of constructing a model of higher-order computation from a first-order one. As this model validates call-by-name, it is natural to consider a call-by-name CPS translation; we use a variant of the one of Hofmann and Streicher [11].

To give an outline of how CPS translation, defunctionalization and the interpretation in an interactive model are related, we consider the very simple example of a function that increments a natural number: $\lambda x : \mathbb{N}.\, 1 + x$. We next outline how this function is translated by the two approaches and how the results compare.

1.1 CPS Translation and Defunctionalization

A compiler for PCF might first transform $\lambda x : \mathbb{N}.\, 1 + x$ into continuation passing style, perhaps apply some optimisations, and then use defunctionalization to obtain a first-order intermediate program, ready for compilation to machine language.

Hofmann and Streicher's call-by-name CPS transform [11] translates the source term $\lambda x : \mathbb{N}.\, 1 + x$ to the term $\lambda \langle x, k\rangle .\, (\lambda k.k\ 1)\ (\lambda u.x\ (\lambda n.k\ (u+n))) : \neg(\neg\neg\mathbb{N} \times \neg\mathbb{N})$, where we write $\neg A$ for $A \to \bot$, as usual. The argument in this term is a pair $\langle x, k\rangle$ of a continuation $k\colon \neg\mathbb{N}$ that accepts the result and a variable $x\colon \neg\neg\mathbb{N}$ that supplies the function argument. To obtain the actual function argument, one applies x to a continuation (here $\lambda n.k\ (u+n)$) to ask for the actual argument to be thrown into the supplied continuation.

Defunctionalization [19] translates this higher-order term into a first-order one. The basic idea is to give each function a name and to pass around not the function itself, but only its name and the values of its free variables. To this end, each lambda abstraction is named with a label: $\lambda^{l_1} \langle x, k\rangle .\, (\lambda^{l_2} k.k\ 1)\ (\lambda^{l_3} u.x\ (\lambda^{l_4} n.k\ (u+n)))$. The whole term can be represented simply by the label l_1. The function l_3 has free variables x and k and is represented by the label together with the values of x and k, which we write as $l_3(x,k)$.

Each application $s\ t$ is replaced by a procedure call *apply*(s,t), as s is now only a function name and not the function itself. The procedure *apply* is defined by case distinction on the function name and behaves like the body of the respective λ-abstraction in the original term. In the example we have the following definition of *apply*:

$$\begin{aligned} apply(l_1, \langle x,k\rangle) &= apply(l_2, l_3(x,k)) & \qquad apply(l_2,k) &= apply(k,1) \\ apply(l_3(x,k),u) &= apply(x, l_4(k,u)) & \qquad apply(l_4(k,u),n) &= apply(k,u+n) \end{aligned}$$

To understand concretely how these equations represent the original term, it is perhaps useful see what happens when a concrete argument and a continuation are supplied: $(\lambda^{l_1} \langle x, k\rangle .\, (\lambda^{l_2} k.k\ 1)\ (\lambda^{l_3} u.x\ (\lambda^{l_4} n.k\ (u+n))))\ \langle \lambda^{l_5} k.\, k\ 42, \lambda^{l_6} n.\mathtt{print\_int}(n)\rangle$. Then we get the following cases for l_5 and l_6 in addition to the cases above

$$apply(l_5,k) = apply(k,42), \qquad apply(l_6,n) = \mathtt{print\_int}(n),$$

and the fully applied term defunctionalizes to *apply*$(l_1, \langle l_5, l_6\rangle)$. Executing it results in 43 being printed, as expected.

This outlines a naive defunctionalization method for translating a higher-order language into a first-order language with (tail) recursion. This method can be improved in various ways. The above *apply*-procedure performs a case distinction on the function name each time it is invoked. In this example, the label of the invoked function can be determined statically, however, so that the case distinction is not in fact necessary. Instead, we may define one function $apply_l$ for each label l and choose the appropriate label statically. The label l then does not need to be passed as an argument anymore. A defunctionalization procedure that takes into account control flow information in this way was introduced by Banerjee et al. [2]. If we apply it to this example and moreover simplify the result by removing unneeded function arguments, then we get:

$$\begin{array}{ll} apply_{l_1}() = apply_{l_2}() & apply_{l_2}() = apply_{l_3}(1) \\ apply_{l_3}(u) = apply_{l_5}(u) & apply_{l_4}(u,n) = apply_{l_6}(u+n) \end{array} \tag{1}$$

The term itself simplifies to $apply_{l_1}()$. The interface where these equations interact with the environment consists of the labels l_1, l_4, l_5 and l_6. Applying the term to concrete arguments as above amounts to extending the environment with the following equations:

$$apply_{l_5}(u) = apply_{l_4}(u,42) \qquad apply_{l_6}(n) = \texttt{print\_int}(n)$$

The point of this paper is that the program (1) is the same as what we get from interpreting the source term in a model of computation by interaction.

1.2 Interpretation in an Interactive Computation Model

In computation by interaction the general idea is to study models of computation that interpret programs by interaction dialogues in the style of game semantics and to consider actual implementations of such dialogue interaction. For example, a function of type $\mathbb{N} \to \mathbb{N}$ may be implemented in interactive style by a program that, for a suitable type S, takes as input a value of type $\texttt{unit} + S \times \texttt{nat}$ and gives as output a value of type $\texttt{nat} + S \times \texttt{unit}$. The input $\mathsf{inl}(\langle\rangle)$ to this program is interpreted as a request for the return value of the function. An output of the form $\mathsf{inl}(n)$ means that n is the requested value. If the output is of the form $\mathsf{inr}(s, \langle\rangle)$, however, then this means that the program would like to know the argument of the function. It also requests that the value s is returned along with the answer, as programs here do not have state and s can thus not be stored until the request is answered. To answer the program's request, we can pass a value of the form $\mathsf{inr}(s,m)$, where m is our answer.

The particular function $\lambda x : \mathbb{N}. 1 + x$ is implemented by the program specified in the following diagram, where S is $\texttt{nat}$. This diagram is to be understood so that one may pass a message along any of its input wires. The message must be a value of the type labelling the wire. When a message arrives at an input of box, the box will react by sending a message on one of its outputs. Thus, at any time there is one message in the network. Computation ends when a message is passed along an output wire.

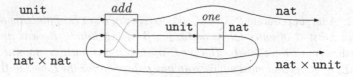

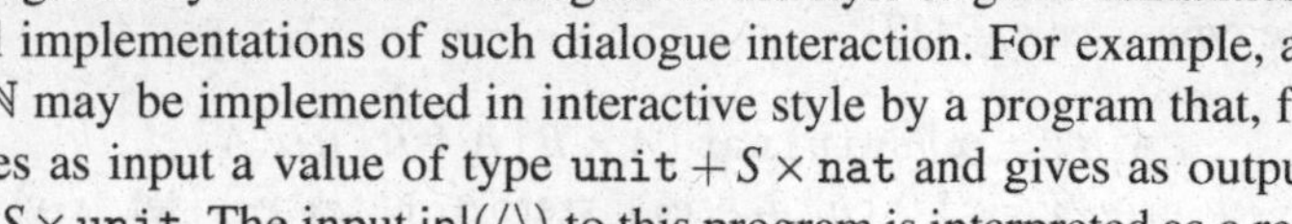

In this diagram, *add* takes as input messages of type $\mathtt{unit} + (\mathtt{nat} + (\mathtt{nat} \times \mathtt{nat}))$; the three input arrows in the figure represent the summands of this type. It outputs a message of type $\mathtt{nat} + (\mathtt{unit} + (\mathtt{nat} \times \mathtt{unit}))$. Its behaviour is given by the mappings $\mathsf{inl}(\langle\rangle) \mapsto \mathsf{inr}(\mathsf{inl}(\langle\rangle))$, $\mathsf{inr}(\mathsf{inl}(n)) \mapsto \mathsf{inr}(\mathsf{inr}(n, \langle\rangle))$ and $\mathsf{inr}(\mathsf{inr}(n,m)) \mapsto \mathsf{inl}(n+m)$. The box labelled *one* maps the request $\langle\rangle$ to the number 1.

This interactive implementation of $\lambda x : \mathbb{N}. 1 + x$ may be described as the interpretation of the term in a semantic model $\mathrm{Int}(\mathbb{T})$ built by applying the general categorical Int construction to a category $\mathbb{T}$ that is constructed from the target language, see Sec. 6.

Compare the above interaction diagram to the definitions in (1) obtained by defunctionalization. The labels l_1, l_3 and l_4 there correspond to the three inputs of the *add*-box (from top to bottom), l_2 is the input of box *one*, and l_5 and l_6 are the destination labels of the two outgoing wires. One may consider the *apply*-definitions in (1) a particular implementation of the diagram, where a call to $apply_l(m)$ means that message m is sent to point l in the diagram. A naive implementation would introduce a label for the end of each arrow in the diagram and implement the message passing accordingly.

This outlines the relation between the translations, which we now describe in detail.

2 Target Language

Programs in the target language consist of mutually (tail-)recursive definitions of first-order functions, such as the examples for *apply*-equations above. The target language is at the same abstraction level as SSA-form compiler intermediate languages, e.g. [3].

The set of *target types* is defined by the grammar below. Sum types and recursive types will be needed at the end of Sec. 8 only. *Target expressions* and *values* are standard terms for these types, see e.g. [17]:

$$\begin{array}{ll}
\text{Types:} & A, B ::= \alpha \mid \mathtt{unit} \mid \mathtt{nat} \mid A \times B \mid A + B \mid \mu\alpha.A \\
\text{Expressions:} & e, e_1, e_2 ::= x \mid \langle\rangle \mid n \mid e_1 + e_2 \mid \mathsf{iszero?}(e) \mid \langle e_1, e_2\rangle \mid \mathsf{let}\ \langle x, y\rangle = e_1\ \mathsf{in}\ e_2 \\
& \qquad \mid \mathsf{inl}(e) \mid \mathsf{inr}(e) \mid \mathsf{case}\ e\ \mathsf{of}\ \mathsf{inl}(x) \Rightarrow e_1; \mathsf{inr}(y) \Rightarrow e_2 \\
& \qquad \mid \mathsf{fold}(e) \mid \mathsf{unfold}(e) \\
\text{Values:} & v, v_1, v_2 ::= \langle\rangle \mid n \mid \langle v_1, v_2\rangle \mid \mathsf{inl}(v) \mid \mathsf{inr}(v) \mid \mathsf{fold}(v)
\end{array}$$

Here, n ranges over natural numbers as constants and $\mathsf{iszero?}(e)$ is intended to have type $\mathtt{unit} + \mathtt{unit}$ with $\mathsf{inl}(\langle\rangle)$ representing true. We assume a standard (non-linear) typing and equality judgement, so that each well-typed closed expression e equals a unique value v of the same type, written as $e = v$.

Target programs consist of a set of first-order function definitions.

Definition 1. *Let $\mathcal{L}$ be an infinite set of program labels.* A definition of a label $f \in \mathcal{L}$ *is given by an equation of the form $f(x) = g(e)$ or the form $f(x) = \mathsf{case}\ e\ \mathsf{of}\ \mathsf{inl}(y) \Rightarrow g(e_1); \mathsf{inr}(z) \Rightarrow h(e_2)$, wherein $g, h \in \mathcal{L}$ and e, e_1 and e_2 range over target expressions.*

Definition 2. A target program $P = (\alpha, E, \beta)$ *consists of a set E of function definitions together with a list α of entry labels and a list β of exit labels. Both α and β must be lists of pairwise distinct labels. The set E of definitions must contain at most one definition for any label and must not contain any definition for the labels in β.*

The list α assigns an order to the function labels that may be used as entry points for the program and β identifies external labels as return points.

We allow ourselves to use syntactic sugar, such as writing $f(x,y) = g(e)$ for $f(z) = g(\mathsf{let}\ \langle x,y\rangle = z\ \mathsf{in}\ e)$ or writing just $f()$ for $f(\langle\rangle)$.

We use an informal graphical notation for target programs, depicting for example the program $(f_1 f_2 f_3, \{f_3(x) = f_4(x), f_4(x) = g_1(x+1)\}, g_1 f_1 g_3)$ as shown below.

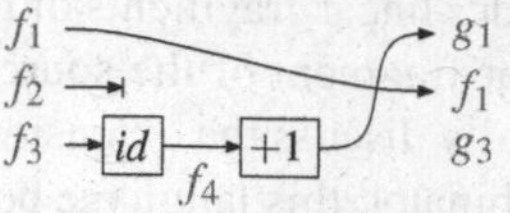

Target programs can be typed in the evident way, so that in each function definition $f(x) = \ldots$ the variable x is assigned a type and function calls must preserve types. If P is the program $(f_1 \cdots f_n, E, g_1 \cdots g_m)$, then we write $P\colon (A_1 \ldots A_n) \to (B_1 \ldots B_m)$ if the functions $f_1, \cdots, f_n, g_1, \cdots, g_m$ in it can be typed such that they accept values of type $A_1, \ldots, A_n, B_1, \ldots, B_m$ respectively.

We define a simple reduction semantics for programs. A *function call* is an expression of the form $f(v)$, where f is a function label and v is a value. A relation $\to_P$ formalises the function calls as they happen during the execution of a program P. It is the smallest relation satisfying the following conditions: if P contains a definition $f(x) = g(e)$ then $f(v) \to_P g(w)$ for all values v and w with $e[v/x] = w$; and if P contains a definition $f(x) = \mathsf{case}\ e\ \mathsf{of}\ \mathsf{inl}(y) \Rightarrow g(e_1);\ \mathsf{inr}(z) \Rightarrow h(e_2)$ then $f(v) \to_P g(w)$ for all values v and w with $\exists u.\, e[v/x] = \mathsf{inl}(u) \wedge e_1[u/y] = w$, and $f(v) \to_P h(w)$ for all values v and w with $\exists u.\, e[v/x] = \mathsf{inr}(u) \wedge e_2[u/z] = w$.

A *call-trace* of program P is a sequence $f_1(v_1) f_2(v_2) \ldots f_n(v_n)$, such that $f_i(v_i) \to_P f_{i+1}(v_{i+1})$ holds for all $i \in \{1, \ldots, n-1\}$.

Two programs $P, Q\colon (A_1 \ldots A_n) \to (B_1 \ldots B_m)$ are *equal* if, for any input, they give the same output, that is, if the entry labels of P and Q are $f_1, \cdots, f_n$ and $g_1, \cdots, g_n$ respectively and the exit labels are $h_1, \ldots, h_m$ and $k_1, \ldots, k_m$ respectively, then, for any v, w, i and j, P has a call-trace of the form $f_i(v) \ldots g_j(w)$ if and only if Q has a call-trace of the form $h_i(v) \ldots k_j(w)$.

The following notation is used in Sec. 7. For any list of target types $X = B_1 \ldots B_n$ and any target type A, we write $A \cdot X$ for the list $(A \times B_1) \ldots (A \times B_n)$. Given a program $P\colon X \to Y$, we write $A \cdot P\colon A \cdot X \to A \cdot Y$ for the program that passes on the value of type A unchanged and otherwise behaves like P. It may be defined by replacing each definition $f(x) = g(e)$ with $f(z,x) = g(z,e)$ for fresh z, and likewise for branching definitions.

Lemma 1. *Target programs can be organised into a category $\mathbb{T}$ whose objects are finite lists of target types and whose morphisms from X to Y are given by an equivalence classes of programs $P\colon X \to Y$ with respect to program equality.*

Lemma 2. *The category $\mathbb{T}$ has finite coproducts, such that the initial object 0 is given by the empty list and the object $X + Y$ is given by the concatenation of the lists X and Y. Moreover, $\mathbb{T}$ has a uniform trace [10] with respect to these coproducts.*

These lemmas show that $\mathbb{T}$ has enough structure so that we can apply the Int construction [13,10] (with respect to coproducts) to it and obtain a category $\mathrm{Int}(\mathbb{T})$ that models interactive computation. We shall describe the interpretation of terms in $\mathrm{Int}(\mathbb{T})$ concretely and refer to loc. cit. for the categorical structure.

3 Source Language

Our source language is $\lambda^{\to,\mathbb{N}}$, the simply-typed λ-calculus with a basic type $\mathbb{N}$ of natural numbers and associated terms for numeral constants $n\colon \mathbb{N}$, addition $s+t$ and case distinction $\mathsf{if0}\ s\ \mathsf{then}\ t_1\ \mathsf{else}\ t_2$. The typing rules are straightforward and can be found in the CPS translation below. There they are formulated with an explicit contraction rule in order to make it easy to consider linear fragments of the source language.

We shall first consider a linear fragment of the source language in Sec. 6, then add the base type $\mathbb{N}$ in Sec. 7 and finally discuss the extension to the full language in Sec. 8.

If one adds a fixed point combinator, this language becomes as expressive as PCF.

4 CPS Translation

We use a variant of the *call-by-name CPS translation* [11], which translates the source language into $\lambda^{\to,\times,\mathbb{N},\bot}$, the calculus that extends $\lambda^{\to,\mathbb{N}}$ with with product types and an empty type $\bot$.

For each type X, the type $\underline{X}$ of its continuations is defined by $\underline{\mathbb{N}} = \neg\mathbb{N}$ and $\underline{X \to Y} = \neg\underline{X} \times \underline{Y}$. We write $\overline{X}$ for $\neg\underline{X}$. A continuation for type $X \to Y$ is thus a pair consisting of a continuation $\underline{Y}$, where the result can be returned, and a function $\neg\underline{X}$ to access the argument. A function can request its argument by applying this function to a continuation of type $\underline{X}$. The argument will then be provided to this continuation.

The CPS translation takes any sequent $x_1 : X_1, \dots, x_n : X_n \vdash t\colon Y$ derivable in $\lambda^{\to,\mathbb{N}}$ to a sequent $x_1 : \overline{X_1}, \dots, x_n : \overline{X_n} \vdash \underline{t}\colon \overline{Y}$ derivable in $\lambda^{\to,\times,\mathbb{N},\bot}$. It is given by the following translation of typing rules of $\lambda^{\to,\mathbb{N}}$ on the left to derived rules of $\lambda^{\to,\times,\mathbb{N},\bot}$ on the right.

$$\frac{}{\Gamma, x\colon X \vdash x\colon X} \quad\Longrightarrow\quad \frac{}{\overline{\Gamma}, x\colon \overline{X} \vdash \eta(x,\overline{X})\colon \overline{X}}$$

$$\frac{\Gamma, x\colon X \vdash t\colon Y}{\Gamma \vdash \lambda x : X.t\colon X \to Y} \quad\Longrightarrow\quad \frac{\overline{\Gamma}, x\colon \overline{X} \vdash \underline{t}\colon \overline{Y}}{\overline{\Gamma} \vdash \lambda\langle x,k\rangle.\underline{t}\,k\colon \overline{X \to Y}}$$

$$\frac{\Gamma \vdash s\colon X \to Y \qquad \Delta \vdash t\colon X}{\Gamma, \Delta \vdash s\,t\colon Y} \quad\Longrightarrow\quad \frac{\overline{\Gamma} \vdash \underline{s}\colon \overline{X \to Y} \qquad \overline{\Delta} \vdash \underline{t}\colon \overline{X}}{\overline{\Gamma}, \overline{\Delta} \vdash \lambda k.\underline{s}\,\langle \underline{t}, k\rangle\colon \overline{Y}}$$

$$\frac{}{\Gamma \vdash n\colon \mathbb{N}} \quad\Longrightarrow\quad \frac{}{\overline{\Gamma} \vdash \lambda k.k\,n\colon \overline{\mathbb{N}}}$$

$$\frac{\Gamma \vdash s\colon \mathbb{N} \qquad \Delta \vdash t\colon \mathbb{N}}{\Gamma, \Delta \vdash s+t\colon \mathbb{N}} \quad\Longrightarrow\quad \frac{\overline{\Gamma} \vdash \underline{s}\colon \overline{\mathbb{N}} \qquad \overline{\Delta} \vdash \underline{t}\colon \overline{\mathbb{N}}}{\overline{\Gamma}, \overline{\Delta} \vdash \lambda k.\underline{s}\,(\lambda x.\underline{t}\,(\lambda y.k\,(x+y)))\colon \overline{\mathbb{N}}}$$

$$\frac{\Gamma \vdash s\colon \mathbb{N} \quad \Delta_1 \vdash t_1\colon \mathbb{N} \quad \Delta_2 \vdash t_2\colon \mathbb{N}}{\Gamma, \Delta_1, \Delta_2 \vdash \mathsf{if0}\ s\ \mathsf{then}\ t_1\ \mathsf{else}\ t_2\colon \mathbb{N}} \quad\Longrightarrow\quad \frac{\overline{\Gamma} \vdash \underline{s}\colon \overline{\mathbb{N}} \qquad \overline{\Delta_1} \vdash \underline{t_1}\colon \overline{\mathbb{N}} \qquad \overline{\Delta_2} \vdash \underline{t_2}\colon \overline{\mathbb{N}}}{\overline{\Gamma}, \overline{\Delta_1}, \overline{\Delta_2} \vdash \lambda k.\underline{s}\,(\lambda x.\mathsf{if}\ x\ \mathsf{then}\ \underline{t_1}\,(\lambda y.k\,y)\ \mathsf{else}\ \underline{t_2}\,(\lambda y.k\,y))\colon \overline{\mathbb{N}}}$$

$$\frac{\Gamma, x\colon X, y\colon X \vdash t\colon Y}{\Gamma, x\colon X \vdash t[x/y]\colon Y} \quad\Longrightarrow\quad \frac{\overline{\Gamma}, x\colon \overline{X}, y\colon \overline{X} \vdash \underline{t}\colon \overline{Y}}{\overline{\Gamma}, x\colon \overline{X} \vdash \underline{t}[x/y]\colon \overline{Y}}$$

This CPS translation differs from the standard call-by-name CPS translation [11] in the translation of variables. Instead of letting $\underline{x}$ be just x, we take it to be an η-expansion $\eta(t,X)$ of x. The term $\eta(t,X)$, is defined by induction on X as follows: $\eta(t,\mathbb{N}) := t$ and

$\eta(t, X \to Y) := \lambda x.\eta(t\ \eta(x,X),Y)$, where x is a fresh variable. For example, we have $\eta(f,(\mathbb{N}\to\mathbb{N})\to\mathbb{N}) = \lambda x_1.f\ (\lambda x_2.x_1\ x_2)$ and $\eta(f,\mathbb{N}\to(\mathbb{N}\to\mathbb{N})) = \lambda x_1.\lambda x_2.f\ x_1\ x_2$.

The use of η-expansion allows us to use compositional reasoning in Sec. 7. In the example in the Introduction, we have not applied this η-expansion for better readability.

5 Defunctionalization

After the CPS transform, the term is annotated with control flow information and then translated into the target language by a defunctionalization procedure that takes the control flow information into account. In this section we define a particularly simple variant of such a procedure, which suffices to show the relation to the Int construction. It is a special case of the flow-based defunctionalization described by Banerjee et al. [2].

The input of the defunctionalization procedure is a term of the labelled λ-calculus $\lambda_\ell^{\to,\times,\mathbb{N},\bot}$. Its syntax differs from that of $\lambda^{\to,\times,\mathbb{N},\bot}$ only in that abstractions, applications and function types are each annotated with a label from $\mathscr{L}$. Thus, the terms $\lambda x : X.t$ and $s\ t$ are replaced by $\lambda^l x{:}X.t$ and $s@_l t$. The type $X \to Y$ becomes $X \xrightarrow{l} Y$.

We require that each abstraction is uniquely identified by its label, i.e. we allow only terms in which no two abstractions have the same label. In the application $s@_l t$ the label l expresses that the function s applied here will be defined by an abstraction with label l. Note that each application can be annotated with a single label l only, which means that for each application the label of the function that is being applied is statically known. In general, one needs to allow a set of labels for more than one possible definition site, as in e.g. [2]. We discuss this in Sec. 8, but until then the variant with a single label suffices and much simplifies the exposition.

The typing rules of $\lambda_\ell^{\to,\times,\mathbb{N},\bot}$ enforce that terms are annotated with correct control flow information. An abstraction $\lambda^l x{:}X.t$ has type $X \xrightarrow{l} Y$. If s has type $X \xrightarrow{l} Y$ and t has type X then $s@_l t$ has type Y.

In the rest of this section we explain how the terms of $\lambda_\ell^{\to,\times,\mathbb{N},\bot}$ are defunctionalized into target terms. We defer the question of how to annotate terms with labels to Sec. 6.

The defunctionalization of a term t in $\lambda_\ell^{\to,\times,\mathbb{N},\bot}$ is defined by the following judgement, which has the form $t \Downarrow t^*; D_t$, where t^* is the defunctionalized term in the target language and D_t is a set of equations. In general, the set D_t need not be function definitions in the sense of Def. 1. We shall however use defunctionalizion only for terms t for which D_t consists only of function definitions.

$$\frac{}{x \Downarrow x; \emptyset} \qquad \frac{}{n \Downarrow n; \emptyset} \qquad \frac{s \Downarrow s^*; D_s \qquad t \Downarrow t^*; D_t}{s+t \Downarrow s^* + t^*; D_s \cup D_t}$$

$$\frac{s \Downarrow s^*; D_s \qquad t \Downarrow t^*; D_t \qquad u \Downarrow u^*; D_u}{\mathsf{if0}\ s\ \mathsf{then}\ t\ \mathsf{else}\ u \Downarrow \mathsf{case\ iszero?}(s^*)\ \mathsf{of\ inl}(\langle\rangle) \Rightarrow t^*;\ \mathsf{inr}(\langle\rangle) \Rightarrow u^*; D_s \cup D_t \cup D_u}$$

$$\frac{s \Downarrow s^*; D_s \qquad t \Downarrow t^*; D_t}{\langle s,t\rangle \Downarrow \langle s^*,t^*\rangle; D_s \cup D_t} \qquad \frac{s \Downarrow s^*; D_s \qquad t \Downarrow t^*; D_t}{\mathsf{let}\ \langle x,y\rangle = t\ \mathsf{in}\ s \Downarrow \mathsf{let}\ \langle x,y\rangle = t^*\ \mathsf{in}\ s^*; D_s \cup D_t}$$

$$\frac{s \Downarrow s^*; D_s \qquad t \Downarrow t^*; D_t}{s@_l t \Downarrow apply_l(s^*,t^*); D_s \cup D_t} \qquad \frac{t \Downarrow t^*; D_t \qquad fv(t) = \{x_1,\dots,x_n\}}{\lambda^l x{:}A.t \Downarrow \langle x_1,\dots,x_n\rangle; D_t \cup \{apply_l(\langle x_1,\dots,x_n\rangle,x) = t^*\}}$$

In the rule for abstraction we assume a global ordering on all variables, so that the order of the tuple is well-defined.

Note that for closed terms of function type, such as the closed terms the form $\underline{t}$ obtained by CPS translation, t^* is just $\langle\rangle$. We therefore concentrate on the set D_t.

With annotations the example from the Introduction becomes the term $\underline{t}$ given by $\lambda^{l_1} z.\,\mathsf{let}\ \langle x,k\rangle = z\ \mathsf{in}\ (\lambda^{l_2} k'.k' @_{l_3} 1) @_{l_2} (\lambda^{l_3} u.x @_{l_5} (\lambda^{l_4} n.k @_{l_6} (u+n)))$, whose type is $\vdash \underline{t}\colon \neg_{l_1}(\neg_{l_5}\neg_{l_4}\mathbb{N} \times \neg_{l_6}\mathbb{N})$. The set $D_{\underline{t}}$ of definitions consists of

$$apply_{l_1}(\langle\rangle, \langle x,k\rangle) = apply_{l_2}(\langle\rangle, \langle x,k\rangle), \qquad apply_{l_2}(\langle\rangle, k') = apply_{l_3}(k',1),$$
$$apply_{l_3}(\langle x,k\rangle, u) = apply_{l_5}(x, \langle k\rangle), \qquad apply_{l_4}(\langle k\rangle, n) = apply_{l_6}(k, u+n).$$

Compared with the Introduction, it appears that more data is being passed around in these *apply*-equations. However, consider once again the application of $\underline{t}$ to the concrete arguments from the Introduction. Then one gets the additional equations

$$apply_{l_5}(\langle\rangle, k) = apply_{l_4}(k, 42), \qquad apply_{l_6}(\langle\rangle, n) = \texttt{print\_int}(n),$$

and the fully applied term defunctionalizes to $apply_{l_1}(\langle\rangle, \langle\langle\rangle, \langle\rangle\rangle)$. Thus, all the variables in the *apply*-equations only ever store the value $\langle\rangle$ or tuples thereof, and these arguments may just as well be omitted.

Note that the defunctionalization procedure yields a set of equations, but it does not specify an interface of entry and exit labels. When one applies defunctionalization to closed source programs of base type, as is usually done in compilation, choosing an interface is not important. For the entry labels one would typically just choose a single entry point `main`, for example. For open terms or terms of higher type, however, one needs to fix the interface that matches the type. In the above example of $\underline{t}$, a suitable choice of entry and exit labels would be $l_1 l_4$ and $l_5 l_4$ respectively. We shall explain how to define an interface for terms in the image of the CPS translation in the next section.

Of course, the defunctionalization procedure described above is quite simple. In actual applications one would certainly want to apply optimisations, not least to remove unnecessary functions arguments. An example of such an optimisation is *lightweight defunctionalization* of Banerjee et al. [2]. We shall see that the Int construction captures one such optimisation of the defunctionalization procedure.

6 The Linear Fragment

To explain the basic idea of how the interpretation in a model of interactive computation (namely $\mathrm{Int}(\mathbb{T})$) relates to CPS translation and defunctionalization, we first consider the simplest non-trivial case. Consider the linear fragment of the source language and instead of the natural number type $\mathbb{N}$ just a type o without any term constructors:

$$X,Y ::= o \mid X \multimap Y \qquad\qquad s,t ::= x \mid s\,t \mid \lambda x{:}X.t$$

The standard typing rules AX, $\multimap$I and $\multimap$E for this calculus are shown below.

First we describe directly what the interpretation of this source language in $\mathrm{Int}(\mathbb{T})$ amounts to. A type X is modelled by an interface (X^-, X^+), which consists of two finite lists X^- and X^+ of target types. Closed terms of type X will be modelled by programs of type $P\colon X^- \to X^+$. The interfaces are defined by induction on the type: both o^- and o^+ are the singleton list `unit`, $(X \multimap Y)^-$ is $Y^- X^+$, i.e. the concatenation of Y^- and X^+, and $(X \multimap Y)^+$ is $Y^+ X^-$. For a context $\Gamma = x_1 : X_1, \dots, x_n : X_n$, we write Γ^- and Γ^+ for the concatenations $X_n^- \dots X_1^-$ and $X_n^+ \dots X_1^+$.

The interpretation of a term $\Gamma \vdash t\colon X$ in Int($\mathbb{T}$) is a morphism $[\![\Gamma \vdash t\colon X]\!]\colon X^-\Gamma^+ \to X^+\Gamma^-$ in $\mathbb{T}$, i.e. an equivalence class of programs. This interpretation is defined by induction on the derivation; as depicted below.

$$\text{AX}\ \frac{}{\Gamma, x\colon X \vdash x\colon X} \Longrightarrow$$

$$\multimap\text{I}\ \frac{\Gamma, x\colon X \vdash t\colon Y}{\Gamma \vdash \lambda x{:}X.t\colon X \multimap Y} \Longrightarrow$$

$$\multimap\text{E}\ \frac{\Gamma \vdash s\colon X \multimap Y \qquad \Delta \vdash t\colon X}{\Gamma, \Delta \vdash s\,t\colon Y} \Longrightarrow$$

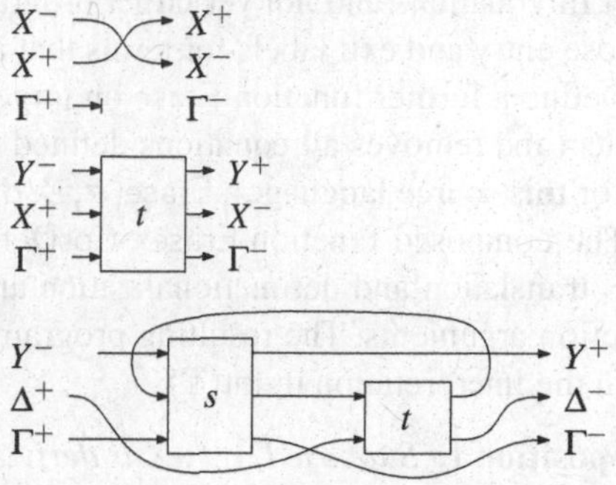

The aim is now to show that this interpretation in Int($\mathbb{T}$) is closely related to CPS translation followed by defunctionalization.

Our flow-based defunctionalization depends on suitable labellings of terms and types. We introduce notation for labellings of types of the form $\overline{X}$. For any type X and any $x^-, x^+ \in \mathcal{L}^*$ with $length(x^-) = length(X^-)$ and $length(x^+) = length(X^+)$, we define a type $\overline{X}[x^-, x^+]$ in the labelled λ-calculus as follows: $\overline{o}[q,a]$ is $\neg_q\neg_a\mathtt{unit}$ and $\overline{(X \multimap Y)}[y^- x^+, y^+ x^-]$ is $\neg_q(\neg_r X' \times Y')$ if $\overline{X}[x^-, x^+]$ is $\neg_r X'$ and $\overline{Y}[y^-, y^+]$ is $\neg_q Y'$.

For example, $\overline{(o \multimap o)}[qa', aq']$ denotes $\neg_q(\neg_{q'}\neg_{a'}\mathtt{unit} \times \neg_a\mathtt{unit})$.

If Γ is $x_1 : X_1, \dots, x_n : X_n$, then we write short $\overline{\Gamma}[x_n^- \dots x_1^-, x_n^+ \dots x_1^+]$ for the context $x_1 \colon \overline{X_1}[x_1^-, x_1^+], \dots, x_n \colon \overline{X_n}[x_n^-, x_n^+]$. We say that a sequent $\overline{\Gamma}[\gamma^-, \gamma^+] \vdash t\colon \overline{X}[x^-, x^+]$ is *well-labelled* if the labels in $\gamma^-, \gamma^+, x^-, x^+$ are pairwise distinct.

Although defined as an abbreviation for a labelled type, one may alternatively think of $\overline{X}[x^-, x^+]$ as the type X together with a labelling of the ports of the interface (X^-, X^+).

Lemma 3. *If $\Gamma \vdash t\colon X$ is derivable in the linear type system, then the sequent $\overline{\Gamma} \vdash \underline{t}\colon \overline{X}$ obtained by CPS transform can be annotated with labels such that the sequent $\overline{\Gamma}[\gamma^-, \gamma^+] \vdash \underline{t}\colon \overline{X}[x^-, x^+]$ is well-labelled and derivable, for some $\gamma^-, \gamma^+, x^-, x^+ \in \mathcal{L}^*$.*

The proof is a straightforward induction on derivations. We note that the case for variables depends on the η-expanded form of the CPS translation. With the expansion a well-labelled $x\colon \overline{\mathbb{N}}[q,a] \vdash \underline{x}\colon \overline{\mathbb{N}}[q',a']$ is derivable; without it this would only be possible if $q = q'$ and $a = a'$. The defunctionalization of $\underline{x}$ consists of definitions of $apply_{q'}$ and $apply_a$, which just forward their arguments to $apply_q$ and $apply_{a'}$ respectively. We believe that it is simpler to consider the case with these indirections first and study their removal (which is non-compositional, due to renaming) in a second step.

We now define a function CpsDefun that combines CPS transformation and defunctionalization. Given any judgement $\Gamma \vdash t\colon X$ derivable in the linear fragment of the source language, let $\overline{\Gamma}[\gamma^-,\gamma^+] \vdash \underline{t} \colon \overline{X}[x^-,x^+]$ be the judgement from the above lemma for a suitable choice of labels. Let $D_{\underline{t}}$ be the set of equations determined by the defunctionalization judgement $\underline{t} \Downarrow \underline{t}^* \,; D_{\underline{t}}$. The function CpsDefun maps the source judgement $\Gamma \vdash t \colon X$ to the target program $(x^-\gamma^+, D_{\underline{t}}, x^+\gamma^-)$. It is not hard to see the set $D_{\underline{t}}$ is such that this indeed a target program.

We use a single function CpsDefun rather than a composition of two functions general Cps and Defun, as we do not have a canonical choice of entry and exit labels for defunctionalization in general. Thus, the composition Defun $\circ$ Cps would only return a set of equations and not yet target program. With a combined function, it suffices to choose entry and exit labels for terms that are in the image of the CPS translation.

Define a further function Erase on target programs, which erases all function arguments (and removes all equations defined by case distinction, which cannot appear in $D_{\underline{t}}$ for this source language): $\mathsf{Erase}(\alpha, E, \beta) := (\alpha, \{f() = g() \mid f(x) = g(e) \in E\}, \beta)$.

The composed function Erase $\circ$ CpsDefun takes a source program, first applies the CPS translation and defunctionalization and then 'optimises' the result by erasing all function arguments. The resulting program is in fact correct and it is what one obtains from the interpretation in $\mathrm{Int}(\mathbb{T})$:

Proposition 1. *Suppose $\Gamma \vdash t \colon X$ is derivable in the linear type system. Then the target program* $\mathsf{Erase}(\mathsf{CpsDefun}(\Gamma \vdash t \colon X))$ *has type $X^-\Gamma^+ \to X^+\Gamma^-$ and is an element of the equivalence class of programs obtained by Int interpretation of $\Gamma \vdash t \colon X$.*

7 Base Types

We now work towards extending the result to a more expressive source language, starting with non-trivial base types. We replace the type o by the type of natural numbers $\mathbb{N}$ and add terms for constant numbers, addition and case distinction.

$$X, Y ::= \mathbb{N} \mid X \multimap Y \qquad s, t ::= n \mid s + t \mid \mathsf{if0}\ s\ \mathsf{then}\ t_1\ \mathsf{else}\ t_2 \mid x \mid s\,t \mid \lambda x{:}X.\,t$$

The example from the Introduction illustrates that for this fragment of the source language it is not possible to remove all arguments from the *apply*-functions, as we have done above. At least certain natural numbers must be passed as arguments.

Again, we first consider the interpretation of the fragment in $\mathrm{Int}(\mathbb{T})$. To this end we define $\mathbb{N}^- = \mathtt{unit}$ (a request to compute the number) and $\mathbb{N}^+ = \mathtt{nat}$ (the actual number as an answer). It is however not completely straightforward to extend the Int interpretation described in the previous section. Consider for example the case of an addition $s+t$ for two closed terms $\vdash s \colon \mathbb{N}$ and $\vdash t \colon \mathbb{N}$. Suppose we already have programs (q_s, E_s, a_s) and (q_t, E_t, a_t) for s and t. For $s+t$ it would be natural to use the program (q, E, a) with equations $apply_q() = apply_{q_s}()$, $apply_{a_s}(x) = apply_{q_t}(x, \langle\rangle)$, $apply_{a_t}(x,y) = apply_a(x+y)$, the equations from E_s, and the equations from $\mathtt{nat} \cdot E_t$ (recall the notation $\mathtt{nat} \cdot -$ from Sec. 2). We use $\mathtt{nat} \cdot E_t$ instead of E_t in order to keep the value x available until the end when we want to compute the sum. The difficulty is to decide which values must, like x, must be preserved in which equations.

One solution to this issue was proposed by Dal Lago and the author in the form of IntML [4]. We consider here a simple special case of this system. The basic idea is to annotate the each function type $X \multimap Y$ with a *subexponential* A, which is a target type:

$$X, Y ::= \mathbb{N} \mid A \cdot X \multimap Y$$

The subexponential annotation may be explained such that a term s of type $A \cdot X \multimap Y$ is a function that uses its argument within an environment that contains an additional value of type A. The function s may be applied to any argument t of type X. In the interactive interpretation of the application $s\ t$, whenever s sends a query to t it needs to preserve a value of type A. It does so by sending the value along with the query, expecting it to be returned unmodified along with a reply. For example, addition naturally gets the type $\mathtt{unit} \cdot \mathbb{N} \multimap \mathtt{nat} \cdot \mathbb{N} \multimap \mathbb{N}$, as it needs to remember the already queried value of the first argument (having type $\mathtt{nat}$) when it queries the second one.

In the type system, subexponential annotations are integrated by letting contexts consist of variable declarations of the form $x \colon A \cdot X$. The typing rules are shown below.

$$\text{AX}\ \frac{}{\Gamma, x\colon \mathtt{unit} \cdot X \vdash x : X} \qquad \text{CONST}\ \frac{}{\Gamma \vdash n : \mathbb{N}}$$

$$\multimap\text{I}\ \frac{\Gamma, x\colon A \cdot X \vdash t : Y}{\Gamma \vdash \lambda x : X.t : A \cdot X \multimap Y} \qquad \multimap\text{E}\ \frac{\Gamma \vdash t : A \cdot X \multimap Y \qquad \Delta \vdash s : X}{\Gamma, A \cdot \Delta \vdash t\ s : Y}$$

$$\text{ADD}\ \frac{\Gamma \vdash s : \mathbb{N} \qquad \Delta \vdash t : \mathbb{N}}{\Gamma, \mathtt{nat} \cdot \Delta \vdash s + t : \mathbb{N}} \qquad \text{IF}\ \frac{\Gamma \vdash s : \mathbb{N} \qquad \Delta_1 \vdash t_1 : X \qquad \Delta_2 \vdash t_2 : X}{\Gamma, \Delta_1, \Delta_2 \vdash \mathsf{if0}\ s\ \mathsf{then}\ t_1\ \mathsf{else}\ t_2 : X}$$

In these rules, we write $A \cdot \Gamma$ for the context obtained by replacing each $x\colon B \cdot X$ with $x\colon (A \times B) \cdot X$. We note that rule IF enforces more linearity than usual. We have chosen to treat this linear version here, as for the defunctionalization of the non-linear version with $\Delta_1 = \Delta_2$, we would need to extend the labelled type system, and this case is already covered when we allow contraction in the next section.

With subexponential annotations, it is straightforward to define the Int interpretation. Define $(A \cdot X \multimap Y)^- = Y^-(A \cdot X^+)$ and $(A \cdot X \multimap Y)^+ = Y^+(A \cdot X^-)$. We write Γ^- and Γ^+ for $A_n \cdot X_n^- \dots A_1 \cdot X_1^-$ and $A_n \cdot X_n^+ \dots A_1 \cdot X_1^+$ if $\Gamma = x_1 \colon A_1 \cdot X_1, \dots, x_n \colon A_n \cdot X_n$.

The interpretation of rule AX changes from the linear case only by insertion of isomorphisms of the form $A \simeq \mathtt{unit} \times A$. The interpretation of rule CONST is given by the program $(apply_q \gamma^+, \{apply_q() = apply_a(n)\}, apply_a \gamma^-)$. The interpretation of the other rules is shown graphically in Fig. 7. In the cases for $\multimap$E and ADD, the boxes labelled with A and $\mathbb{N}^+$ respectively denote the program obtained by applying the operations $A \cdot (-)$ and $\mathbb{N}^+ \cdot (-)$ respectively to the contents of the box. For ADD, CONST and IF we do not show the contexts for brevity. In the case for IF, we write 0? for the program given by $apply_{a_s}(x) = \mathsf{case}\ \mathsf{iszero?}(x)\ \mathsf{of}\ \mathsf{inl}(y) \Rightarrow apply_{q_1}(y); \mathsf{inr}(z) \Rightarrow apply_{q_2}(z)$.

Let us now consider how the above interpretation relates to the one obtained by CPS translation and defunctionalization, where the subexponential annotations are ignored.

A CPS translation of n is given by $\lambda^q k. k@_a n \colon \overline{\mathbb{N}}[q,a]$, where $\overline{\mathbb{N}}[q,a] = \neg_q \neg_a \mathtt{nat}$, which defunctionalizes to $apply_q(\langle\rangle, k) = apply_a(k, n)$. The Int interpretation yields the definition $apply_q() = apply_a(n)$, which differs only in the removal of arguments.

For addition $s + t$ a CPS translation is $\lambda^q k. \underline{s} @_{q_s} (\lambda^{a_s} x. \underline{t} @_{q_t} (\lambda^{a_t} y. k @_a (x+y)))$. Defunctionalization leads to the following set of equations: $D_{\underline{s}} \cup D_{\underline{t}} \cup \{apply_q(f,k) = apply_{q_s}(\underline{s}^*, \langle k \rangle), apply_{a_s}(\langle k \rangle, x) = apply_{q_t}(\underline{t}^*, \langle k, x \rangle), apply_{a_t}(\langle k, x \rangle, y) = apply_a(k, x+y)\}$. Comparing this to the Int interpretation, we can see that this set of equations has the same shape, albeit with different arguments.

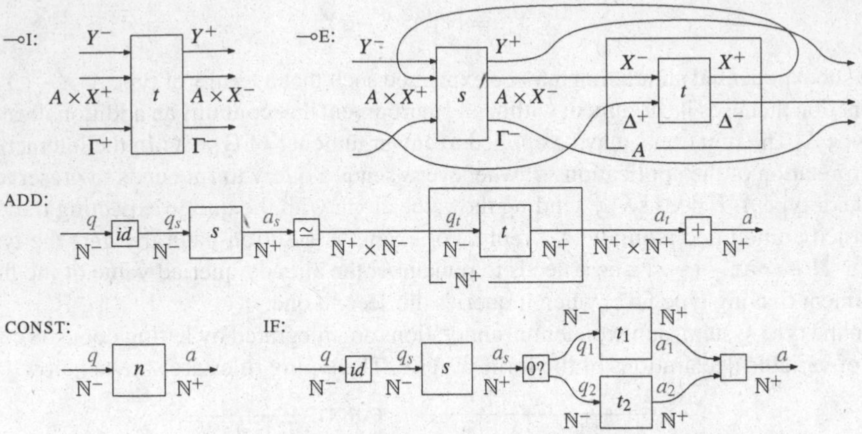

Fig. 1. Int Interpretation of Rules with Subexponentials

Similarly, the source term $\mathsf{if0}\ s\ \mathsf{then}\ t_1\ \mathsf{else}\ t_2$ is CPS translated to the labelled term $\lambda^q k.\, \underline{s} @_{q_s} (\lambda^{a_s} x.\, \mathsf{if0}\ x\ \mathsf{then}\ \underline{t_1} @_{q_1} (\lambda^{a_1} y.\, k @_a y)\ \mathsf{else}\ \underline{t_2} @_{q_2} (\lambda^{a_2} y.\, k @_a y))$. The equations obtained by defunctionalization have the same shape as those of the Int interpretation.

The observation that the programs obtained by Int interpretation and CPS translation followed by defunctionalization have the same shape can be made precise as follows. We say that two programs have the same *skeleton* whenever they have the same interface and the following holds: if one of the programs contains the definition $f(x) = g(e)$, then the other contains $f(x) = g(e')$ for some e'; and if one of the programs contains $f(x) = \mathsf{case}\ e\ \mathsf{of}\ \mathsf{inl}(x) \Rightarrow g(e_1);\ \mathsf{inr}(y) \Rightarrow h(e_2)$, then the other contains $f(x) = \mathsf{case}\ e'\ \mathsf{of}\ \mathsf{inl}(x) \Rightarrow g(e_1');\ \mathsf{inr}(y) \Rightarrow h(e_2')$ for some e', e_1' and e_2'.

We note that Lemma 3 continues to hold and that CpsDefun can be defined as above.

Proposition 2. *For any $\Gamma \vdash t \colon X$ there exists a program $I_{\underline{t}}$ that is a representative of the Int interpretation of the derivation of the sequent and that has the same skeleton as* $\mathsf{CpsDefun}(\Gamma \vdash t \colon X)$.

The proposition establishes a simple connection between the general shape of the programs. Let us now compare the values that are being passed around in them. We show that the values appearing in call traces of the program obtained by Int interpretation can be seen as simplifications of the values appearing at the same time in the traces of the program obtained by defunctionalization.

For any value v, we define a multiset of $\mathscr{V}(v)$ of the numbers it contains as follows: $\mathscr{V}(v) = \{n\}$ if $v = n$, $\mathscr{V}(v) = \mathscr{V}(v_1) \cup \mathscr{V}(v_2)$ if $v = \langle v_1, v_2 \rangle$ and $\mathscr{V}(v) = \emptyset$ otherwise (values of recursive types or sum types cannot appear). We say that a value v *simplifies* a value w if $\mathscr{V}(v) \subseteq \mathscr{V}(w)$. For example, the value $\langle 2, \langle 3, 3 \rangle \rangle$ simplifies $\langle 1, \langle \langle 2, \langle \rangle \rangle, \langle 3, \langle 2, 3 \rangle \rangle \rangle \rangle$, but not $\langle 2, 3 \rangle$. We say that a call trace $f_1(v_1) \dots f_n(v_n)$ *simplifies* the call trace $g(w_1) \dots g_n(w_n)$ if, for any $i \in \{1, \dots, n\}$, $f_i = g_i$ and v_i simplifies w_i.

With this terminology, we can express that the Int interpretation of any term simplifies its CPS translation and defunctionalization in the sense that it differs only in that unused function arguments are removed and function arguments are rearranged.

Proposition 3. *Let* $\vdash t\colon \mathbb{N}$, *let* $(q, D_{\underline{t}}, a) := \mathsf{CpsDefun}(\Gamma \vdash t\colon X)$ *and let* $I_{\underline{t}}$ *be the program from Prop. 2. Then, any call-trace of* $I_{\underline{t}}$ *beginning with* $apply_q()$ *simplifies the call-trace of* $\mathsf{CpsDefun}(\Gamma \vdash t\colon X)$ *of the same length that begins with* $apply_q(\langle\rangle, \langle\rangle)$.

This proposition allows us to consider the Int interpretation as a simplification of the program obtained by defunctionalization. This simplification seems quite similar to other optimisations of defunctionalization, in particular lightweight defunctionalization [2]. However, we do not know any variant of defunctionalization in the literature that gives exactly the same result. One may consider the Int interpretation as a new approach to optimising the defunctionalization of programs in continuation passing style.

8 Simple Types and Recursion

In this section, we strengthen the source language, explain how the Int interpretation can be extended to translate this language to the target language and relate this translation to CPS transform and defunctionalization. Since with increasing expressiveness of the source language it becomes harder to keep track of the syntactic details of defunctionalization, we investigate the relation less formally than in the previous section. We argue that a type system with subexponential annotations, adapted from IntML, offers a simple and conceptually clear way of managing the details.

First, we consider contraction in the source language. The CPS translation will remain unchanged, of course. The defunctionalization procedure described in Sec. 5, however, is too simple to handle this case. The control-flow annotations used therein do not suffice for the simply-typed case; they would need to be extended so that functions and applications are annotated with sets of labels. For instance, $s @_{\{l_1,\dots,l_n\}} t$ would mean that the label of s may not be uniquely determined, but that it is known to be among $l_1, \dots, l_n$. Such an application is transformed by the defunctionalization into a case distinction on the function that actually appears for s during evaluation: $(s @_{\{l_1,\dots,l_n\}} t)^* = \mathsf{case}\ s^*\ \mathsf{of}\ l_1(\vec{x}) \Rightarrow apply_{l_1}(l_1(\vec{x}), t^*); \dots ; l_n(\vec{y}) \Rightarrow apply_{l_n}(l_n(\vec{y}), t^*)$. Details appear in [2]. Note that such a case distinction is only possible if labels are actually passed as values; they cannot be omitted as in Sec. 5. To encode labels, one typically uses algebraic data types whose constructors correspond to the function labels. To handle the full λ-calculus, one must allow for recursive algebraic data types.

Let us now consider how the Int translation can be extended to the simply-typed λ-calculus and how it relates to defunctionalization. To this end, we can extend the type system from the previous section with a rule COPY for explicit copying. We also add a rule STRUCT for weakening of subexponential annotations, which is needed for Prop. 4.

$$\textsc{copy}\ \frac{\Gamma \vdash s : X \qquad \Delta, x\colon A \cdot X, y\colon B \cdot X \vdash t : Z}{\Delta, (A+B)\cdot\Gamma \vdash \mathsf{copy}\ s\ \mathsf{as}\ x, y\ \mathsf{in}\ t : Z} \qquad \textsc{struct}\ \frac{\Gamma, x\colon A \cdot X \vdash t : Y}{\Gamma, x\colon B \cdot X \vdash t : Y}\ A \lhd B$$

The side condition $A \lhd B$ means that any value of type A can be encoded into one of type B. Formally, $A \lhd B$ if and only if there are target expressions $x\colon A \vdash s\colon B$ and $y\colon B \vdash r\colon A$, such that $r[s[v/x]/y] = v$ holds for any value v of type A.

Recall the explanation of subexponentials as making explicit the value environment in which a variable is being used. The sequent in the premise of COPY tells us that x_1 and x_2 are used in environments with an additional value of type A and B respectively.

The subexponential $A+B$ in the conclusion tell us that x may be used in two ways: first in an environment that contains an additional variable of type A and second in one with an additional variable of type B. The coproduct identifies the two copies of x.

In the Int translation, copy is implemented by case distinction. Depending on whether the subexponential value is $\mathsf{inl}(a)$ or $\mathsf{inr}(b)$, the message is sent to x_1 or x_2 respectively. This case distinction mirrors the case distinction used in defunctionalization.

Let us outline the relation concretely by modifying the example from the Introduction. Consider the term $\lambda x : \mathbb{N}. x+x$. Its CPS transform is (with slight simplification) the term $\lambda^{l_1}\langle x,k\rangle. x @_{l_4} (\lambda^{l_2} m. x @_{l_4} (\lambda^{l_3} n. k @_{l_5} (m+n)))$. Applying this term to the argument $\langle \lambda^{l_4} k. k @_{\{l_2,l_3\}} 42, \lambda^{l_5} n. \mathtt{print\_int}(n)\rangle$ gives us an example of an application that needs to be annotated with a set of labels.

If we apply the defunctionalization procedure of Banerjee et al. [2] and then manually remove unneeded arguments, then we get the following equations, in which we consider labels as constructors of an algebraic data type.

$$\begin{aligned} \mathit{apply}_{l_1}() &= \mathit{apply}_{l_4}(l_2()) & \mathit{apply}_{l_4}(k) &= \mathsf{case}\ k\ \mathsf{of}\ l_2() \Rightarrow \mathit{apply}_{l_2}(42) \\ \mathit{apply}_{l_2}(m) &= \mathit{apply}_{l_4}(l_3(m)) & &\mid l_3(m) \Rightarrow \mathit{apply}_{l_3}(m,42) \\ \mathit{apply}_{l_3}(m,n) &= \mathit{apply}_{l_5}(m+n) & \mathit{apply}_{l_5}(n) &= \mathtt{print\_int}(n) \end{aligned}$$

We compare these equations to what we obtain by applying the Int interpretation to the term $\lambda x.\, \mathsf{copy}\ x\ \mathsf{as}\ x_1, x_2\ \mathsf{in}\ x_1 + x_2$ of type $(\mathtt{unit}+\mathtt{nat}) \cdot \mathbb{N} \multimap \mathbb{N}$:

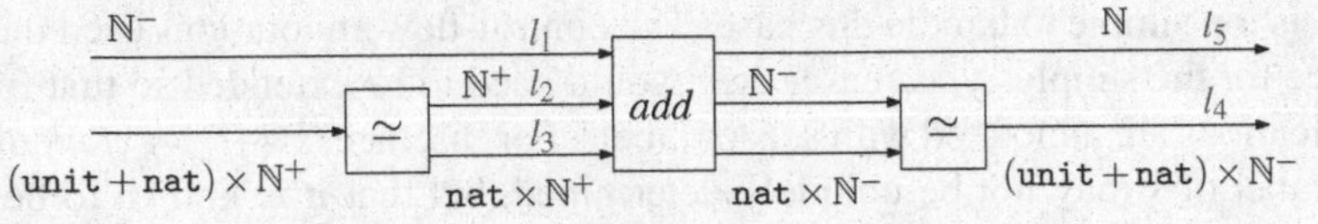

The interpretation of COPY inserts the boxes labelled $\simeq$, which denote the canonical isomorphism of their type.

To apply the program given by the diagram to the actual argument 42, one connects the output of type $(\mathtt{unit}+\mathtt{nat}) \times \mathbb{N}^-$ to the input of type $(\mathtt{unit}+\mathtt{nat}) \times \mathbb{N}^+$ such that when the value $\langle k', \langle\rangle\rangle$ arrives at the output port, then the value $\langle k', 42\rangle$ is fed back to the input port. This is what the equation for apply_{l_4} in the defunctionalized program does. The two possible cases of k, namely $\mathsf{inl}(\langle\rangle)$ or $\mathsf{inr}(m)$, correspond to $l_2()$ and $l_3(m)$ in the equation. Thus, in the interactive implementation of $\lambda x : \mathbb{N}. x+x$, duplication is treated just as in defunctionalization above. The points corresponding to the apply_{l_i}-equations are indicated in the diagram.

We note that with rules COPY and STRUCT the type system is as expressive as the simply-typed λ-calculus, if the target language has recursive types. This mirrors the use of recursive types in defunctionalization. Write $|t|$ for the term obtained by replacing in it any subterm of the form $\mathsf{copy}\ s_1\ \mathsf{as}\ x,y\ \mathsf{in}\ s_2$ with $s_2[s_1/x, s_1/y]$. Write $|X|$ and $|\Gamma|$ for the type and context of the simply-typed λ-calculus obtained by replacing any $A \cdot Y \multimap Z$ with $Y \to Z$ and removing subexponentials in the context.

Proposition 4. *If $\Gamma \vdash t \colon X$ is derivable in $\lambda^{\to,\mathbb{N}}$, then there exist Δ, s and Y with $\Gamma = |\Delta|$, $t = |s|$ and $X = |Y|$, such that $\Delta \vdash s \colon Y$ is derivable.*

The proof goes by using the simple type inference procedure from [5] and noting that the constraints generated therein can be solved easily in the presence of recursive types. Thus, there exists a simple type inference algorithm that finds the derivation of $\Delta \vdash s\colon Y$.

Finally, we mention that recursion can be accounted for by a fixed point combinator of type $\mathsf{fix}_{A,X} : (A\ \mathsf{list}) \cdot (A \cdot X \multimap X) \multimap X$, where A list abbreviates $\mu\alpha.\mathtt{unit} + A \times \alpha$.

9 Conclusion

We have observed that the non-standard compilation methods based on computation by interaction are closely related to CPS translation and defunctionalization. The interpretation in an interactive model may be regarded as a simple direct description of the combination of CPS translation, defunctionalization and a final optimisation of arguments. Subexponential types, in this form originally introduced in IntML, provide a logical account for the issues of managing value environments inherent to defunctionalization. The use of recursive algebraic data types in defunctionalization is explained by type theoretic means in rule COPY, Prop. 4 and the type of the fixed point combinator.

Subexponentials refine the exponentials in AJM games [1], where $!X$ is implemented using $\mathbb{N} \cdot X$. If we had used full exponentials in the Int interpretation above, then we would have obtained a compilation that encodes function values as numbers in $\mathbb{N}$, which is akin to storing closures on the heap. Subexponentials give us more control to avoid such encodings where unnecessary.

The subexponential type system bears resemblance to the linear logic with subexponentials of [16], which is where the terminology of subexponentials was first used. The type system is also quite similar to the type system for Syntactic Control of Concurrency (SCC) in [9]. A main difference appears to be that while SCC controls the number of program threads, subexponentials account for both the threads and their local data.

The observation that there is a connection between game models and continuations is not new. It appears for example in Levy's work on a jump-with-argument calculus [14]. Connections of game models to compilation have been made in [15], for example. It is also well-known that continuation passing is related to message passing, see e.g. [22]. However, we are not aware of work that makes explicit a connection to defunctionalization. We believe that the connection between game models and machine languages deserves to be better known and studied further. The call traces in this paper, for example, should have the same status as plays in a game semantic model.

In further work, we should like to understand if there are connections to Danvy's work on defunctionalized interpreters [7], or more generally to work on abstract machines, e.g. [21,6]. A relation is not obvious: Danvy considers the defunctionalization of particular implementations of interpreters, while here we show that the whole compilation itself may be described extensionally by the Int construction.

In another direction, an interactive view of CPS transform and defunctionalization may also give insight into issues that are often seen as problematic in the context of defunctionalization, such as compositional compilation and polymorphism: the interpretation in Int($\mathbb{T}$) is compositional and polymorphism can also be accounted for [20].

References

1. Abramsky, S., Jagadeesan, R., Malacaria, P.: Full abstraction for PCF. Inf. Comput. 163(2), 409–470 (2000)
2. Banerjee, A., Heintze, N., Riecke, J.G.: Design and correctness of program transformations based on control-flow analysis. In: Kobayashi, N., Pierce, B.C. (eds.) TACS 2001. LNCS, vol. 2215, pp. 420–447. Springer, Heidelberg (2001)
3. Cejtin, H., Jia, L., Weeks, S.: Flow-directed closure conversion for typed languages. In: Smolka, G. (ed.) ESOP 2000. LNCS, vol. 1782, pp. 56–71. Springer, Heidelberg (2000)
4. Dal Lago, U., Schöpp, U.: Functional programming in sublinear space. In: Gordon, A.D. (ed.) ESOP 2010. LNCS, vol. 6012, pp. 205–225. Springer, Heidelberg (2010)
5. Dal Lago, U., Schöpp, U.: Type inference for sublinear space functional programming. In: Ueda, K. (ed.) APLAS 2010. LNCS, vol. 6461, pp. 376–391. Springer, Heidelberg (2010)
6. Danos, V., Herbelin, H., Regnier, L.: Game semantics and abstract machines. In: Proceedings of the 11th IEEE Symposium on Logic in Computer Science (LICS 1996), pp. 394–405. IEEE, Los Alamitos (1996)
7. Danvy, O.: Defunctionalized interpreters for programming languages. In: Hook, J., Thiemann, P. (eds.) ICFP, pp. 131–142. ACM (2008)
8. Ghica, D.R.: Geometry of synthesis: a structured approach to VLSI design. In: Hofmann, M., Felleisen, M. (eds.) POPL, pp. 363–375. ACM (2007)
9. Ghica, D.R., Murawski, A.S., Ong, C.H.L.: Syntactic control of concurrency. Theor. Comput. Sci. 350(2-3), 234–251 (2006)
10. Hasegawa, M.: On traced monoidal closed categories. Mathematical Structures in Computer Science 19(2), 217–244 (2009)
11. Hofmann, M., Streicher, T.: Continuation models are universal for lambda-mu-calculus. In: LICS, pp. 387–395. IEEE Computer Society (1997)
12. Hyland, J.M.E., Ong, C.H.L.: On full abstraction for PCF: I, II, and III. Inf. Comput. 163, 285–408 (2000)
13. Joyal, A., Street, R., Verity, D.: Traced monoidal categories. Math. Proc. Cambridge Philos. Soc. 119(3), 447–468 (1996)
14. Levy, P.B.: Call-By-Push-Value: A Functional/Imperative Synthesis, Semantics Structures in Computation, vol. 2. Springer, Heidelberg (2004)
15. Melliès, P.A., Tabareau, N.: An algebraic account of references in game semantics. Electr. Notes Theor. Comput. Sci. 249, 377–405 (2009)
16. Nigam, V., Miller, D.: Algorithmic specifications in linear logic with subexponentials. In: Proceedings of the 11th International ACM SIGPLAN Conference on Principles and Practice of Declarative Programming (PPDP 2009), pp. 129–140. ACM, New York (2009)
17. Pierce, B.C.: Types and programming languages. MIT Press (2002)
18. Plotkin, G.D.: Call-by-name, call-by-value and the lambda-calculus. Theor. Comput. Sci. 1(2), 125–159 (1975)
19. Reynolds, J.C.: Definitional interpreters for higher-order programming languages. In: Proceedings of the ACM Annual Conference, ACM 1972, vol. 2, pp. 717–740. ACM, New York (1972)
20. Schöpp, U.: Stratified bounded affine logic for logarithmic space. In: Proceedings of the 22nd IEEE Symposium on Logic in Computer Science (LICS 2007), pp. 411–420. IEEE (2007)
21. Streicher, T., Reus, B.: Classical logic, continuation semantics and abstract machines. J. Funct. Program. 8(6), 543–572 (1998)
22. Thielecke, H.: Categorical Structure of Continuation Passing Style. Ph.D. thesis, University of Edinburgh (1997)

Completeness of Conversion between Reactive Programs for Ultrametric Models

Paula Severi and Fer-Jan de Vries

Computer Science, University of Leicester
{ps56,fdv1}@mcs.le.ac.uk

Abstract. In 1970 Friedman proved completeness of beta eta conversion in the simply-typed lambda calculus for the set-theoretical model. Recently Krishnaswami and Benton have captured the essence of Hudak's reactive programs in an extension of simply typed lambda calculus with causal streams and a temporal modality and provided this typed lambda calculus for reactive programs with a sound ultrametric semantics.

We show that beta eta conversion in the typed lambda calculus of reactive programs is complete for the ultrametric model.

1 Introduction

Krishnaswami and Benton have recently introduced a typed lambda calculus for reactive programs [1, 2]. Their basic idea was to have "a lambda calculus with types not only for data, but also indexed with time." This led them to extend simply typed lambda calculus with causal streams and a temporal modality and secondly, to define an ultrametric semantics for reactive programs. In the ultrametric model, types are interpreted as ultrametric spaces and terms as non-expansive maps [1, 3, 4]. They demonstrated the soundness of this extension for the ultrametric semantics.

This raises the natural question of *completeness.* In this paper we show that two terms typable in the calculus of reactive programs are $\beta\eta$-convertible if and only if they have the same interpretation in the model of ultrametric spaces.

Completeness has been well studied for simply typed lambda calculus. It has been proved for the set-theoretical model [5], the model of CPOs and the model of modest sets [6, 7]. Towards completeness for the *ultrametric semantics*, we introduce the notions of *step-indexed applicative structure* and *Henkin model for reactive programs.* We show that the term model (consisting of reactive programs modulo conversion) and the ultrametric model can be seen as step-indexed applicative structures and also as Henkin models for reactive programs. Since for the ultrametric model, a stream is a function on natural numbers, we need a strong notion of extensionality that requires that two streams are equal if all their components are equal. Strong extensionality of the term model is not so easy to prove. It does not follow immediately from the η-rule but from the fact that our calculus is confluent and strongly normalising.

Actually, we show two completeness results. The first one, called *completeness (of $\beta\eta$-conversion) for Henkin models*, says that there exists a Henkin model for

M. Hasegawa (Ed.): TLCA 2013, LNCS 7941, pp. 221–235, 2013.

reactive programs satisfying exactly the theory of $\beta\eta$-conversion. The second one, mentioned before, is about *completeness (of $\beta\eta$-conversion) for the ultrametric model.* The latter is proved by constructing a partial surjective step-indexed logical relation between the ultrametric model and the term model.

One interesting aspect of our paper is that we consider (ultra) metric spaces in a proof of completeness. We show that on the term model of the typed lambda calculus for reactive programs an ultrametric d can be defined for which the equivalence classes of terms of type $\sigma \to \tau$ are *non-expansive*, i.e. for M of type $\sigma \to \tau$ we have that $d([MP],[MQ]) \leq d([P],[Q])$.

This paper is organised as follows. Section 2 defines the typed lambda calculus for reactive programs. Section 3 introduces the notions of (step-indexed) applicative structure, Henkin model and (step-indexed) logical relation for reactive programs. Section 4 proves strong normalisation and confluence. Section 5 defines the term model and proves completeness for Henkin models. Section 6 shows that the model of ultrametric spaces is a Henkin model. Section 7 defines an ultrametric on the term model and shows that this metric is well-behaved. Section 8 shows completeness for the ultrametric model.

2 Typed Lambda Calculus for Reactive Programs

We recall the typed lambda calculus λ^{RP} for reactive programs as defined in [1]. It comes with a syntax, rewriting rules and typing rules.

Definition 1 (Syntax for reactive programs). *We define the set* P *of reactive programs (or terms) and the set* T *of types as follows.*

$$\begin{array}{l} \mathsf{P} \ni M ::= x \mid \mathsf{hd}(M) \mid \mathsf{tl}(M) \mid \mathsf{cons}(M,M) \mid \mathsf{await}(M) \mid \circ(M) \mid \lambda x{:}\sigma.M \mid MM \\ \mathsf{T} \ni \sigma \ ::= \mathsf{b} \mid (\sigma \to \sigma) \mid \bullet\sigma \mid \mathsf{S}(\sigma) \end{array}$$

where the parameter x ranges over a set V *of variables,* b *over a set* B *of basic types. A type declaration is a statement of the form $x :_i \tau$. A context is a finite set of type declarations with only distinct variables as subjects.*

Definition 2 (Reduction for reactive programs). *The β-rule is defined by:*

$$\begin{array}{llll} (\lambda x{:}\sigma.N)M & \to N[x:=M]\ (\beta) & \mathsf{await}(\circ(M)) & \to M\ (\beta) \\ \mathsf{hd}(\mathsf{cons}(M,N)) & \to M \qquad\quad (\beta) & \mathsf{tl}(\mathsf{cons}(M,N)) & \to N\ (\beta) \end{array}$$

The η-rule is defined by:

$$\begin{array}{lll} \lambda x{:}\sigma.Mx & \to M,\ \textit{if}\ x \notin \mathsf{fv}(M)\ (\eta) & \circ(\mathsf{await}(M)) \to M\ (\eta) \\ \mathsf{cons}(\mathsf{hd}(M),\mathsf{tl}(M)) & \to M \qquad\qquad\qquad (\eta) & \end{array}$$

Let $\rho \in \{\beta,\eta,\beta\eta\}$. The relation $\to_\rho$ is defined as the smallest relation on P *that is closed under under contexts and the ρ-rule(s). The reflexive and transitive closure of $\to_\rho$ is denoted by $\twoheadrightarrow_\rho$. The reflexive, symmetric and transitive closure of $\to_\rho$ is denoted by $=_\rho$, called* ρ-conversion. *The ρ-normal form of a term M is N, if $M \twoheadrightarrow_\rho N$ and N is in ρ-normal form. If the ρ-normal form of M exists, we denote it by $\mathsf{nf}_\rho(M)$.*

Definition 3 (Typing rules for reactive programs). *A type declaration is a statement of the form $x :_i \tau$. A context is a finite set of type declarations with only distinct variables as subjects. A type declaration $M :_i \tau$ is derivable from the context Γ, if the typing judgement $\Gamma \vdash M :_i \sigma$ can be derived from the following typing rules:*

$$\frac{x{:_i}\sigma \in \Gamma \quad j \geq i}{\Gamma \vdash x :_j \sigma}\,(\mathsf{var}) \qquad \frac{\Gamma \vdash M :_i \mathsf{S}(\sigma)}{\Gamma \vdash \mathsf{hd}(M) :_i \sigma}\,(\mathsf{head})$$

$$\frac{\Gamma \vdash M :_i \mathsf{S}(\sigma)}{\Gamma \vdash \mathsf{tl}(M) :_{i+1} \mathsf{S}(\sigma)}\,(\mathsf{tail}) \qquad \frac{\Gamma \vdash M :_i \sigma \quad \Gamma \vdash N :_{i+1} \mathsf{S}(\sigma)}{\Gamma \vdash \mathsf{cons}(M, N) :_i \mathsf{S}(\sigma)}\,(\mathsf{cons})$$

$$\frac{\Gamma, x{:_i}\sigma \vdash N :_i \tau}{\Gamma \vdash \lambda x{:}\sigma.N :_i (\sigma \to \tau)}\,(\to \mathsf{I}) \qquad \frac{\Gamma \vdash M :_{i+1} \sigma}{\Gamma \vdash \circ(M) :_i \bullet\sigma}\,(\bullet\mathsf{I})$$

$$\frac{\Gamma \vdash N :_i (\sigma \to \tau) \quad \Gamma \vdash M :_i \sigma}{\Gamma \vdash NM :_i \tau}\,(\to \mathsf{E}) \qquad \frac{\Gamma \vdash M :_i \bullet\sigma}{\Gamma \vdash \mathsf{await}(M) :_{i+1} \sigma}\,(\bullet\mathsf{E})$$

If a judgement $\Gamma \vdash M :_i \sigma$ is derivable from these rules, we call M a typable term.

The intuition is that $M :_i \sigma$ expresses that at time stamp i we know about the existence of a term M with type σ. If time stamp 0 represents 'now', then time stamp i represents 'i steps from now into the future.' To observe the tail N of a stream $\mathsf{cons}(M, N)$ of which we can see the head M now, we must wait one time step. We cannot force the future into the present.

Lemma 1 (Time adjustment [1]). *If $\Gamma, \Delta \vdash M :_i \sigma$ then $\Gamma, \Delta_{+n} \vdash M :_{i+n} \sigma$, where Δ_{+n} is obtained from Δ by raising the indexing time by n in all type declarations in Δ. Moreover, the derivations of $\Gamma, \Delta \vdash M :_i \sigma$ and $\Gamma, \Delta_{+n} \vdash M :_{i+n} \sigma$ have the same size.*

The following lemma is proved by induction on the derivation.

Lemma 2 (Subject reduction for reactive programs). *If $\Gamma \vdash M :_i \sigma$ and $M \twoheadrightarrow_{\beta\eta} N$ then $\Gamma \vdash N :_i \sigma$.*

3 Applicative Structures, Henkin Models and Logical Relations for Reactive Programs

In this section, we extend the notions of applicative structure, Henkin model and logical relations as defined for the simply typed lambda calculus, e.g., [7], to reactive programs. The time indices $i \in \mathbb{N}$ will play a similarly crucial role as they did in the typing rules for reactive programs.

Definition 4 (Applicative structures for reactive programs). *A (step-indexed) applicative structure for reactive programs is a tuple*

$$\mathcal{A} = \langle \{\mathcal{A}_i^\sigma\}, \{\delta_{ij}^\sigma\}, \{\mathit{hd}_i^\sigma, \mathit{tl}_i^\sigma, \mathit{await}_i^\sigma, \mathit{app}_i^{\sigma\tau}\} \rangle$$

of families of sets and functions indexed by types from T *such that for all* $\sigma, \tau \in \mathsf{T}$ *and all* $i, j \in \mathbb{N}$ *with* $i \leq j$ *we have:*

1. $\mathcal{A}_i^\sigma$ *is a set,*
2. $\delta_{ij}^\sigma \in \mathcal{A}_i^\sigma \to \mathcal{A}_j^\sigma$ *(expressing "delay"),*
3. $\delta_{ii}^\sigma = id : \mathcal{A}_i^\sigma \to \mathcal{A}_i^\sigma$ *and* $\delta_{jk}^\sigma \circ \delta_{ij}^\sigma = \delta_{ik}^\sigma$*,*
4. $\mathit{hd}_i^\sigma \in \mathcal{A}_i^{\mathsf{S}(\sigma)} \to \mathcal{A}_i^\sigma$, $\mathit{tl}_i^\sigma \in \mathcal{A}_i^{\mathsf{S}(\sigma)} \to \mathcal{A}_{i+1}^{\mathsf{S}(\sigma)}$,
   $\mathit{await}_i^\sigma \in \mathcal{A}_i^{\bullet\sigma} \to \mathcal{A}_{i+1}^\sigma$, $\mathit{app}_i^{\sigma\tau} \in \mathcal{A}_i^{\sigma\to\tau} \to \mathcal{A}_i^\sigma \to \mathcal{A}_i^\tau$.
5. $\delta_{ij}^\sigma(\mathit{hd}_i^\sigma(a)) = \mathit{hd}_j^\sigma(\delta_{ij}^\sigma(a))$ *for all* $a \in \mathcal{A}_i^{\mathsf{S}(\sigma)}$,
   $\delta_{(i+1)j}^\sigma(\mathit{tl}_i^\sigma(a)) = \mathit{tl}_j^\sigma(\delta_{ij}^\sigma(a))$ *for all* $a \in \mathcal{A}_i^{\mathsf{S}(\sigma)}$ *and* $i+1 \leq j$,
   $\delta_{(i+1)j}^\sigma(\mathit{await}_i^\sigma(a)) = \mathit{await}_j^\sigma(\delta_{ij}^\sigma(a))$ *for all* $a \in \mathcal{A}_i^{\bullet\sigma}$ *and* $i+1 \leq j$,
   $\delta_{ij}^\tau(\mathit{app}_i^{\sigma\tau}(f, b)) = \mathit{app}_j^{\sigma\tau}((\delta_{ij}^{\sigma\to\tau} f), (\delta_{ij}^\sigma b))$ *for all* $f \in \mathcal{A}_i^{\sigma\to\tau}$, $b \in \mathcal{A}_i^\sigma$.

To define extensional applicative structures, we have to define when the element of all three kind of types $\mathcal{A}_i^{\sigma\to\tau}$, $\mathcal{A}_i^{\bullet\sigma}$ and $\mathcal{A}_i^{\mathsf{S}(\sigma)}$ are extensional. For the first two this is straightforward. However for the latter we consider a strong version of extensionality that views streams as functions from natural numbers: two streams are equal if all their components are equal.

Definition 5 (Extensional applicative structure for reactive programs). *We say that an applicative structure for reactive programs is extensional if it satisfies the following conditions:*

1. **Extensionality on** $\sigma \to \tau$**.** *For all* $j \geq i$ *and all* $a, b \in \mathcal{A}_j^{\sigma\to\tau}$, *if* $\mathit{app}_j^{\sigma\tau}(\delta_{ij}^{\sigma\to\tau}(a), d) = \mathit{app}_j^{\sigma\tau}(\delta_{ij}^{\sigma\to\tau}(b), d)$ *for all* $d \in \mathcal{A}_j^\sigma$, *then* $a = b$.
2. **Extensionality on** $\bullet\sigma$**.** *For all* $a, b \in \mathcal{A}_i^{\bullet\sigma}$, *if* $\mathit{await}_i^\sigma(a) = \mathit{await}_i^\sigma(b)$ *then* $a = b$.
3. **Extensionality on** $\mathsf{S}(\sigma)$**.** *For all* $a, b \in \mathcal{A}_i^{\mathsf{S}(\sigma)}$, *if for all* $n \in \mathbb{N}$ *we have that* $\mathit{hd}_{i+n}^\sigma(\mathit{tl}_{i+n}^\sigma(\ldots(\mathit{tl}_i^\sigma(a)))) = \mathit{hd}_{i+n}^\sigma(\mathit{tl}_{i+n}^\sigma(\ldots(\mathit{tl}_i^\sigma(b))))$ *then* $a = b$.

Extensionality for the arrow type requires that the applications are equal for all $j \geq i$. This is clearly stronger than having the same condition for just i. However, for the other two cases, the formulations with $j \geq i$ and just i are equivalent.

It is easy to show that extensionality implies the next weaker notion.

Definition 6 (Weak extensional applicative structure for reactive programs). *We say that an applicative structure for reactive programs is weakly extensional if extensionality on* $\mathsf{S}(\sigma)$ *is replaced by the weaker condition:*

3'. For all $MN \in \mathcal{A}_i^{\mathsf{S}(\sigma)}$, *if* $\mathit{hd}_i^\sigma(M) = \mathit{hd}_i^\sigma(N)$ *and* $\mathit{tl}_i^\sigma(M) = \mathit{tl}_i^\sigma(N)$ *then* $M = N$.

Let $\mathcal{A}$ be an applicative structure for reactive programs and Γ be a context. An environment ρ is a function from the set of variables V to the union of all $\mathcal{A}_i^\sigma$. For $a \in \mathcal{A}_i^\sigma$, the update environment $\rho[x \leftarrow a]$ is the environment mapping x to a and all other variables $y \neq x$ to $\rho(y)$. We write $\rho \models \Gamma$ if $\rho(x) \in \mathcal{A}_i^\sigma$ holds for all $x :_i \sigma \in \Gamma$. A *meaning function* for an applicative structure $\mathcal{A}$ is a (total) function that maps any derivation $\Gamma \vdash M :_i \sigma$ and any environment ρ, to an element $[\![\Gamma \vdash M :_i \sigma]\!]_\rho^{\mathcal{A}}$ in $\mathcal{A}_i^\sigma$.

Definition 7 (Henkin models for reactive programs). *Let $\rho \models \Gamma$. An extensional applicative structure $\mathcal{A}$ for reactive programs is called a* Henkin model *if there exists a meaning function satisfying the following conditions (all together called the* environment model condition*):*

- $\delta^{\sigma}_{ij}([\![\Gamma \vdash M :_i \sigma]\!]^{\mathcal{A}}_{\rho}) = [\![\Gamma \vdash M :_j \sigma]\!]^{\mathcal{A}}_{\rho}$
- $[\![\Gamma \vdash x :_j \sigma]\!]^{\mathcal{A}}_{\rho} = \delta^{\sigma}_{ij}(\rho(x))$ *for all* $x :_i \sigma \in \Gamma$
- $[\![\Gamma \vdash MN :_i \tau]\!]^{\mathcal{A}}_{\rho} = \mathit{app}^{\sigma\tau}_i([\![\Gamma \vdash M :_i \sigma \to \tau]\!]^{\mathcal{A}}_{\rho},\ [\![\Gamma \vdash N :_i \sigma]\!]^{\mathcal{A}}_{\rho})$
- $[\![\Gamma \vdash \lambda x{:}\sigma.M :_i \sigma \to \tau]\!]^{\mathcal{A}}_{\rho} = \begin{cases} \text{the unique } f \in \mathcal{A}^{\sigma\to\tau}_i \text{ such that} \\ \text{for all } j \geq i,\ d \in \mathcal{A}^{\sigma}_j \\ \mathit{app}^{\sigma\tau}_j(\delta^{\sigma\to\tau}_{ij}(f), d) = [\![\Gamma, x :_j \sigma \vdash M :_j \tau]\!]_{\rho[x:=d]} \end{cases}$
- $[\![\Gamma \vdash \mathsf{await}(M) :_{i+1} \sigma]\!]^{\mathcal{A}}_{\rho} = \mathit{await}^{\sigma}_i([\![\Gamma \vdash M :_i \sigma]\!]^{\mathcal{A}}_{\rho})$
- $[\![\Gamma \vdash \circ(M) :_i \bullet\sigma]\!]^{\mathcal{A}}_{\rho} = \begin{cases} \text{the unique } a \in \mathcal{A}^{\bullet\sigma}_i \text{ such that} \\ \mathit{await}^{\sigma}_i(a) = [\![\Gamma \vdash M :_{i+1} \sigma]\!]^{\mathcal{A}}_{\rho} \end{cases}$
- $[\![\Gamma \vdash \mathsf{hd}(M) :_i \sigma]\!]^{\mathcal{A}}_{\rho} = \mathit{hd}^{\sigma}_i([\![\Gamma \vdash M :_i \mathsf{S}(\sigma)]\!]^{\mathcal{A}}_{\rho})$
- $[\![\Gamma \vdash \mathsf{tl}(M) :_{i+1} \mathsf{S}(\sigma)]\!]^{\mathcal{A}}_{\rho} = \mathit{tl}^{\sigma}_i([\![\Gamma \vdash M :_i \mathsf{S}(\sigma)]\!]^{\mathcal{A}}_{\rho})$
- $[\![\Gamma \vdash \mathsf{cons}(M, N) :_i \mathsf{S}(\sigma)]\!]^{\mathcal{A}}_{\rho} = \begin{cases} \text{the unique } s \in \mathcal{A}^{\mathsf{S}(\sigma)}_i \text{ such that} \\ \mathit{hd}^{\sigma}_i(s) = [\![\Gamma \vdash M :_i \sigma]\!]^{\mathcal{A}}_{\rho} \text{ and} \\ \mathit{tl}^{\sigma}_i(s) = [\![\Gamma \vdash N :_{i+1} \mathsf{S}(\sigma)]\!]^{\mathcal{A}}_{\rho} \end{cases}$

We will use the notation

$$\mathcal{A} \models \Gamma \vdash M = N :_i \sigma$$

if $\Gamma \vdash M :_i \sigma$, $\Gamma \vdash N :_i \sigma$ *and* $[\![\Gamma \vdash M :_i \sigma]\!]^{\mathcal{A}}_{\rho} = [\![\Gamma \vdash N :_i \sigma]\!]^{\mathcal{A}}_{\rho}$ *for all* ρ *with* $\rho \models \Gamma$.

Lemma 3 (Soundness of Henkin models for reactive programs)

1. *If* $\Gamma \vdash M :_i \sigma$, *then* $[\![\Gamma \vdash M :_i \sigma]\!]^{\mathcal{A}}_{\rho} \in \mathcal{A}^{\sigma}_i$ *for all* $\rho \models \Gamma$.
2. *If* $\Gamma \vdash M :_i \sigma$ *and* $\Gamma \vdash N :_i \sigma$ *then* $M =_{\beta\eta} N$ *implies* $\mathcal{A} \models \Gamma \vdash M = N :_i \sigma$.

Both items of the lemma can be proved by induction on the size of the derivation. It is enough to consider one step $\to_{\beta\eta}$ in the proof of the second item.

Definition 8 (Logical relations for reactive programs). A (step-indexed) logical relation for reactive programs $\mathcal{R}$ *between two applicative structures for reactive programs* $\mathcal{A}$ *and* $\mathcal{B}$ *is a family* $\{\mathcal{R}^{\sigma}_i\}$ *of indexed relations such that*

- $\mathcal{R}^{\sigma}_i \subseteq \mathcal{A}^{\sigma}_i \times \mathcal{B}^{\sigma}_i$ *for each* σ *and* i,
- *if* $\mathcal{R}^{\sigma}_i(a, b)$ *then* $\mathcal{R}^{\sigma}_j(\delta^{\sigma}_{ij}(a), \delta^{\sigma}_{ij}(b))$ *for all* $j \geq i$,
- $\mathcal{R}^{\bullet\sigma}_i(a, b)$ *iff* $\mathcal{R}^{\sigma}_{i+1}(\mathit{await}^{\sigma}_i(a), \mathit{await}^{\sigma}_i(b))$,
- $\mathcal{R}^{\mathsf{S}(\sigma)}_i(a, b)$ *iff* $\mathcal{R}^{\sigma}_{i+n}(\mathit{hd}^{\sigma}_{i+n}(\mathit{tl}^{\sigma}_{i+n} \cdots \mathit{tl}^{\sigma}_i(a)), \mathit{hd}^{\sigma}_{i+n}(\mathit{tl}^{\sigma}_{i+n} \cdots \mathit{tl}^{\sigma}_i(b)))$ *for all* n,
- $\mathcal{R}^{\sigma\to\tau}_i(f, g)$ *iff* $\forall j \geq i. \forall a \in \mathcal{A}^{\sigma}_j. \forall b \in \mathcal{B}^{\sigma}_j. \mathcal{R}^{\sigma}_j(a, b) \Rightarrow$
$$\mathcal{R}^{\tau}_j(\mathit{app}^{\sigma\tau}_j(\delta^{\sigma\to\tau}_{ij} f)\ a, \mathit{app}^{\sigma\tau}_j(\delta^{\sigma\to\tau}_{ij} g)\ b).$$

A logical relation $\mathcal{R}$ *between* $\mathcal{A}$ *and* $\mathcal{B}$ *is called a* logical partial (surjective) function *from* $\mathcal{A}$ *to* $\mathcal{B}$ *if each* $\mathcal{R}^{\sigma}_i$ *is a partial (surjective) function.*

The definition of binary logical relations generalises easily to any arity. In this paper we will define a logical relation of arity one (a logical predicate) and one of arity two.

Lemma 4 (Basic lemma on logical relations for reactive programs). *Let $\mathcal{R}$ be a logical relation for reactive programs between two Henkin models $\mathcal{A}$ and $\mathcal{B}$. Let $\rho_{\mathcal{A}}$ and $\rho_{\mathcal{B}}$ be environments for $\mathcal{A}$ and $\mathcal{B}$ respectively, such that $\rho_{\mathcal{A}} \models \Gamma$, $\rho_{\mathcal{B}} \models \Gamma$ and $\mathcal{R}^{\tau}_{j}(\rho_{\mathcal{A}}(x), \rho_{\mathcal{B}}(x))$ for all $x :_j \tau$ in Γ.*
If $\Gamma \vdash M :_i \sigma$ then $\mathcal{R}^{\sigma}_{i}([\![\Gamma \vdash M :_i \sigma]\!]^{\mathcal{A}}_{\rho_{\mathcal{A}}}, [\![\Gamma \vdash M :_i \sigma]\!]^{\mathcal{B}}_{\rho_{\mathcal{B}}})$.

The above lemma is proved by induction on the size of the derivation.

The theory induced by a Henkin model $\mathcal{A}$, denoted by $\mathsf{Th}(\mathcal{A})$ is the set $\{(M, N) \mid \mathcal{A} \models \Gamma \vdash M = N :_i \sigma\}$.

Lemma 5 (Theory inclusion). *Let $\mathcal{A}, \mathcal{B}$ be Henkin models for reactive programs. If there is a logical partial function from $\mathcal{A}$ to $\mathcal{B}$, then $\mathsf{Th}(\mathcal{A}) \subseteq \mathsf{Th}(\mathcal{B})$.*

This lemma is proved similarly as [7, Lemma 8.2.17].

4 Confluence and Strong Normalisation

In this section, we prove confluence and strong normalisation of $\beta\eta$ for the typed lambda calculus of reactive programs.

Failure of confluence of $\beta\eta$ on untypable terms has several causes. One cause is the presence of explicit types in the abstractions. Nederpelt's term $\lambda x{:}\sigma.(\lambda y{:}\tau.y)x$ provides a counterexample [8]. Another cause is the non-left linear η-rule for streams. This is shown through a variation of Klop's counterexample on surjective pairs [9, 10]. Define

$$\mathsf{D} = \lambda x{:}\sigma.\lambda y{:}\tau.(\mathsf{cons}(\mathsf{hd}(\lambda z.zx), \mathsf{tl}(\lambda z.zy))\lambda z.u).$$

Then, we have that $\mathsf{D}MM \twoheadrightarrow_{\beta\eta} u$ for any M. Note that the η-step creates a β-redex that cannot be performed earlier (this shows that η cannot be postponed over β on untypable terms). Next, we define $\mathsf{E} = Y\ (\lambda f{:}\sigma'.\lambda x{:}\tau'.\mathsf{D}\ x\ (f\ x))$ and $\mathsf{F} = Y\ (\lambda f{:}\sigma''.\mathsf{E}\ f)$. We have that $\mathsf{E} \twoheadrightarrow_{\beta} \lambda x{:}\tau'.\mathsf{D}\ x\ (\mathsf{E}\ x)$ and $\mathsf{F} \twoheadrightarrow_{\beta} \mathsf{E}\ \mathsf{F}$. So, $\mathsf{F} \twoheadrightarrow_{\beta\eta} u$ and $\mathsf{F} \twoheadrightarrow_{\beta\eta} \mathsf{E}\ u$. But u and $\mathsf{E}\ u$ do not have a common reduct.

We will show that the typable terms are $\beta\eta$-strongly normalising using a logical predicate similar to the ones used in strong normalisation proofs of the simply typed lambda calculus (e.g., see Section 8.3.2 of [7]). For this proof, we use an applicative structure $\mathcal{T}$ constructed from typable terms.

Notation 1. *From now on, we assume the existence of a family $\{\mathsf{V}^{\sigma}_{i}\}$ of pairwise disjoint, infinite sets of pairwise distinct variables. We define Γ_{∞} to be the infinite context consisting of all type declarations of the form $x :_i \sigma$ with $x \in \mathsf{V}^{\sigma}_{i}$ for some type σ and index i.*

Definition 9 (Term applicative structure). *For each type σ and index i, let $\mathcal{T}_i^\sigma$ be the set $\{M \mid \Gamma \vdash M :_i \sigma$, for some σ and finite $\Gamma \subset \Gamma_\infty\}$ of all terms that can be typed with σ at "stage" i with some finite subcontext of Γ_∞. We define* the term applicative structure *as the applicative structure*

$$\mathcal{T} = \langle \{\mathcal{T}_i^\sigma\}, \{\delta_{ij}^\sigma\}, \{\mathit{hd}_i^\sigma, \mathit{tl}_i^\sigma, \mathit{await}_i^\sigma, \mathit{app}_i^{\sigma\tau}\}\rangle$$

where $\mathit{app}_i^{\sigma\tau}(M, N) = MN$, $\mathit{hd}_i^\sigma(M) = \mathsf{hd}(M)$, $\mathit{await}_i^\sigma(M) = \mathsf{await}(M)$ *and* $\mathit{tl}_i^\sigma(M) = \mathsf{tl}(M)$. *We take set inclusion for δ_{ij}^σ when $i \leq j$. This is well-defined by the Time Adjustment Lemma. As meaning function, we take $[\![\Gamma \vdash M :_i \sigma]\!]_\rho^\mathcal{T} = \rho(M)$ where $\rho(M)$ is the result of performing the substitution ρ to M.*

Note that $[\![\Gamma \vdash M :_i \sigma]\!]_\rho^\mathcal{T} \in \mathcal{T}_i^\sigma$. This meaning function does not satisfy the environment model condition.

Definition 10 (Logical predicate of strongly normalizing terms). *Let $\mathcal{SN}$ be the set of $\beta\eta$-strongly normalising terms. We define the family of predicates $\mathcal{P}_i^\sigma \subseteq \mathcal{T}_i^\sigma$ by induction on σ:*

$$\begin{array}{ll}
\mathcal{P}_i^{\mathsf{b}} & = \{M \in \mathcal{T}_i^{\mathsf{b}} \mid M \in \mathcal{SN}\} \\
\mathcal{P}_i^{\sigma\to\tau} & = \{M \in \mathcal{T}_i^{\sigma\to\tau} \mid \forall j \geq i, N \in \mathcal{P}_j^\sigma, MN \in \mathcal{P}_j^\tau\} \\
\mathcal{P}_i^{\bullet\sigma} & = \{M \in \mathcal{T}_i^{\bullet\sigma} \mid \mathsf{await}(M) \in \mathcal{P}_{i+1}^\sigma\} \\
\mathcal{P}_i^{\mathsf{S}(\sigma)} & = \{M \in \mathcal{T}_i^{\mathsf{S}(\sigma)} \mid \forall n \in \mathbb{N}, \mathsf{hd}(\mathsf{tl}^n(M)) \in \mathcal{P}_{i+n}^\sigma\}
\end{array}$$

where $\mathsf{tl}^n(M)$ is the term $\mathsf{tl}(\mathsf{tl}(\ldots(\mathsf{tl}(M)))$ consisting of n applications of tl.

It is easy to see that $\mathcal{P}$ is a step-indexed logical relation for reactive programs of arity one. Note that $\mathcal{P}_i^\sigma \subseteq \mathcal{P}_j^\sigma$ if $i \leq j$. We define the elimination contexts $\mathcal{E}[\,]$ with one hole with the grammar $\mathcal{E}[\,] := [\,] \mid \mathcal{E}[\,]M \mid \mathsf{hd}(\mathsf{tl}^n(\mathcal{E}[\,])) \mid \mathsf{await}(\mathcal{E}[\,])$. We write $\mathcal{E}[\,] \in \mathcal{SN}$ if all terms used in the construction of $\mathcal{E}[\,]$ are $\beta\eta$-strongly normalising.

Lemma 6. *1. $\mathcal{P}_i^\sigma \subseteq \mathcal{SN}$. 2. If $\mathcal{E}[\,] \in \mathcal{SN}$ and $\mathcal{E}[x] \in \mathcal{T}_i^\sigma$ then $\mathcal{E}[x] \in \mathcal{P}_i^\sigma$.*

The two statements are proved simultaneously by induction on the type σ. For the base case $\sigma = \mathsf{b}$, it is easy to see that $\mathcal{E}[x] \in \mathcal{SN}$ because $\mathcal{E}[\,] \in \mathcal{SN}$. The cases in the next lemma all follow by induction on σ.

Lemma 7 (Closure under β-expansion inside context $\mathcal{E}$)

- *If $\mathcal{E}[(\lambda x.M)N] \in \mathcal{T}_i^\sigma$ and $\mathcal{E}[M[x := N]] \in \mathcal{P}_i^\sigma$ then $\mathcal{E}[(\lambda x.M)N] \in \mathcal{P}_i^\sigma$.*
- *If $\mathcal{E}[\mathsf{await}(M)] \in \mathcal{T}_i^{\bullet\sigma}$ and $\mathcal{E}[M] \in \mathcal{P}_i^\sigma$ then $\mathcal{E}[\mathsf{await}(M)] \in \mathcal{P}_i^{\bullet\sigma}$.*
- *If $\mathcal{E}[\mathsf{hd}(\mathsf{cons}(M, N))] \in \mathcal{T}_i^\sigma$ and $\mathcal{E}[M] \in \mathcal{P}_i^\sigma$ then $\mathcal{E}[\mathsf{hd}(\mathsf{cons}(M, N))] \in \mathcal{P}_i^\sigma$.*
- *If $\mathcal{E}[\mathsf{tl}(\mathsf{cons}(M, N))] \in \mathcal{T}_i^\sigma$ and $\mathcal{E}[N] \in \mathcal{P}_i^\sigma$ then $\mathcal{E}[\mathsf{tl}(\mathsf{cons}(M, N))] \in \mathcal{P}_i^\sigma$.*

Lemma 8 (Soundness for logical predicate $\mathcal{P}$). *Let $\Gamma \vdash M :_i \sigma$ and $\rho(x) \in \mathcal{P}_i^\sigma$ for all $x :_i \sigma \in \Gamma$. Then, $[\![\Gamma \vdash M :_i \sigma]\!]_\rho^\mathcal{T} \in \mathcal{P}_i^\sigma$.*

Proof. By induction on the derivation using Lemma 7. □

Remark 1 (Alternative proof of Lemma 8). It is possible to give a general version of the basic lemma for logical relations instead of Lemma 8. For that we would have to introduce more notational machinery like the notions of acceptable meaning function and admissible relation, as on pages 540-541 in [7].

Theorem 1 (Strong $\beta\eta$-Normalisation on typable terms). *If $\Gamma \vdash M :_i \sigma$ then M is $\beta\eta$-strongly normalising.*

Proof. Suppose $\Gamma \vdash M :_i \sigma$. As environment ρ, we take identity. Now $\rho \models \Gamma$, since $\rho(x) = x \in \mathcal{P}_i^\tau$ for all $x :_i \tau \in \Gamma$ by the second item of Lemma 6. From Lemma 8 we obtain $[\![\Gamma \vdash M :_i \sigma]\!]_\rho^{\mathcal{T}} = M \in \mathcal{P}_i^\sigma$. Using the first item of Lemma 6 we find $\mathcal{P}_i^\sigma \subseteq \mathcal{SN}$. Hence M is $\beta\eta$-strongly normalising. □

Theorem 2 (Confluence of $\beta\eta$ on typable terms). *The $\beta\eta$-reduction is confluent on typable terms.*

Proof. We apply Newman's Lemma [11, Theorem 1.2.1]. Since, by Theorem 1, $\beta\eta$ is strongly normalising, it is sufficient to verify that $\beta\eta$-reduction is locally confluent. This is straightforward. □

5 Term Model

In this section, we construct the term model $\mathcal{T}/{=_{\beta\eta}}$ from the term applicative structure $\mathcal{T}$ by quotienting over $\beta\eta$ conversion. We prove that $\mathcal{T}/{=_{\beta\eta}}$ is extensional. This gives us our first completeness result, i.e. there exists a Henkin model for reactive programs satisfying exactly the theory of $\beta\eta$-conversion.

We write $[M]$ to denote the set of terms that are $\beta\eta$ convertible to M.

Definition 11 (Term model). *For each type σ and index i, let $(\mathcal{T}/{=_{\beta\eta}})_i^\sigma = \{[M] \mid M \in \mathcal{T}_i^\sigma\}$. We define the applicative structure $\mathcal{T}/{=_{\beta\eta}}$ as*

$$\langle\{(\mathcal{T}/{=_{\beta\eta}})_i^\sigma\}, \{\delta_{ij}^\sigma\}, \{\mathit{hd}_i^\sigma, \mathit{tl}_i^\sigma, \mathit{await}_i^\sigma, \mathit{app}_i^{\sigma\tau}\}\rangle$$

with $\mathit{app}_i^{\sigma\tau}([M],[N]) = [MN]$, $\mathit{hd}_i^\sigma([M]) = [\mathsf{hd}(M)]$, $\mathit{await}_i^\sigma([M]) = [\mathsf{await}(M)]$ and $\mathit{tl}_i^\sigma([M]) = [\mathsf{tl}(M)]$. We take set inclusion for δ_{ij}^σ when $i \le j$, and define as meaning function $[\![\Gamma \vdash M :_i \sigma]\!]_\rho = [M[x_1{:=}N_1, \ldots, x_n{:=}N_n]]$ where $\rho(x_i) = [N_i]$ for all $1 \le i \le n$ and all x_i occur in Γ.

We write $\mathsf{size}(M)$, $\mathsf{size}(\sigma)$ and $\mathsf{size}(\Gamma)$ for the number of symbols of M, σ and all types in Γ, respectively.

Lemma 9 (Shape of β-normal forms). *Let M be a typable β-normal form. Then, M is of the form $\lambda x_1{:}\sigma_1 \ldots x_n{:}\sigma_n.N$ where N satisfies one of the clauses:*

1. *N is of the form $\mathsf{cons}(P,Q)$ where P and Q are in β-normal form.*
2. *N is of the form $\circ P$ where P is in β-normal form.*

3. N belongs to the grammar:

$$\mathsf{X} \ni X := x \mid XP \mid \mathsf{hd}(X) \mid \mathsf{tl}(X) \mid \mathsf{await}(X)$$

where P is in β-normal form and $x \in \mathsf{V}$. Instead of a variable X ranging over X we may occasionally write $X[P_1, \ldots, P_n]$ to list explicitly all arguments P used in the construction of X. For such X we have that if $\Gamma \vdash X :_j \tau$ then $\mathsf{size}(\tau) \leq \mathsf{size}(\Gamma)$ and $\mathsf{size}(P_i) < \mathsf{size}(\Gamma)$ for all $1 \leq i \leq n$.

The previous lemma can be proved by induction on the derivation.

Note that weak extensionality of the term model is a direct consequence of the η-rule. To prove extensionality we need more, namely that λ^{RP} is confluent and strongly normalising. The assumption made for Γ_∞ in Notation 1 is important in the proof of extensionality on $\sigma \to \tau$, as it allows us to pick an $x \in \mathcal{T}_i^\sigma$.

Lemma 10 (Extensionality of term model). *The applicative structure for the term model is extensional.*

Proof. We only prove extensionality on $\mathsf{S}(\sigma)$ and leave the other cases to the reader. Let $M, N \in \mathcal{T}_i^{\mathsf{S}(\sigma)}$ be in $\beta\eta$-normal form. We now analyse the shape of these $\beta\eta$-normal forms. Suppose that $\mathsf{hd}(\mathsf{tl}^n(M)) =_{\beta\eta} \mathsf{hd}(\mathsf{tl}^n(N))$ for all $n \in \mathbb{N}$. We prove that $M =_\eta N$ by induction on the number of cons that appear in M and N. We distinguish cases depending on the shape of M and N by Lemma 9.

1. Case $M, N \in \mathsf{X}$. Then $\mathsf{hd}(M)$ and $\mathsf{hd}(N)$ are in $\beta\eta$-normal form. By Confluence of $\beta\eta$ (Theorem 2) we find $\mathsf{hd}(M) = \mathsf{hd}(N)$. Hence $M = N$.
2. Case $M = \mathsf{cons}(P, Q)$ and $N = \mathsf{cons}(P', Q')$. We have $P =_{\beta\eta} \mathsf{hd}(M) =_{\beta\eta} \mathsf{hd}(N) =_{\beta\eta} P'$. Since P and P' are in $\beta\eta$-normal form, we have $P = P'$ by confluence of $\beta\eta$. We also have $\mathsf{hd}(\mathsf{tl}^n(Q)) =_{\beta\eta} \mathsf{hd}(\mathsf{tl}^n(Q'))$ for all $n \in \mathbb{N}$. Since Q and Q' have fewer number of cons than M and N, $Q =_\eta Q'$ by induction hypothesis.
3. Case $M = \mathsf{cons}(P, Q)$ and $N \in \mathsf{X}$. Then P, $\mathsf{hd}(N)$ and $\mathsf{tl}(N)$ are all in $\beta\eta$-normal form. We get $P =_{\beta\eta} \mathsf{hd}(M) =_{\beta\eta} \mathsf{hd}(N)$. By confluence of $\beta\eta$ we conclude $P = \mathsf{hd}(N)$. We also have $\mathsf{hd}(\mathsf{tl}^n(Q)) =_{\beta\eta} \mathsf{hd}(\mathsf{tl}^n(\mathsf{tl}(N)))$ for all $n \in \mathbb{N}$. Applying the induction hypothesis to Q and $\mathsf{tl}(N)$ we get $Q =_\eta \mathsf{tl}(N)$. □

Theorem 3 (Completeness for Henkin models of reactive programs). *There exists a Henkin model for reactive programs satisfying exactly the theory of $\beta\eta$-conversion.*

Proof. It is routine to show that the meaning function of $\mathcal{T}/{=_{\beta\eta}}$ satisfies the environment condition. The term model trivially satisfies the theory of $\beta\eta$-conversion, i.e., $M =_{\beta\eta} N$ iff $\mathcal{T}/{=_{\beta\eta}} \models \Gamma \vdash M = N ;_i \sigma$ whenever $\Gamma \vdash M, N :_i \sigma$. □

6 Ultrametric Model for Reactive Programs

In this section, we present the ultrametric model of [1] as a Henkin Model for reactive programs.

A *complete 1-bounded ultrametric space* is a tuple (U, d_U), where U is a set and the distance function $d_U : U \times U \to [0,1]$ satisfies: 1. $d_U(u,v) = 0$ iff $u = v$, 2. $d_U(u,v) = d_U(v,u)$, 3. $d_U(u,z) \le \max(d_U(u,v), d_U(v,z))$, 4. every Cauchy sequence in U has a limit in X. A function $f : U \to V$ between ultrametric spaces is *non-expansive* if $d_V(f(u_1), f(u_2)) \le d_U(u_1, u_2)$. It is well-known that the complete 1-bounded ultrametric spaces and nonexpansive functions form a cartesian-closed category. The *shrink functor* $\frac{1}{2}$ maps (U, d_U) to $(U, \frac{1}{2}d_U)$ and a non-expansive function $f \in U \to V$ to the non-expansive function $\frac{1}{2}(f) \in \frac{1}{2}U \to \frac{1}{2}V$ where $\frac{1}{2}(f)(u) = f(u)$.

Definition 12 (Ultrametric applicative structure). *An* ultrametric applicative structure *is an applicative structure*

$$\mathcal{U} = \langle \{\mathcal{U}_i^\sigma\}, \{\delta_{ij}^\sigma\}, \{\mathit{hd}_i^\sigma, \mathit{tl}_i^\sigma, \mathit{await}_i^\sigma, \mathit{app}_i^{\sigma\tau}\} \rangle$$

1. $\mathcal{U}_i^\sigma = \frac{1}{2}^i \mathcal{U}_0^\sigma$ *where* $\mathcal{U}_0^\sigma$ *is defined by induction on* σ*:*
   (a) $\mathcal{U}_0^{\mathsf{b}}$ *is some ultrametric space* (U, d_U).
   (b) $\mathcal{U}_0^{\sigma \to \tau}$ *is the set of nonexpansive maps from* $\mathcal{U}_0^\sigma$ *to* $\mathcal{U}_0^\tau$*, equipped with the supremum metric:* $d_{\mathcal{U}_0^\sigma \to \mathcal{U}_0^\tau}(f,g) = \sup\{d_{\mathcal{U}_0^\tau}(f(x), g(x)) \mid x \in \mathcal{U}_0^\sigma\}$.
   (c) $\mathcal{U}_0^{\bullet\sigma} = \frac{1}{2}\mathcal{U}_0^\sigma$.
   (d) $\mathcal{U}_0^{\mathsf{S}(\sigma)}$ *is the set of total functions from* $\mathbb{N}$ *to* $\mathcal{U}_0^\sigma$*, equipped with the stream metric:* $d_{\mathcal{U}_0^{\mathsf{S}(\sigma)}}(f,g) = \sup\{\frac{1}{2}^n d_{\mathcal{U}_0^\sigma}(f(n), g(n)) \mid n \in \mathbb{N}\}$.
2. $\delta_{ij}^\sigma \in \mathcal{U}_i^\sigma \to \frac{1}{2}^{j-i}\mathcal{U}_j^\sigma$ *is defined by* $\delta_{ij}^\sigma(u) = u$.
3. $\mathit{app}_i^{\sigma\tau}(f,a) = f(a)$, $\mathit{await}_i^\sigma(a) = a$, $\mathit{hd}_i^\sigma(f) = f(0)$ *and* $\mathit{tl}_i^\sigma(f)(n) = f(n+1)$ *for all* $n \ge 0$.

It is easy to see that an ultrametric applicative structure is extensional. We define $\mathit{cons}_i^\sigma(a, f) = g$ where $g(0) = a$ and $g(n+1) = f(n)$ for $n \ge 0$.

Lemma 11 (Ultrametric model). *Let* $\rho \models \Gamma$*. The ultrametric applicative structure together with the meaning function defined as*

- $[\![\Gamma \vdash x :_j \sigma]\!]_\rho = \delta_{ij}^\sigma(\rho(x))$ *if* $x{:}_i\sigma \in \Gamma$,
- $[\![\Gamma \vdash MN :_i \tau]\!]_\rho = \mathit{app}_i^{\sigma\tau}([\![\Gamma \vdash M :_i \sigma \to \tau]\!]_\rho, [\![\Gamma \vdash N :_i \tau]\!]_\rho)$,
- $[\![\Gamma \vdash \lambda x{:}\sigma.M :_i \sigma \to \tau]\!]_\rho = \{(a, [\![\Gamma, x :_i \sigma \vdash M :_i \tau]\!]_{\rho[x:=a]}) \mid a \in \mathcal{U}_i^\sigma\}$,
- $[\![\Gamma \vdash \mathsf{await}(M) :_{i+1} \sigma]\!]_\rho = [\![\Gamma \vdash M :_i \bullet\sigma]\!]_\rho$,
- $[\![\Gamma \vdash \circ(M) :_i \bullet\sigma]\!]_\rho = [\![\Gamma \vdash M :_{i+1} \sigma]\!]_\rho$,
- $[\![\Gamma \vdash \mathsf{hd}(M) :_i \mathsf{S}(\sigma)]\!]_\rho = \mathit{hd}_i^\sigma([\![\Gamma \vdash M :_i \mathsf{S}(\sigma)]\!]_\rho)$,
- $[\![\Gamma \vdash \mathsf{tl}(M) :_{i+1} \mathsf{S}(\sigma)]\!]_\rho = \mathit{tl}_i^\sigma([\![\Gamma \vdash M :_i \mathsf{S}(\sigma)]\!]_\rho)$,
- $[\![\Gamma \vdash \mathsf{cons}(M,N) :_i \mathsf{S}(\sigma)]\!]_\rho = \mathit{cons}_i^\sigma([\![\Gamma \vdash M :_i \sigma]\!]_\rho, [\![\Gamma \vdash N :_{i+1} \sigma]\!]_\rho)$,

is a Henkin model for reactive programs called the ultrametric model.

7 Metric on the Term Model

In this section, we define an ultrametric d on the term model for which the equivalence classes of terms of type $\sigma \to \tau$ are *non-expansive*, i.e. for M of type $\sigma \to \tau$ we have that $d([MP],[MQ]) \leq d([P],[Q])$.

We recall the notions of depth, truncation and metric on terms of [12]. The depth of N in argument positions in $\mathsf{cons}(M,N)$ and $\circ(N)$ is counted one deeper than the depth of the terms $\mathsf{cons}(M,N)$ and $\circ(N)$ themselves. To define truncation, we extend the syntax with a constant $\perp$.

Definition 13 (Truncation). *The truncation of M at depth n, denoted by M^n, is defined by induction as follows.*

$$
\begin{aligned}
M^0 &= \perp & x^{n+1} &= x \\
(\lambda x.M)^{n+1} &= \lambda x.M^{n+1} & (M\ N)^{n+1} &= (M^{n+1}\ N^{n+1}) \\
(\circ(M))^{n+1} &= \circ(M^n) & (\mathsf{await}(M))^{n+1} &= \mathsf{await}(M^{n+1}) \\
(\mathsf{hd}(M))^{n+1} &= \mathsf{hd}(M^{n+1}) & (\mathsf{tl}(M))^{n+1} &= \mathsf{tl}(M^{n+1}) \\
(\mathsf{cons}(M,N))^{n+1} &= \mathsf{cons}(M^{n+1}, N^n)
\end{aligned}
$$

Definition 14 (Metric on terms). *Define $d : \mathsf{P} \times \mathsf{P} \to [0,1]$ as $d(M,N) = 0$, if $M = N$ and $d(M,N) = 2^{-m}$ otherwise, where $m = \max\{n \in \mathbb{N} \mid M^n = N^n\}$.*

Note that d is not invariant under $\beta\eta$-conversion. The metric on equivalence classes defined by $d([M],[N]) = d(\mathsf{nf}_{\beta\eta}(M), \mathsf{nf}_{\beta\eta}(N))$ is not the right one since there may be elements in $(\mathcal{T}/{=_{\beta\eta}})_i^{\sigma\to\tau}$ that are not non-expansive. For example, $d([MP],[MQ]) = 1$ and $d([P],[Q]) = 1/2$ if $M = \lambda x{:}\mathsf{S}(\sigma).\mathsf{cons}(\mathsf{hd}(y),\mathsf{tl}(x))$, $P = y$ and $Q = z$. The distance between $[MP] = [y]$ and $[MQ] = [\mathsf{cons}(\mathsf{hd}(y),\mathsf{tl}(z))]$ should be $\frac{1}{2}$ and not 1 for $[M]$ to be non-expansive.

To define the right metric, we introduce the notions of infinite term and extensional long normal form. The notion of extensional long normal form does not coincide with the notion of eta long normal form. In order to define the notion of extensional long normal, we express a function f on natural numbers as an infinite term $\mathsf{cons}(M_1, \mathsf{cons}(M_2, \ldots))$ where M_1 corresponds to $f(1)$, M_2 to $f(2)$, etc. For example, the extensional long normal form of a variable x of type $\mathsf{S}(\sigma)$ is the infinite term $\mathsf{cons}(\mathsf{hd}(x), \mathsf{cons}(\mathsf{hd}(\mathsf{tl}(x)), \ldots))$.

Definition 15 (Infinitary terms). *We define the set P^∞ of infinitary terms as the metric completion of (P, d).*

Definition 16 (Extensional long normal form). *Let $\Gamma \vdash M :_i \sigma$. The (extensional) long normal form of M is a term in P^∞ denoted by $\mathsf{L}(M)$ and defined as follows. If M is not in β-normal form, we define $\mathsf{L}(M) = \mathsf{L}(\mathsf{nf}_\beta(M))$. If M is in β-normal form then we define it by induction on the pair $(\mathsf{size}(\Gamma) + \mathsf{size}(\sigma), \mathsf{size}(M))$ with the lexicographic order as follows.*

1. *Case σ is a base type. Then $M = X[M_1, \ldots, M_n] \in \mathsf{X}$. We define $\mathsf{L}(M) = X[\mathsf{L}(M_1), \ldots, \mathsf{L}(M_n)]$.*

2. *Case* σ *is* $\sigma_1 \to \sigma_2$. *If* $M = \lambda x.N$ *then we define* $\mathsf{L}(M) = \lambda x.\mathsf{L}(N)$.
   Otherwise, $M \in \mathsf{X}$ *and we define* $\mathsf{L}(M) = \lambda y.\mathsf{L}(M\ y)$.
3. *Case* σ *is* $\bullet\tau$. *If* $M = \circ(P)$, *we define* $\mathsf{L}(M) = \circ(\mathsf{L}(P))$.
   Otherwise, $M \in \mathsf{X}$ *and we define* $\mathsf{L}(M) = \circ(\mathsf{L}(\mathsf{await}(M)))$.
4. *Case* $\sigma = \mathsf{S}(\tau)$. *Either* $M = \mathsf{cons}(P, Q)$ *and we put* $\mathsf{L}(M) = \mathsf{cons}(\mathsf{L}(P), \mathsf{L}(Q))$, *or* $M \in \mathsf{X}$ *and we define* $\mathsf{L}(M) = \mathsf{cons}(\mathsf{L}(\mathsf{hd}(M)), \mathsf{cons}(\mathsf{L}(\mathsf{hd}(\mathsf{tl}(M))), \ldots))$.

Lemma 12. *1. Let* M *be in* β*-normal form such that* $\Gamma \vdash M :_i \sigma$. *If* $M \to_\eta N$, *then* $\mathsf{L}(M) = \mathsf{L}(N)$.
2. If $\Gamma \vdash M :_i \sigma$ *and* $M =_{\beta\eta} N$, *then* $\mathsf{L}(M) = \mathsf{L}(N)$.

The first statement is proved by induction on $(\mathsf{size}(\Gamma) + \mathsf{size}(\sigma), \mathsf{size}(M))$. The second is proved using confluence, strong normalisation and the first one.

Definition 17 (Metric on equivalence classes of typable terms). *We define a metric* $d : (\mathcal{T}/{=_{\beta\eta}})^\sigma_i \times (\mathcal{T}/{=_{\beta\eta}})^\sigma_i \to [0, 1]$ *as* $d([M], [N]) = d(\mathsf{L}(M), \mathsf{L}(N))$.

We define $\Gamma \vdash^n M :_i \sigma$ by adding to the typing rules of Definition 3 the rule $\Gamma \vdash^n \bot :_i \sigma$ if $n \geq i$ for all $i \in \mathbb{N}$ and all types σ. It is easy to show that this typing rules satisfy subject reduction, confluence and strong normalisation. The notion of extensional long normal form is extended to terms with $\bot$ and typable in $\vdash^n$. The follow lemma is proved by induction on the derivation.

Lemma 13. *If* $\Gamma \vdash^{n+i} M :_i \sigma$ *then the* $\bot$*'s in* M *all occur at depth greater or equal than* n.

The following three lemmas are proved by induction on $(\mathsf{size}(\Gamma) + \mathsf{size}(\sigma), n)$.

Lemma 14. *If* $\Gamma \vdash M :_i \sigma$ *then* $\Gamma \vdash^{n+i} (\mathsf{L}(M))^n :_i \sigma$.

Lemma 15. *Let* M *be in* β*-normal form such that* $\Gamma \vdash M :_i \sigma$. *Then, there exists* N *in* β*-normal form such that* $N \twoheadrightarrow_\eta M$ *and* $(\mathsf{L}(M))^n = N^n$.

Lemma 16. *Let* M, N *be in* β*-normal form such that* $\Gamma \vdash M, N :_i \sigma$. *If* $M^n = N^n$ *then* $(\mathsf{L}(M))^n = (\mathsf{L}(N))^n$.

We write $M \prec N$ if M is the result of replacing some subterms of N by $\bot$.

Lemma 17. *If* $M \prec N$ *and* $M \to_\beta M'$ *then* $M' \prec N'$ *and* $N \to_\beta N'$ *for some* N'.

Theorem 4 (Non-expansiveness). *Let* $M \in \mathcal{T}^{\sigma\to\tau}_i$ *and* $P, Q \in \mathcal{T}^\sigma_i$. *Then:*

1. $(\mathsf{L}(MP))^n = (\mathsf{L}(M(\mathsf{L}(P))^n))^n$. *2.* $d([MP], [MQ]) \leq d([P], [Q])$.

Proof. (1) Assume M, P are in β-normal form. By Lemma 15, there exists P' in β-normal form such that $P' \twoheadrightarrow_\eta P$ and $(\mathsf{L}(P))^n = (P')^n$. Then, $M(\mathsf{L}(P))^n \prec MP'$. By Lemma 14, $\Gamma \vdash^{n+i} (\mathsf{L}(P))^n :_i \sigma$ and hence, $\Gamma \vdash^{n+i} M(\mathsf{L}(P))^n :_i \sigma$. By Subject reduction, $\Gamma \vdash^{n+i} N :_i \sigma$ where $N = \mathsf{nf}_\beta(M(\mathsf{L}(P))^n)$. By Lemma 17, there exists N' such that $N \prec N'$ and $MP' \twoheadrightarrow_\beta N'$. By Lemma 13, the $\bot$'s in N occur at depth greater than n. Hence, $N^n = (N')^n = (\mathsf{nf}_\beta(MP'))^n$ since N is in β-normal form. By Lemma 16, $(\mathsf{L}(N))^n = (\mathsf{L}(\mathsf{nf}_\beta(MP')))^n$. By Lemma 12, $(\mathsf{L}(MP))^n = (\mathsf{L}(M(\mathsf{L}(P))^n))^n$.
(2) Suppose $(\mathsf{L}(P))^n = (\mathsf{L}(Q))^n$. By the first part, $(\mathsf{L}(MP))^n = (\mathsf{L}(M(\mathsf{L}(P))^n))^n = (\mathsf{L}(M(\mathsf{L}(Q))^n))^n = (\mathsf{L}(MQ))^n$. □

Theorem 5 (Well-behaviour of metric)

1. $d([M],[N]) = \sup\{d([MP],[NP]) \mid P \in \mathcal{T}_i^\sigma\}$.
2. $d([M],[N]) = \frac{1}{2} d([\mathsf{await}(M)],[\mathsf{await}(N)])$.
3. $d([M],[N]) = \sup\{\frac{1}{2}^n d([\mathsf{hd}(\mathsf{tl}^n(M))],[\mathsf{hd}(\mathsf{tl}^n(N))]) \mid n \in \mathbb{N}\}$.

The proof of this theorem is similar to the one for Theorem 4.

8 Completeness for the Ultrametric Model

In this section we show that $\beta\eta$-conversion captures semantic equality between reactive programs, i.e. two terms typable in the calculus of reactive programs are $\beta\eta$-convertible if and only if they have the same interpretation in the ultrametric model. Our proof follows closely the proof of completeness of the simply typed lambda calculus given in [7, Section 8].

An ultrametric frame is an ultrametric applicative structure where $\mathcal{U}_i^{\sigma\to\tau}$ is a subset of the set of non-expansive maps from $\mathcal{U}_i^\sigma$ to $\mathcal{U}_i^\tau$.

Theorem 6. *There exists an ultrametric frame isomorphic to the term model.*

Proof. We define $\mathcal{U}_i^\sigma = \{\phi_i^\sigma([M]) \mid M \in \mathcal{T}_i^\sigma\}$ where ϕ_i^σ is a function from $(\mathcal{T}/=_{\beta\eta})_i^\sigma$ to $\mathcal{U}_i^\sigma$ defined by induction on σ.

$$\phi_i^{\mathsf{b}}([M]) = [M] \qquad \phi_i^{\sigma\to\tau}([M]) = \{(a, \phi_i^\tau([M(\phi_i^\sigma)^{-1}(a)])) \mid a \in \mathcal{U}_i^\sigma\}$$
$$\phi_i^{\bullet\sigma}([M]) = \phi_{i+1}^\sigma([\mathsf{await}(M)]) \qquad \phi_i^{\mathsf{S}(\sigma)}([M]) = \{(n, \phi_i^\sigma([\mathsf{hd}(\mathsf{tl}^n(a))])) \mid n \in \mathbb{N}\}$$

Surjectivity of ϕ_i^σ is trivial. Injectivity follows by induction on σ using Lemma 10. That ϕ_i^σ and its inverse are non-expansive follows by induction on σ using Theorem 5. It remains to prove that $\phi_i^{\sigma\to\tau}([M])$ is non-expansive. This follows from Theorem 4 and the fact that ϕ_i^τ and the inverse of ϕ_i^σ are non-expansive. □

Definition 18 (Embedding). *Let (U, d_U), (V, d_V) be ultrametric spaces. An embedding from V to U is a pair (ϕ, ψ) of non-expansive maps $\phi : V \to U$ and $\psi : U \to V$ such that $\psi \circ \phi = id_V$.*

Lemma 18 (Partial Surjections for ultrametric spaces). *Let $\mathcal{U}$ be an ultrametric applicative structure and $\mathcal{V}$ be an ultrametric frame. If there exists an embedding from $\mathcal{V}_i^{\mathsf{b}}$ to $\mathcal{U}_i^{\mathsf{b}}$ for each constant type b and $i \in \mathbb{N}$, then there exists a partial logical surjective function $\mathcal{R}$ from $\mathcal{U}$ to $\mathcal{V}$ and two families of non-expansive maps ϕ_i^σ, ψ_i^σ such that*

1. $\mathcal{R}_i^\sigma(\phi_i^\sigma(v), v)$ for all $v \in \mathcal{V}_i^\sigma$. 2. $\mathcal{R}_i^\sigma(u, \psi_i^\sigma(u))$ for all $u \in dom(\mathcal{R}_i^\sigma)$.

Proof. By induction on σ. We do only the case $\mathsf{S}(\sigma)$ for streams. Then we define:

$$\phi_i^{\mathsf{S}(\sigma)}(v) = \{(n, \phi_{i+n}^\sigma(v(n))) \mid n \in \mathbb{N}\} \quad \forall v \in \mathcal{V}_i^{\mathsf{S}(\sigma)}$$
$$\psi_i^{\mathsf{S}(\sigma)}(u) = \{(n, \psi_{i+n}^\sigma(u(n))) \mid n \in \mathbb{N}\} \quad \forall u \in dom(\mathcal{R}_i^{\mathsf{S}(\sigma)})$$
$$\mathcal{R}_i^{\mathsf{S}(\sigma)} = \{(f, g) \mid (f(n), g(n)) \in \mathcal{R}_{i+n}^\sigma\}$$

Statements (1) and (2) follow from induction hypothesis. Surjectivity follows from (1). That $\mathcal{R}_i^{\mathsf{S}(\sigma)}$ is a function follows by extensionality. Surjectivity plays a role only in the arrow type case, for proving that $\mathcal{R}_i^{\sigma\to\tau}$ is a function. □

Theorem 7 (Completeness for the ultrametric model). *Suppose there exists an embedding from* $(\mathcal{T}/{=_{\beta\eta}})_i^{\mathsf{b}}$ *to* $\mathcal{U}_i^{\mathsf{b}}$ *for each constant type* b *and* $i \in \mathbb{N}$. *Let* $\Gamma \vdash M, N :_i \sigma$. *Then,* $M =_{\beta\eta} N$ *if and only if* $\mathcal{U} \models \Gamma \vdash M = N{:_i}\sigma$.

Proof. ($\Rightarrow$) [1, Theorem 4]. Alternatively, it also follows from Soundness for Henkin Models (Lemma 3) and the fact that the ultrametric model is a Henkin one (Lemma 11). ($\Leftarrow$). Let $\mathcal{V}$ be the ultrametric frame isomorphic to $\mathcal{T}/{=_{\beta\eta}}$ of Theorem 6. By Lemma 18, there exists a logical partial function from $\mathcal{U}$ to $\mathcal{V} \sim \mathcal{T}/{=_{\beta\eta}}$. By Lemma 11 and Theorem 3, the ultrametric model and the term model are Henkin models. Hence, $\mathsf{Th}(\mathcal{U}) \subseteq \mathsf{Th}(\mathcal{T}/{=_{\beta\eta}})$ by Lemma 5. □

9 Conclusions and Future Work

As a natural sequel, we are currently studying the theory induced by the ultrametric model for the typed lambda calculus of reactive programs extended with the fixpoint operator of [1].

Statman's 1-Section Theorem [13, 14, 15] generalises Friedman's result by giving necessary and sufficient conditions for a model to satisfy completeness of $\beta\eta$-conversion on terms typable in the simply typed lambda calculus. It will be interesting to prove a similar result to Statman's 1-Section Theorem for the typed lambda calculus of reactive programs.

Our step-indexed applicative structures are in fact Kripke lambda models over the partial order $(\mathbb{N}, \leq)$ in the terminology of Mitchell and Moggi [16]. By using natural numbers, the additional operators for streams such as tl_i^σ can move from time i to $i+1$. However, the notion of Henkin model for reactive programs is not a particular case of Kripke model as defined in [16]. Our environment and meaning function do not have the natural number i as argument since that information is provided by the judgement $\Gamma \vdash M :_i \sigma$.

The notion of step indexed logical relation for recursive types in [17, 18, 19] use the index in a different way from ours. In our definition of logical relation on the type $\sigma \to \tau$ we quantify over $j \geq i$ for some given i. While in the definition of logical relation for recursive types, the quantification is over $j \leq i$ for some given i. The choice to quantify over $j \geq i$ is essential for our proofs to go through. The logical predicate of strongly normalising terms should satisfy $\mathcal{P}_i^{\sigma\to\tau} \subseteq \mathcal{P}_j^{\sigma\to\tau}$ for $i \leq j$ and similarly, the logical surjective function defined in Lemma 18 should satisfy $\mathcal{R}_i^{\sigma\to\tau} \subseteq \mathcal{R}_j^{\sigma\to\tau}$ for $i \leq j$. This holds trivially when we quantify over $j \geq i$ but it does not hold if the quantification is done reversing the order.

Our step-indexed notion of applicative structure can be described as families of *covariant* functors from $(\mathbb{N}, \leq)$ to the category **Set** of sets and functions. The topos of trees [20] that Birkedal and coworkers use for step-indexing models of various programming languages consists of the *contravariant* functors from $(\mathbb{N}, \leq)$ to **Set**. These are functors from $(\mathbb{N}, \leq)^{op}$, that is $(\mathbb{N}, \geq)$, to **Set**.

Acknowledgements. We thank Lars Birkedal for asking us a question during ICFP 2012 that led to this paper.

References

[1] Krishnaswami, N.R., Benton, N.: Ultrametric semantics of reactive programs. In: LICS, pp. 257–266 (2011)
[2] Krishnaswami, N.R., Benton, N.: A semantic model for graphical user interfaces. In: ICFP, pp. 45–57 (2011)
[3] Birkedal, L., Schwinghammer, J., Støvring, K.: A metric model of lambda calculus with guarded recursion. Presented at FICS 2010 (2010)
[4] Escardó, M.: A metric model of PCF. Presented at the Workshop on Realizability Semantics and Applications, June 30-July 1 (1999)
[5] Friedman, H.: Equality between functionals. In: Logic Colloquium. Springer Lecture Notes, vol. 453, pp. 22–37 (1975)
[6] Plotkin, G.: Notes on completeness of the full continuous hierarchy (1982) (unpublished manuscript)
[7] Mitchell, J.C.: Foundations for programming languages. Foundation of Computing Series. MIT Press (1996)
[8] Nederpelt, R.P.: Strong normalization in a typed lambda calculus. Ph.D. dissertation, Technische Universiteit Eindhoven (1973)
[9] Klop, J.W.: Combinatory reduction systems. Ph.D. dissertation, Rijksuniversiteit Utrecht (1980)
[10] Barendregt, H.P.: The Lambda Calculus: Its Syntax and Semantics, 2nd edn. North-Holland, Amsterdam (1984)
[11] Terese (ed.): Term Rewriting Systems. Cambridge Tracts in Theoretical Computer Science, vol. 55. Cambridge University Press (2003)
[12] Severi, P., de Vries, F.-J.: Pure type systems with corecursion on streams: from finite to infinitary normalisation. In: ICFP, pp. 141–152 (2012)
[13] Statman, R.: Completeness, invariance and λ-definability. Journal of Symbolic Logic (1982)
[14] Statman, R.: Equality between functionals, revisted. In: Harvey Friedman's Research on the Foundations of Mathematics, pp. 331–338 (1985)
[15] Riecke, J.G.: Statman's 1-section theorem. Inf. Comput. 116(2), 294–303 (1995)
[16] Mitchell, J.C., Moggi, E.: Kripke-style models for typed lambda calculus. Ann. Pure Appl. Logic 51(1-2), 99–124 (1991)
[17] Ahmed, A.: Step-indexed syntactic logical relations for recursive and quantified types. In: Sestoft, P. (ed.) ESOP 2006. LNCS, vol. 3924, pp. 69–83. Springer, Heidelberg (2006)
[18] Birkedal, L., Reus, B., Schwinghammer, J., Støvring, K., Thamsborg, J., Yang, H.: Step-indexed Kripke models over recursive worlds. In: POPL, pp. 119–132 (2011)
[19] Dreyer, D., Ahmed, A., Birkedal, L.: Logical step-indexed logical relations. Logical Methods in Computer Science 7(2) (2011)
[20] Birkedal, L., Møgelberg, R.E., Schwinghammer, J., Støvring, K.: First steps in synthetic guarded domain theory: Step-indexing in the topos of trees. In: LICS, pp. 55–64 (2011)

A Constructive Model of Uniform Continuity

Chuangjie Xu and Martín Escardó

University of Birmingham, UK

Abstract. We construct a continuous model of Gödel's system T and its logic HA^ω in which all functions from the Cantor space $\mathbf{2}^{\mathbb{N}}$ to the natural numbers are uniformly continuous. Our development is constructive, and has been carried out in intensional type theory in Agda notation, so that, in particular, we can compute moduli of uniform continuity of T-definable functions $\mathbf{2}^{\mathbb{N}} \to \mathbb{N}$. Moreover, the model has a continuous *Fan functional* of type $(\mathbf{2}^{\mathbb{N}} \to \mathbb{N}) \to \mathbb{N}$ that calculates moduli of uniform continuity. We work with sheaves, and with a full subcategory of *concrete* sheaves that can be presented as sets with structure, which can be regarded as spaces, and whose natural transformations can be regarded as continuous maps.

Keywords. Constructive mathematics, topological models, uniform continuity, Fan functional, intuitionistic type theory, topos theory, sheaves, HA^ω, Gödel's system T.

1 Introduction

Gödel's system T has a well-known topological models in which all integer-valued functions on the Cantor space are uniformly continuous:

$$\forall f\colon \mathbf{2}^{\mathbb{N}} \to \mathbb{N}.\ \exists n \in \mathbb{N}.\ \forall \alpha, \beta \in \mathbf{2}^{\mathbb{N}}.\ \alpha =_n \beta \implies f(\alpha) = f(\beta),$$

where $\alpha =_n \beta$ means $\forall i < n.\ \alpha_i = \beta_i$. These models include Kleene–Kreisel functionals [19], compactly generated spaces [19], limit spaces [20], equilogical spaces [3], sequential spaces [12], and QCB spaces [12]. However, even though these models are introduced for the purposes of computability theory, they are developed within a classical meta-theory.

The purpose of this paper is to develop such a topological model in a weak constructive meta-theory, without explicit reference to computability, but with computability in mind. In fact, we conjecture that our model is classically equivalent to the model of Kleene–Kreisel spaces. Our continuous model of system T consists of certain C-spaces, which can be seen as sheaves (see below), and is developed in Section 2. Like the above models, this model has a *Fan functional* of type $(\mathbf{2}^{\mathbb{N}} \to \mathbb{N}) \to \mathbb{N}$ that continuously calculates moduli of uniform continuity (Section 3). Importantly, we do not rely on the Fan Theorem [5] or any such principle to construct the Fan functional. In particular, we recover the well-known fact that the T-definable functions $\mathbf{2}^{\mathbb{N}} \to \mathbb{N}$ are uniformly continuous [5] by defining a logical relation between the set-theoretical and continuous models (Section 4). To model the logic HA^ω extended with the above uniform continuity

M. Hasegawa (Ed.): TLCA 2013, LNCS 7941, pp. 236–249, 2013.

axiom, we realize the logical operations by continuous functions in our model (Section 5). We also discuss how our results can be extended to give a continuous model of dependent types (Section 7).

As mentioned above, we develop this in a weak constructive meta-language, which does not incorporate constructively contentious axioms such as continuity principles, Fan Theorem, Bar Induction, or Church's Thesis [5]. In this presentation we deliberately leave the details of the meta-language unexplained, relying on the readers' ability to recognize constructive arguments. One possible formal constructive meta-language for our development is intensional Martin-Löf type theory (MLTT), and, in fact, we have developed the main results of this paper in Agda [6] (Section 6). Because MLTT has a computational interpretation, our model can be used to compute moduli of uniform continuity of system T definable functions $\mathbf{2}^{\mathbb{N}} \to \mathbb{N}$. More importantly, our model can be used to extract computational content from proofs in HA^{ω} extended with the above uniform continuity axiom.

Our model is a sheaf topos, with a full subcategory of *concrete* sheaves [2] that can be presented as sets with structure, which can be regarded as spaces, and whose natural transformations can be regarded as continuous maps. The underlying category of our site is the monoid of uniformly continuous endomaps of the Cantor space, with a natural coverage consisting of families of concatenation maps. The coverage axiom amounts precisely to the uniform continuity of the elements of our monoid. Our development of this topos is fairly standard, but we have taken care of making sure the arguments are presented in a form suitable for a formalization in a predicative type theory, and this is one of the main contributions of this work.

Our work builds upon Johnstone's paper *On a topological topos* (1979), Fourman's papers *Continuous truth* and *Notions of choice sequence* (1982), van der Hoeven and Moerdijk's paper *Sheaf models for choice sequences* (1984), Bauer and Simpson's unpublished work *Continuity begets continuity* (2006), and Coquand and Jaber's paper *A note on forcing and type theory* (2010) and *A computational interpretation of forcing in type theory* (2012).

Johnstone, Fourman, van der Hoeven and Moerdijk's work with sheaf toposes over different sites. Johnstone's site is the monoid of continuous endo-functions of the one-point compactification of the discrete natural numbers with the canonical Grothendieck topology. A full subcategory of Johnstone's topological topos is that of sequential topological spaces, which is cartesian closed. A bigger full subcategory, which is locally cartesian closed, is that of Kuratowski limit spaces. The concrete sheaves, or spaces, in our model correspond to the Kuratowski limit spaces in Johnstone's construction, and are also related to Spanier's quasi-topological spaces [22], as we discuss in the body of the paper. Bauer and Simpson's work can be seen as taking place in the topological topos.

Fourman works with a site whose underlying category is the semilattice of finite sequences of natural numbers under the prefix order, and van der Hoeven and Moerdijk work with a site whose underlying category is the monoid of continuous endomaps of the Baire space, and they relate their work to Fourman's.

Coquand and Jaber's forcing model instead uses the semilattice of finite binary sequences under the prefix order as the underlying category of the site, modelling the idea of a generic infinite binary sequence. They iterate their construction in order to be able to model the Fan functional, and our model can be regarded as accomplishing this iteration directly (personal communication with Coquand).

2 Sheaves and Spaces

2.1 Sheaves and Natural Transformations

Recall that a presheaf on a small category $\mathbf{C}$ is a functor $\mathbf{C}^{\mathrm{op}} \to \mathbf{Set}$. When $\mathbf{C}$ is a one-object category, i.e. a monoid, this can be formulated in terms of monoid actions [18, §I.1]. A *presheaf* on a monoid $(\mathbf{C}, \circ, \mathrm{id})$ amounts to a set P with an *action*

$$((p,t) \mapsto p \cdot t) \colon P \times \mathbf{C} \to P$$

such that for all $p \in P$ and $t, u \in \mathbf{C}$

$$p \cdot \mathrm{id} = p, \qquad p \cdot (t \circ u) = (p \cdot t) \cdot u.$$

A *natural transformation* of presheaves $(P, \cdot)$ and $(Q, \cdot)$ amounts to a function $\phi \colon P \to Q$ that preserves the action, i.e.

$$\phi(p \cdot t) = (\phi\ p) \cdot t.$$

We work with the monoid C of uniformly continuous endo-maps on the Cantor space $\mathbf{2}^{\mathbb{N}}$, that is, the functions $t \colon \mathbf{2}^{\mathbb{N}} \to \mathbf{2}^{\mathbb{N}}$ such that

$$\forall m \in \mathbb{N}.\ \exists n \in \mathbb{N}.\ \forall \alpha, \beta \in \mathbf{2}^{\mathbb{N}}.\ \alpha =_n \beta \implies t(\alpha) =_m t(\beta).$$

Notice that any continuous function $\mathbf{2}^{\mathbb{N}} \to \mathbf{2}^{\mathbb{N}}$ is uniformly continuous, assuming classical logic or the Fan Theorem. Because we do not assume such principles, we need to explicitly require uniform continuity in the definition of the monoid C.

Our site is the monoid C equipped with the countable coverage $\mathcal{J}$ consisting of the finite covering families

$$\langle \mathrm{cons}_s \rangle_{s \in \mathbf{2}^n}$$

for all natural numbers n, where $\mathbf{2}^n$ is the set of binary sequences of length n and $\mathrm{cons}_s \colon \mathbf{2}^{\mathbb{N}} \to \mathbf{2}^{\mathbb{N}}$ is the concatenation map:

$$\mathrm{cons}_s(\alpha) = s\alpha.$$

It is easy to verify that, for any $n \in \mathbb{N}$ and for any $s \in \mathbf{2}^n$, the concatenation map cons_s is uniformly continuous and so $\mathrm{cons}_s \in \mathrm{C}$.

The coverage axiom specialized to our situation amounts to saying that for all $t \in \mathrm{C}$,

$$\forall m \in \mathbb{N}.\ \exists n \in \mathbb{N}.\ \forall s \in \mathbf{2}^n.\ \exists t' \in \mathrm{C}.\ \exists s' \in \mathbf{2}^m.\ t \circ \mathrm{cons}_s = \mathrm{cons}_{s'} \circ t'. \qquad (\dagger)$$

It is routine to show that:

Lemma 1. *A map* $t\colon \mathbf{2}^{\mathbb{N}} \to \mathbf{2}^{\mathbb{N}}$ *satisfies* (†) *iff it is uniformly continuous.*

Thus, not only does the coverage axiom hold, but also it amounts to the fact that the elements of the monoid C are the uniformly continuous functions. Notice that every covering family is jointly surjective. Because the maps in each covering family have disjoint images, we do not need to consider the compatibility condition in the definition of sheaf:

Lemma 2. *A presheaf* $(P,\cdot)$ *is a* sheaf *iff for every* $n \in \mathbb{N}$ *and every family* $\langle p_s \in P\rangle_{s\in\mathbf{2}^n}$*, there is a unique* amalgamation $p \in P$ *such that, for all* $s \in \mathbf{2}^n$,

$$p \cdot \mathrm{cons}_s = p_s.$$

Notice also that, by induction, it is enough to consider the case $n = 1$:

Lemma 3. *A presheaf* $(P,\cdot)$ *is a* sheaf *iff for any two* $p_0, p_1 \in P$*, there is a unique* $p \in P$ *such that*

$$p \cdot \mathrm{cons}_0 = p_0 \qquad \text{and} \qquad p \cdot \mathrm{cons}_1 = p_1.$$

This construction gives a full subcategory $\mathbf{Shv}(\mathrm{C}, \mathcal{J})$ of the category of presheaves, consisting of the sheaves over the site $(\mathrm{C}, \mathcal{J})$.

2.2 Spaces and Continuous Maps

An important example of a sheaf is the monoid C itself with function composition as the action. Given $t_0, t_1 \in \mathrm{C}$, the amalgamation $t\colon \mathbf{2}^{\mathbb{N}} \to \mathbf{2}^{\mathbb{N}}$ is simply

$$t(\alpha) = t_{\alpha_0}(\lambda n.\alpha_{n+1}).$$

We say a presheaf is *concrete* if its action is function composition. Then all the elements in a concrete presheaf $(P, \circ)$ must be maps from the Cantor space to some set X. Concrete sheaves admit a more concrete description as the set X with the additional structure given by the maps in P. We denote the full subcategory of concrete sheaves by $\mathbf{CShv}(\mathrm{C}, \mathcal{J})$.

Concrete sheaves can be regarded as spaces, and their natural transformations as continuous maps. More precisely, they are analogous to Spanier's quasi-topological spaces [22], which have the category of topological spaces and continuous maps as a *full* subcategory. One advantage of quasi-topological spaces over topological spaces, which is the main reason for Spanier's introduction of the notion of quasi-space, is that continuous maps of quasi-spaces form a cartesian closed category. This category serves as a model of system T and HA^ω that validates the uniform continuity principle, assuming classical logic in the meta-language. Our concrete sheaves can be seen as analogues of quasi-topological spaces, admitting a constructive treatment.

A *quasi-topology* on a set X assigns to each compact Hausdorff space K a set $P(K, X)$ of functions $K \to X$ such that:

(1) All constant maps are in $P(K, X)$.

(2) If $t\colon K' \to K$ is continuous and $p \in P(K, X)$, then $p \circ t \in P(K', X)$.
(3) If $\langle t_i\colon K_i \to K\rangle_{i\in I}$ is a finite, jointly surjective family and $p\colon K \to X$ is a map with $p \circ t_i \in P(K_i, X)$ for every $i \in I$, then $p \in P(K, X)$.

A *quasi-topological space* is a set endowed with a quasi-topology, and a *continuous map* of quasi-spaces (X, P) and (Y, Q) is a function $f\colon X \to Y$ such that $f \circ p \in Q(K, Y)$ whenever $p \in P(K, X)$. For example, every topological space X is a quasi-topological space with the quasi-topology P such that $P(K, X)$ is the set of continuous maps $K \to X$, and this construction gives the full embedding of topological spaces into quasi-topological spaces.

This definition can be modified by considering just one compact Hausdorff space, the Cantor space, rather than all compact Hausdorff spaces, and by restricting the jointly surjective finite families of continuous maps to the covering families $\langle \mathrm{cons}_s\rangle_{s\in\mathbf{2}^n}$ considered in the previous section. We call the resulting objects C-*spaces*.

Definition 1. *A* C-space *is a set X equipped with a* C-topology *P, i.e. a collection of maps $\mathbf{2}^{\mathbb{N}} \to X$, called* probes, *satisfying the following conditions:*

(1) *All constant maps are in P.*
(2) (Presheaf condition) *If $p \in P$ and $t \in$ C, then $p \circ t \in P$.*
(3) (Sheaf condition) *For any $n \in \mathbb{N}$ and any family $\langle p_s \in P\rangle_{s\in\mathbf{2}^n}$, the unique map $p\colon \mathbf{2}^{\mathbb{N}} \to X$ defined by $p(s\alpha) = p_s(\alpha)$ is in P.*

A continuous map *of* C-*spaces (X, P) and (Y, Q) is a map $f\colon X \to Y$ with $f \circ p \in Q$ whenever $p \in P$. We write* C-**Space** *for the category of* C-*spaces and continuous maps. The above three conditions are called the* probe axioms.

Notice that the sheaf condition is equivalent to

(3′) *If $p\colon \mathbf{2}^{\mathbb{N}} \to X$ is a map such that there exists $n \in \mathbb{N}$ with $p \circ \mathrm{cons}_s \in P$ for all $s \in \mathbf{2}^n$, then $p \in P$.*

The idea is that we "topologize" the set X by choosing a designated set P of maps $\mathbf{2}^{\mathbb{N}} \to X$ that we want, and hence declare, to be continuous. For example, if X already has some form of topology, e.g. a metric, we can take P to be the set of continuous functions $\mathbf{2}^{\mathbb{N}} \to X$ with respect to this topology and the natural topology of the Cantor space. Of course we have to make sure the sheaf condition is satisfied.

As mentioned earlier, C-spaces provide a more concrete description of concrete sheaves in the following sense. Given a C-space (X, P), the C-topology P together with function composition is a concrete sheaf. Conversely, if $(P, \circ)$ is a concrete sheaf, then all maps in P should have the same codomain.

Proposition 1. *The two categories* C-**Space** *and* **CShv**(C, $\mathcal{J}$) *are naturally equivalent.*

By virtue of this equivalence, C-**Space** can also be viewed as a full subcategory of **Shv**(C, $\mathcal{J}$). Moreover, C-spaces are closed under products and form an exponential ideal.

To improve the readability, we abbreviate X for the space $(|X|, \text{Probe}(X))$ where $|X|$ stands for the underlying set and $\text{Probe}(X)$ for the collection of probes, i.e. the C-topology on $|X|$, and we often write X to mean $|X|$ by an abuse of notation.

2.3 The Cartesian Closed Structure of C-Space

C-spaces have several convenient categorical properties, the first of which is cartesian closedness.

Theorem 1. *The category* C-**Space** *is cartesian closed.*

Proof. Any singleton set $\mathbf{1} = \{\star\}$ with the unique map $\mathbf{2}^{\mathbb{N}} \to \mathbf{1}$ as the only probe is a C-space as well as a terminal object in C-**Space**.

Given C-spaces (X, P) and (Y, Q), their *product* is the cartesian product $X \times Y$ equipped with the collection R of probes defined by the condition that $r \colon \mathbf{2}^{\mathbb{N}} \to X \times Y$ is in R iff $\pi_0 \circ r \in P$ and $\pi_1 \circ r \in Q$, where π_0 and π_1 are the projections. We have to verify that R satisfies the probe axioms and that this has the universal property of a categorical product in C-**Space**, i.e. continuity of projection functions and its universal property, but this is routine.

Given C-spaces (X, P) and (Y, Q), their *exponential* is the set Y^X of continuous maps $X \to Y$ equipped with the collection R of probes defined by the condition that $r \colon \mathbf{2}^{\mathbb{N}} \to Y^X$ is in R iff for any $t \in \mathrm{C}$ and $p \in P$ the map $\lambda\alpha.r(t\alpha)(p\alpha)$ is in Q. Again, we have to verify that the probe axioms are satisfied and that this has the universal property of an exponential in C-**Space**, which involves some subtleties regarding the coverage axiom. □

Theorem 2. *The category* C-**Space** *has finite coproducts.*

Proof. The empty set equipped with the empty collection of probes is a C-space and an initial object in C-**Space**.

Binary coproducts can be constructed as follows: given C-spaces (X, P) and (Y, Q), their *coproduct* is the disjoint union $X + Y$ equipped with the collection R of probes defined by the condition that $r \colon \mathbf{2}^{\mathbb{N}} \to X + Y$ is in R iff there exists $n \in \mathbb{N}$ such that for all $s \in \mathbf{2}^n$ either there exists $p \in P$ with $r(\text{cons}_s \alpha) = \text{in}_0(p\alpha)$ for all $\alpha \in \mathbf{2}^{\mathbb{N}}$ or there exists $q \in Q$ with $r(\text{cons}_s \alpha) = \text{in}_1(q\alpha)$ for all $\alpha \in \mathbf{2}^{\mathbb{N}}$. We have to verify that the probe axioms are satisfied and that this has the required universal property. □

2.4 Discrete C-Spaces and Natural Numbers Object

We say that a C-space X is *discrete* if for every C-space Y, all functions $X \to Y$ are continuous. A map $p \colon \mathbf{2}^{\mathbb{N}} \to X$ into a set X is called *locally constant* iff

$$\exists n \in \mathbb{N}.\ \forall \alpha, \beta \in \mathbf{2}^{\mathbb{N}}.\ \alpha =_n \beta \implies p(\alpha) = p(\beta).$$

Lemma 4. *Let X be any set.*

(1) *The locally constant functions $\mathbf{2}^{\mathbb{N}} \to X$ form a* C*-topology on X.*
(2) *For any* C*-topology P on X, every locally constant function $\mathbf{2}^{\mathbb{N}} \to X$ is in P.*

In other words, the locally constant maps $\mathbf{2}^{\mathbb{N}} \to X$ form the smallest C-topology on the set X. Moreover:

Lemma 5. *A* C*-space is discrete iff the probes on it are precisely the locally constant functions.*

We thus refer to the collection of locally constant maps $\mathbf{2}^{\mathbb{N}} \to X$ as the *discrete* C-topology on X. In particular, when the set X is $\mathbf{2}$ or $\mathbb{N}$, the locally constant functions amount to the uniformly continuous functions. Hence we have a discrete two-point space $\mathbf{2}$ and a discrete space $\mathbb{N}$ of natural numbers, which play an important role in our model:

Theorem 3. *In the category* **C-Space***:*

1. *The coproduct of two copies of the terminal space $\mathbf{1}$ is the discrete two-point space $\mathbf{2}$.*
2. *The discrete space of natural numbers is the natural numbers object.*

Proof. The universal properties of $\mathbf{2}$ and $\mathbb{N}$ can be constructed in the same way as in the category **Set**, because the unique maps g and h in the diagrams below are continuous by the discreteness of $\mathbb{N}$ and $\mathbf{2}$:

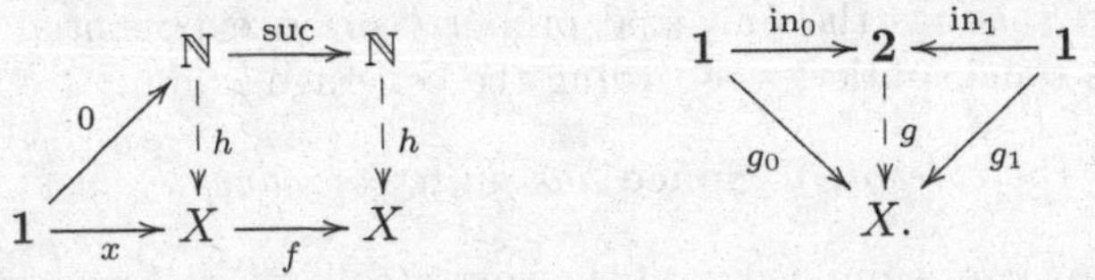

3 The Fan Functional

The monoid C can be regarded as a one-object category **C** with the object $\mathbf{2}^{\mathbb{N}}$ and the morphisms all uniformly continuous maps $\mathbf{2}^{\mathbb{N}} \to \mathbf{2}^{\mathbb{N}}$. The *Yoneda embedding* $y \colon \mathbf{C} \to$ **C-Space** satisfies

$$y\left(\mathbf{2}^{\mathbb{N}}\right) = (\mathbf{2}^{\mathbb{N}}, \mathrm{C}),$$

where $(\mathbf{2}^{\mathbb{N}}, \mathrm{C})$ is the C-space corresponding to the concrete sheaf $(\mathrm{C}, \circ)$ given as an example in the previous section.

In a cartesian closed category with a natural numbers object $\mathbb{N}$ and a finite coproduct $\mathbf{2} = \mathbf{1} + \mathbf{1}$, call their exponential $\mathbf{2}^{\mathbb{N}}$ the *Cantor space*. With this terminology, we have that $y(\mathbf{2}^{\mathbb{N}})$ is precisely the Cantor space in **C-Space**, i.e.

$$y\left(\mathbf{2}^{\mathbb{N}}\right) = \mathbf{2}^{\mathbb{N}},$$

where in the left-hand side $\mathbf{2}^{\mathbb{N}}$ is the only object of the monoid C and in the right-hand side $\mathbf{2}^{\mathbb{N}}$ is the exponential of the two discrete spaces $\mathbb{N}$ and $\mathbf{2}$. Since all maps $\mathbb{N} \to \mathbf{2}$ are continuous by the discreteness of $\mathbb{N}$, the underlying set of the exponential $\mathbf{2}^{\mathbb{N}}$ is precisely the Cantor space $\mathbf{2}^{\mathbb{N}}$ (space of all maps $\mathbb{N} \to \mathbf{2}$). Of course one has to verify that

$$r \in \mathrm{C} \iff \forall t \in C.\ \forall p \in \mathrm{Probe}(\mathbb{N}).\ \lambda\alpha.r(t\alpha)(p\alpha) \in \mathrm{Probe}(\mathbf{2}),$$

i.e. the two C-topologies are the same, which is routine.

As the category **C** has only one object $\mathbf{2}^{\mathbb{N}}$, the Yoneda Lemma amounts to the following.

Lemma 6 (Yoneda). *For any* C*-space* X*, a map* $\mathbf{2}^{\mathbb{N}} \to X$ *is a probe on* X *iff it is continuous.*

By the Yoneda Lemma, we get that the continuous maps from the Cantor space in C-**Space** to the natural numbers object are in natural bijection with the uniformly continuous maps $\mathbf{2}^{\mathbb{N}} \to \mathbb{N}$ of the meta-language used to define the model:

Corollary 1. *Writing* $[\mathbf{2}^{\mathbb{N}}, \mathbb{N}]_{\text{C-}\mathbf{Space}}$ *for the set of continuous maps* $\mathbf{2}^{\mathbb{N}} \to \mathbb{N}$*, and* $\mathrm{cts}(\mathbf{2}^{\mathbb{N}}, \mathbb{N})$ *for the set of uniformly continuous maps* $\mathbf{2}^{\mathbb{N}} \to \mathbb{N}$*, we have*

$$[\mathbf{2}^{\mathbb{N}}, \mathbb{N}]_{\text{C-}\mathbf{Space}} \cong \mathrm{cts}(\mathbf{2}^{\mathbb{N}}, \mathbb{N}).$$

Moreover, the topology on $[\mathbf{2}^{\mathbb{N}}, \mathbb{N}]_{\text{C-}\mathbf{Space}}$ is discrete:

Lemma 7. *The exponential* $\mathbb{N}^{\mathbf{2}^{\mathbb{N}}}$ *is a discrete* C*-space.*

Proof. Given a probe $p \colon \mathbf{2}^{\mathbb{N}} \to \mathbb{N}^{\mathbf{2}^{\mathbb{N}}}$, we want to show that it is locally constant. By the construction of exponentials in Section 2, we know that for all $t, r \in \mathrm{C}$,

$$\lambda\alpha.p(t\alpha)(r\alpha) \in \mathrm{Probe}(\mathbb{N}),$$

i.e. $\lambda\alpha.p(t\alpha)(r\alpha)$ it is uniformly continuous. In particular, we can take

$$t(\alpha)(i) = \alpha_{2i} \qquad \text{and} \qquad r(\alpha)(i) = \alpha_{2i+1},$$

which are both uniformly continuous, and define $q(\alpha) = p(t\alpha)(r\alpha)$. From the proof of uniform continuity of q, we get its modulus n. (NB. Here we are implicitly using choice, but this is not a problem in intensional type theory. In a setting without choice, we would need to define uniform continuity by explicitly requiring a modulus.) Now define a map join: $\mathbf{2}^{\mathbb{N}} \times \mathbf{2}^{\mathbb{N}} \to \mathbf{2}^{\mathbb{N}}$ by

$$\begin{aligned} \mathrm{join}(\alpha, \beta)(2i) &= \alpha_i \\ \mathrm{join}(\alpha, \beta)(2i+1) &= \beta_i. \end{aligned}$$

Given $\alpha, \alpha', \beta \in \mathbf{2}^{\mathbb{N}}$ with $\alpha =_n \alpha'$, we have

$$\begin{aligned} p(\alpha)(\beta) &= p(t(\mathrm{join}(\alpha, \beta)))(r(\mathrm{join}(\alpha, \beta))) && \text{(by the definitions of } t, r, \mathrm{join}) \\ &= q(\mathrm{join}(\alpha, \beta)) && \text{(by the definition of } q) \\ &= q(\mathrm{join}(\alpha', \beta)) && (\mathrm{join}(\alpha, \beta) =_{2n} \mathrm{join}(\alpha', \beta), 2n \geq n) \\ &= p(\alpha')(\beta). \end{aligned}$$

Hence p is locally constant and therefore $\mathbb{N}^{\mathbf{2}^{\mathbb{N}}}$ is discrete. □

Theorem 4. *There is a* Fan functional

$$\text{fan}\colon \mathbb{N}^{\mathbf{2}^{\mathbb{N}}} \to \mathbb{N}$$

in C-**Space** *that continuously calculates moduli of uniform continuity.*

Proof. Given a continuous map $f\colon \mathbf{2}^{\mathbb{N}} \to \mathbb{N}$, i.e. an element of $\mathbb{N}^{\mathbf{2}^{\mathbb{N}}}$, we know f is uniformly continuous as $f = f \circ \mathrm{id}_{\mathbf{2}^{\mathbb{N}}} \in \mathrm{Probe}(\mathbb{N})$ by the continuity of f. Then we can get a modulus n from the proof of its uniform continuity. From this n we can compute the smallest modulus of f as follows. We define a function $\mathrm{lmod}\colon (\mathbf{2}^{\mathbb{N}} \to \mathbb{N}) \to \mathbb{N} \to \mathbb{N}$ by induction on its second argument:

$$\begin{aligned} \mathrm{lmod}\ f\ 0 &= 0 \\ \mathrm{lmod}\ f\ (n+1) &= \text{if } (\forall s \in \mathbf{2}^n.\ f(s0^\omega) \equiv f(s1^\omega)) \text{ then } (\mathrm{lmod}\ f\ n) \text{ else } (n+1). \end{aligned}$$

With a proof by induction, we can show that $\mathrm{lmod}\ f\ n$ is the smallest modulus if n is a modulus of f. Hence, we define

$$\mathrm{fan}(f) = \mathrm{lmod}\ f\ n.$$

According to the previous lemma, the space $\mathbb{N}^{\mathbf{2}^{\mathbb{N}}}$ is discrete and hence this functional is continuous. □

4 Uniform Continuity of T-Definable Functions

Now we recover a well known result, using a logical relation between the set-theoretical and the C-**Space** models of Gödel's system T.

Recall that system T is a simply typed lambda calculus with a ground type N for natural numbers and a primitive recursor $\mathsf{rec}\colon \sigma \to \sigma \to \mathsf{N} \to \sigma$ for every type σ. For our purpose of formulating the uniform continuity principle, we add the binary type 2 as another ground type and a case function $\mathsf{if}\colon \sigma \to \sigma \to 2 \to \sigma$ for every T type σ. Such a system can be interpreted in a cartesian closed category with a natural numbers object $\mathbb{N}$ and a coproduct $\mathbf{2}$ (or $\mathbf{1}+\mathbf{1}$) of two copies of the terminal object. Specifically, types are interpreted as objects: N is interpreted as $\mathbb{N}$, the type 2 as $\mathbf{2}$, product types as products, and function types as exponentials. Contexts are interpreted inductively as products. And a term in context is interpreted as a morphism from the interpretation of its context to the one of its type. Finally, rec and if are interpreted using the universal properties of $\mathbb{N}$ and $\mathbf{2}$.

Both the categories **Set** and C-**Space** are cartesian closed and have a natural numbers objects and a coproduct $\mathbf{1}+\mathbf{1}$; thus, they give models of system T. Throughout this paper, we use the semantic braces $[\![-]\!]$ for the interpretation, and add **Set** and C-**Space** as subscripts to distinguish which model we are working with. Now we apply the logical relations technique to understand the relationship between these two models.

Definition 2. *The* logical relation R *over the set-theoretical and* C-**Space** *models is defined by*

1. *If σ is a* T *type, then $\mathrm{R}_\sigma \subset [\![\sigma]\!]_{\mathbf{Set}} \times [\![\sigma]\!]_{\mathbf{C\text{-}Space}}$ is defined by induction on type σ as follows:*
   (a) $\mathrm{R}_\iota(a, a')$ iff $a = a'$, where ι is the ground type 2 *or* N*;*
   (b) $\mathrm{R}_{\sigma\to\tau}(f, f')$ iff, for any $a \in [\![\sigma]\!]_{\mathbf{Set}}$ and any $a' \in [\![\sigma]\!]_{\mathbf{C\text{-}Space}}$, if $\mathrm{R}_\sigma(a, a')$ then $\mathrm{R}_\tau(f(a), f'(a'))$.
2. *If $\Gamma \equiv x_1 : \sigma_1, \ldots, x_n : \sigma_n$ is a context, then $\mathrm{R}_\Gamma \subset [\![\Gamma]\!]_{\mathbf{Set}} \times [\![\Gamma]\!]_{\mathbf{C\text{-}Space}}$ is defined by $\mathrm{R}_\Gamma(\boldsymbol{a}, \boldsymbol{a'})$ iff $\mathrm{R}_{\sigma_i}(a_i, a'_i)$ for all $i \leq n$.*
3. *Given $f \equiv [\![\Gamma \vdash \boldsymbol{t} : \tau]\!]_{\mathbf{Set}}$ and $f' \equiv [\![\Gamma \vdash \boldsymbol{t} : \tau]\!]_{\mathbf{C\text{-}Space}}$, $\mathrm{R}(f, f')$ iff, for any $\boldsymbol{a} \in [\![\Gamma]\!]_{\mathbf{Set}}$ and any $\boldsymbol{a'} \in [\![\Gamma]\!]_{\mathbf{C\text{-}Space}}$, if $\mathrm{R}_\Gamma(\boldsymbol{a}, \boldsymbol{a'})$ then $\mathrm{R}_\tau(f(\boldsymbol{a}), f'(\boldsymbol{a'}))$.*

With a proof by induction on types as usual, we can easily show that the interpretations of any T term in these two models are related.

Lemma 8. *If $\Gamma \vdash \boldsymbol{t} : \tau$, then $\mathrm{R}([\![\Gamma \vdash \boldsymbol{t} : \tau]\!]_{\mathbf{Set}}, [\![\Gamma \vdash \boldsymbol{t} : \tau]\!]_{\mathbf{C\text{-}Space}})$.*

We say that an element $x \in [\![\sigma]\!]_{\mathbf{Set}}$ in the set-theoretical model is T-*definable* if it is the interpretation of some closed T term, i.e. there exists a closed term $t : \sigma$ such that $x = [\![t]\!]_{\mathbf{Set}}$.

Theorem 5. *Any* T*-definable function $\mathbf{2}^{\mathbb{N}} \to \mathbb{N}$ is uniformly continuous.*

Proof. If $f \colon \mathbf{2}^{\mathbb{N}} \to \mathbb{N}$ interprets the term $\boldsymbol{f} : (\mathsf{N}\to 2)\to\mathsf{N}$, then f is related to the (uniformly) continuous map $[\![\boldsymbol{f}]\!]_{\mathbf{C\text{-}Space}} \colon \mathbf{2}^{\mathbb{N}} \to \mathbb{N}$ according to the above lemma. By the definition of the logical relation, f is uniformly continuous. □

5 A Continuous Realizability Semantics of HA$^\omega$

Recall that HA$^\omega$ has equations between system T terms of the same type as atomic propositions, quantifiers that range over elements of (the interpretation of) system T types, and logical connectives $\wedge$, $\Rightarrow$ (the connectives $\vee$ and $\neg$ are definable from the other connectives). For technical convenience, we add a singleton type 1 and binary product types to the inductive definition of system T types. Throughout this section, we use σ, τ to range over T types, bold lower case letters $\boldsymbol{f}, \boldsymbol{x}, \boldsymbol{n}, \boldsymbol{t}, \boldsymbol{u}$ to range over T terms, and φ, ψ to range over HA$^\omega$ formulas.

With the above definition, the uniform continuity principle can be formulated in HA$^\omega$ by the following

$$\forall \boldsymbol{f} : (\mathsf{N}\to 2)\to\mathsf{N}.\ \exists \boldsymbol{n} : \mathsf{N}.\ \forall \boldsymbol{\alpha}, \boldsymbol{\beta} : \mathsf{N}\to 2.\ \boldsymbol{\alpha} =_{\boldsymbol{n}} \boldsymbol{\beta} \Rightarrow \boldsymbol{f}(\boldsymbol{\alpha}) = \boldsymbol{f}(\boldsymbol{\beta}) \qquad \text{(UC)}$$

where $\boldsymbol{\alpha} =_{\boldsymbol{n}} \boldsymbol{\beta}$ is short for $\forall \boldsymbol{i} : \mathsf{N}.\ \boldsymbol{i} < \boldsymbol{n} \Rightarrow \boldsymbol{\alpha}(\boldsymbol{i}) = \boldsymbol{\beta}(\boldsymbol{i})$. Here we can define the relation $\boldsymbol{i} < \boldsymbol{n}$ by $\exists \boldsymbol{m} : \mathsf{N}.\ \mathsf{suc}(\boldsymbol{i} + \boldsymbol{m}) = \boldsymbol{n}$ where addition $+$ is inductively defined in T.

To any HA$^\omega$ formula φ we associate a type $|\varphi|$ of potential realizers. Then a *continuous realizer* of a formula $\Gamma \vdash \varphi$ is a pair

$$(e, \boldsymbol{q}) \in [\![\, |\varphi| \,]\!]_{\mathbf{C\text{-}Space}} \times [\![\Gamma]\!]_{\mathbf{C\text{-}Space}}.$$

We call this a *continuous realizability semantics.* In the following, semantic brackets without explicit decorations refer to the C-space interpretation.

Definition 3 (Continuous realizability). *The* types of potential realizers *of* HA^ω *formulas are given inductively as follows:*

1. $|\boldsymbol{t} = \boldsymbol{u}| \quad = 1,$
2. $|\varphi \wedge \psi| \quad = |\varphi| \times |\psi|,$
3. $|\varphi \Rightarrow \psi| \quad = |\varphi| \to |\psi|,$
4. $|\forall x{:}\sigma.\ \varphi| \quad = \sigma \to |\varphi|,$
5. $|\exists x{:}\sigma.\ \varphi| \quad = \sigma \times |\varphi|.$

Let Γ *be a context and* $\boldsymbol{q} \in [\![\Gamma]\!]$. *The relation*

$$(e, \boldsymbol{q}) \text{ realizes } \Gamma \vdash \varphi$$

is defined by induction on formulas as follows:

1. $(\star, \boldsymbol{q})$ *realizes* $\Gamma \vdash \boldsymbol{t} = \boldsymbol{u}$ *iff* $[\![\Gamma \vdash \boldsymbol{t} : \sigma]\!](\boldsymbol{q}) \equiv [\![\Gamma \vdash \boldsymbol{u} : \sigma]\!](\boldsymbol{q})$, *where* σ *is the type of the terms* $\boldsymbol{t}$ *and* $\boldsymbol{u}$,
2. $(e, \boldsymbol{q})$ *realizes* $\Gamma \vdash \varphi_0 \wedge \varphi_1$ *iff* $(\pi_i(e), \boldsymbol{q})$ *realizes* $\Gamma \vdash \varphi_i$ *for all* $i \in \{0,1\}$, *where* $e \in [\![\,|\varphi_0|\,]\!] \times [\![\,|\varphi_1|\,]\!]$,
3. $(e, \boldsymbol{q})$ *realizes* $\Gamma \vdash \varphi \Rightarrow \psi$ *iff for all* $a \in [\![\,|\varphi|\,]\!]$ *with* $(a, \boldsymbol{q})$ *realizing* $\Gamma \vdash \varphi$, *the pair* $(e(a), \boldsymbol{q})$ *realizes* $\Gamma \vdash \psi$, *where* $e \in [\![\,|\psi|\,]\!]^{[\![\,|\varphi|\,]\!]}$,
4. $(e, \boldsymbol{q})$ *realizes* $\Gamma \vdash \forall x{:}\sigma.\ \varphi$ *iff for all* $a \in [\![\sigma]\!]$, *the pair* $(e(a), (\boldsymbol{q}, a))$ *realizes* $\Gamma, x{:}\sigma \vdash \varphi$, *where* $e \in [\![\,|\varphi|\,]\!]^{[\![\sigma]\!]}$ *and* $(\boldsymbol{q}, a) \in [\![\Gamma, x{:}\sigma]\!]$,
5. $(e, \boldsymbol{q})$ *realizes* $\Gamma \vdash \exists x{:}\sigma.\ \varphi$ *iff* $(\pi_1(e), (\boldsymbol{q}, \pi_0(e)))$ *realizes* $\Gamma, x{:}\sigma \vdash \varphi$, *where* $e \in [\![\sigma]\!] \times [\![\,|\varphi|\,]\!]$.

We say a closed HA^ω *formula* φ *is* realizable *if there exists* $e \in [\![\,|\varphi|\,]\!]$ *such that* $(e, \star)$ *realizes* $\vdash \varphi$.

The main result of this paper is that our model validates the uniform continuity principle in the following sense.

Theorem 6. *The uniform continuity principle* (UC) *can be realized by the Fan functional.*

Proof. If $(e, \star)$ realizes UC, then e is a continuous map

$$\mathbb{N}^{\mathbf{2}^{\mathbb{N}}} \to \mathbb{N} \times (\mathbf{2}^{\mathbb{N}} \to \mathbf{2}^{\mathbb{N}} \to (\mathbb{N} \to (\mathbb{N} \times \mathbf{1}) \to \mathbf{1}) \to \mathbf{1}).$$

By Definition 3, given any continuous $f\colon \mathbf{2}^{\mathbb{N}} \to \mathbb{N}$, the pair $(e(f), (\star, f))$ realizes

$$\boldsymbol{f}{:}(\mathsf{N}{\to}2){\to}\mathsf{N} \ \vdash\ \exists \boldsymbol{n}{:}\mathsf{N}.\ \forall \boldsymbol{\alpha}, \boldsymbol{\beta}{:}\mathsf{N}{\to}2.\ (\boldsymbol{\alpha} =_{\boldsymbol{n}} \boldsymbol{\beta} \Rightarrow \boldsymbol{f}(\boldsymbol{\alpha}) = \boldsymbol{f}(\boldsymbol{\beta})).$$

We define the first component of $e(f)$ to be $\mathrm{fan}(f)$, i.e. the modulus of uniform continuity of f. Given $n = \mathrm{fan}(f)$, we want that $(\pi_1(e(f)), (\star, f, n))$ realizes

$$\boldsymbol{f}{:}(\mathsf{N}{\to}2){\to}\mathsf{N}, \boldsymbol{n}{:}\mathsf{N} \ \vdash\ \forall \boldsymbol{\alpha}, \boldsymbol{\beta}{:}\mathsf{N}{\to}2.\ (\boldsymbol{\alpha} =_{\boldsymbol{n}} \boldsymbol{\beta} \Rightarrow \boldsymbol{f}(\boldsymbol{\alpha}) = \boldsymbol{f}(\boldsymbol{\beta})).$$

Given $\alpha, \beta \in \mathbf{2}^{\mathbb{N}}$ with $\alpha =_n \beta$, it is easy to verify that there exists a continuous map $e'\colon \mathbb{N} \to (\mathbb{N} \times \mathbf{1}) \to \mathbf{1}$ such that $(e', (\star, f, n, \alpha, \beta))$ realizes

$$\boldsymbol{f}{:}(\mathsf{N}{\to}2){\to}\mathsf{N}, \boldsymbol{n}{:}\mathsf{N}, \boldsymbol{\alpha}{:}\mathsf{N}{\to}2, \boldsymbol{\beta}{:}\mathsf{N}{\to}2 \ \vdash\ \boldsymbol{\alpha} =_{\boldsymbol{n}} \boldsymbol{\beta}.$$

According to the definition of $[\![-]\!]$, we have

$$[\![\Gamma \vdash \boldsymbol{f}(\boldsymbol{\alpha}) : \mathsf{N}]\!](\star, f, n, \alpha, \beta) = f(\alpha)$$

and

$$[\![\Gamma \vdash \boldsymbol{f}(\boldsymbol{\beta}) : \mathsf{N}]\!](\star, f, n, \alpha, \beta) = f(\beta)$$

where $\Gamma \equiv \boldsymbol{f}:(\mathsf{N}\to 2)\to\mathsf{N}, \boldsymbol{n}:\mathsf{N}, \boldsymbol{\alpha}:\mathsf{N}\to 2, \boldsymbol{\beta}:\mathsf{N}\to 2$. As n is the modulus of f, we have $f(\alpha) = f(\beta)$ and hence

$$[\![\Gamma \vdash \boldsymbol{f}(\boldsymbol{\alpha}) : \mathsf{N}]\!](\star, f, n, \alpha, \beta) = [\![\Gamma \vdash \boldsymbol{f}(\boldsymbol{\beta}) : \mathsf{N}]\!](\star, f, n, \alpha, \beta).$$

Thus, $(\star, (\star, f, n, \alpha, \beta))$ realizes $\Gamma \vdash \boldsymbol{f}(\boldsymbol{\alpha}) = \boldsymbol{f}(\boldsymbol{\beta})$. □

6 Construction of the Model in Intensional Type Theory

The above results have been deliberately developed in such a way to be routinely formalizable in intensional type theory. However, certain details require a closer look. We work with an intensional type theory with a universe, $\sum$-types, $\prod$-types, identity types and standard base types such as natural numbers, booleans, unit type and empty type.

We considered three approaches, which we developed in Agda notation [6], and are available at [24]. In the first approach, which is probably the simplest and most readable, we assumed the axiom of function extensionality,

$$\prod_{X,Y:\ \mathrm{Type}}\ \prod_{f,g:\ X\to Y} \left(\prod_{x:\ X} fx \equiv gx\right) \to f \equiv g,$$

where $\equiv$ is the identity type. This approach has two drawbacks. One of them is that, because this axiom does not come with a computational interpretation, it is in principle useless for extracting computational content. In ealier stages of this work, we conjectured that the axiom of extensionality occurs only in computationally irrelevant contexts.

In order to attempt to verify that this is indeed the case, in our second approach, we made use of Agda's irrelevant fields [7], and postulated extensionality within such an irrelevant context. With this second approach, the Agda type checker proved our conjecture false. However, by refining the notion of C-topology, we were able to make it true, and our constructions and proofs type-checked. We needed to add the following condition:

(4) *Any map extensionally equivalent to a probe is also a probe.*

And we also needed to add more steps in each case the construction of a space was performed.

However, a second drawback remains in the two approaches considered above: they do not seem to allow a construction of the Fan functional. To define a continuous Fan functional in the model (Section 3), we derive its continuity by showing that its domain, the space $\mathbb{N}^{\mathbf{2}^{\mathbb{N}}}$, is discrete, i.e. by showing that any probe $p\colon \mathbf{2}^{\mathbb{N}} \to \mathbb{N}^{\mathbf{2}^{\mathbb{N}}}$ is locally constant. We can find an $n \in \mathbb{N}$ and show that for any $\alpha, \alpha' \in \mathbf{2}^{\mathbb{N}}$ with $\alpha =_n \alpha'$, the two maps $p(\alpha), p(\alpha')\colon \mathbf{2}^{\mathbb{N}} \to \mathbb{N}$ are equal using functional extensionality. However, this does not allow us to conclude that their proofs of continuity are also equal, in order to conclude $p(\alpha) \equiv p(\alpha')$, and the proofs of continuity cannot be put in an irrelevant field because they are computationally relevant, at least not without further thought.

In our third and last, fully successful, approach, instead of assuming any form of extensionality or irrelevant fields, we slightly adjusted the definition of C-space. Now a C-topology is defined on a set equipped with an equivalence relation, that is, on a setoid. With this, we can define a notion of equality of continuous functions that ignores continuity proofs, and the Fan functional can be implemented as discussed in the previous sections. We remark that the probe axiom (4) mentioned above is still needed.

This third approach works well, and does not need to assume any non-standard axiom for intensional type theory. However, the drawback is that the proofs are much less readable than in the first two approaches. It would be desirable to find an approach that avoids setoids and addresses the equality problem for continuous functions by hiding information in the definition of continuity without losing computational information to obtain a more concise formalization.

At the moment we formalized everything discussed above, including the Fan functional, the set and continuous interpretations of system T, their logical relation, and the proof that the set-theoretical definable functions $\mathbf{2}^{\mathbb{N}} \to \mathbb{N}$ are uniformly continuous, but excluding the definition of HA^{ω} and its interpretation, which is under development.

7 Future Work

Both the category of sheaves and its full subcategory of C-spaces are locally cartesian closed. For the second category, an exponential in a slice category **C-Space**$/X$ is constructed in the same way as the one in the slice category **Set**$/X$, with a suitable construction of the topology on its domain (see [2, Proposition 43]). Thus, rather than giving a realizability interpretation of UC, we can understand its quantifiers as $\prod$ and $\sum$, and interpret them using the locally cartesian closed structure [21,10,15,11]. With a cursory calculation to be fully verified in future work, we can show that the Fan functional (modulo some type isomorphisms) is an element of the interpretation of UC. Hence our development seems to generalize from system T to dependent types. We have not considered the interpretation of universes with our continuous model so far.

As mentioned in the introduction, we conjecture that the system T model consisting of C-spaces is classically equivalent to the model of Kleene–Kreisel functionals, and hence can be seen as a constructive development of that model.

References

1. Awodey, S., Bauer, A.: Propositions as [types]. Journal of Logic and Computation 14(4), 447–471 (2004)
2. Baez, J.C., Hoffnung, A.E.: Convenient categories of smooth spaces (2008)
3. Bauer, A., Birkedal, L., Scott, D.: Equilogical spaces. Theoret. Comput. Sci. 315(1), 35–59 (2004)
4. Bauer, A., Simpson, A.: Continuity begets continuity. Presented at Trends in Constructive Mathematics in Germany (2006)
5. Beeson, M.J.: Foundations of Constructive Mathematics. Springer (1985)
6. Bove, A., Dybjer, P.: Dependent types at work. In: Bove, A., Barbosa, L.S., Pardo, A., Pinto, J.S. (eds.) LerNet 2008. LNCS, vol. 5520, pp. 57–99. Springer, Heidelberg (2009)
7. Community. Agda wiki, `http://wiki.portal.chalmers.se/agda/`
8. Coquand, T., Jaber, G.: A note on forcing and type theory. Fundamenta Informaticae 100(1-4), 43–52 (2010)
9. Coquand, T., Jaber, G.: A computational interpretation of forcing in type theory. In: Epistemology Versus Ontology, vol. 27, pp. 203–213. Springer, Netherlands (2012)
10. Curien, P.-L.: Substitution up to isomorphism. Fundamenta Informaticae 19(1/2), 51–85 (1993)
11. Dybjer, P.: Internal type theory. In: Berardi, S., Coppo, M. (eds.) TYPES 1995. LNCS, vol. 1158, pp. 120–134. Springer, Heidelberg (1996)
12. Escardó, M., Lawson, J., Simpson, A.: Comparing Cartesian closed categories of (core) compactly generated spaces. Topology Appl. 143(1-3), 105–145 (2004)
13. Fourman, M.P.: Notions of choice sequence. In: The L. E. J. Brouwer Centenary Symposium Proceedings of the Conference Held in Noordwijkerhout, vol. 110, pp. 91–105 (1982)
14. Fourman, M.P.: Continuous truth I, non-constructive objects. In: Proceedings of Logic Colloquium, Florence 1982, vol. 112, pp. 161–180. Elsevier (1984)
15. Hofmann, M.: On the interpretation of type theory in locally Cartesian closed categories. In: Pacholski, L., Tiuryn, J. (eds.) CSL 1994. LNCS, vol. 933, pp. 427–441. Springer, Heidelberg (1995)
16. Johnstone, P.T.: On a topological topos. Proceedings of the London Mathematical Society 38(3), 237–271 (1979)
17. Johnstone, P.T.: Sketches of an Elephant: A Topos Theory Compendium. Oxford University Press (2002)
18. Mac Lane, S., Moerdijk, I.: Sheaves in Geometry and Logic: A First Introduction to Topos Theory. Springer (1992)
19. Normann, D.: Recursion on the countable functionals. Lec. Not. Math, vol. 811. Springer (1980)
20. Normann, D.: Computing with functionals—computability theory or computer science? Bull. Symbolic Logic 12(1), 43–59 (2006)
21. Seely, R.A.G.: Locally cartesian closed categories and type theory. Mathematical Proceedings of the Cambridge Philosophical Society 95(1), 33–48 (1984)
22. Spanier, E.H.: Quasi-topologies. Duke Mathematical Journal 30(1), 1–14 (1963)
23. van der Hoeven, G., Moerdijk, I.: Sheaf models for choice sequences. Annals of Pure and Applied Logic 27(1), 63–107 (1984)
24. Xu, C., Escardó, M.H.: A constructive model of uniform continuity, developed in Agda. Available at the Authors' Institutional Web Pages (2012-2013)

Author Index